W9-AOX-529

MATHEMATICAL ASPECTS OF FINITE ELEMENTS IN PARTIAL DIFFERENTIAL EQUATIONS

Publication No. 33
of the Mathematics Research Center
The University of Wisconsin — Madison

ACADEMIC PRESS RAPID MANUSCRIPT REPRODUCTION

Mathematical Aspects of Finite Elements in Partial Differential Equations

Edited by Carl de Boor

Proceedings of a Symposium
Conducted by the Mathematics Research Center
The University of Wisconsin—Madison
April 1–3, 1974

Academic Press
New York • San Francisco • London 1974

A Subsidiary of Harcourt Brace Jovanovich, Publishers

ACADEMIC PRESS, INC.
111 Fifth Avenue, New York, New York 10003

United Kingdom Edition published by
ACADEMIC PRESS, INC. (LONDON) LTD.
24/28 Oval Road, London NW1

Library of Congress Cataloging in Publication Data

Symposium on Mathematical Aspects of Finite Elements in Partial Differential Equations, University of Wisconsin–Madison, 1974.
Mathematical aspects of finite elements in partial differential equations.

Bibliography: p.
Includes index.
1. Differential equations, Partial–Congresses.
2. Finite element method–Congresses. I. De Boor, Carl, ed. II. Wisconsin. University–Madison. Mathematics Research Center. III. Title.
QA374.S934 1974 515'.353 74-23389
ISBN 0-12-208350-4

PRINTED IN THE UNITED STATES OF AMERICA

Contents

CONTENTS

Preface

This book contains the Proceedings of the Symposium on Mathematical Aspects of Finite Elements in Partial Differential Equations held in Madison, Wisconsin, April 1-3, 1974, under the auspices of the Mathematics Research Center, University of Wisconsin-Madison, and with financial support from the United States Army under Contract No. DA-31-124-ARO-D-462. The fourteen speakers had been invited to address mathematical questions raised by the use of finite elements in the numerical solution of partial differential equations. Due to the necessity of meeting the publication schedule, the text of Professor Garth Baker's fine lecture has not been included in this volume. The sessions were chaired by:

Professor G. Birkhoff, *Harvard University*
Professor H. H. Rachford, Jr., *Rice University*
Professor A. H. Schatz, *Cornell University*
Professor R. S. Varga, *Kent State University*
Professor B. Wendroff, *Los Alamos Scientific Laboratory*.

The program committee consisted of Professors J. Bramble, J. Douglas, Jr., J. Nitsche, and B. Noble, with the editor as chairman. The many organizational details were handled by the experienced symposium secretary, Mrs. Gladys G. Moran, with intelligent dedication and inexhaustible enthusiasm. The preparation of the manuscripts for publication was in the able hands of Mrs. Dorothy M. Bowar.

Higher Order Local Accuracy by Averaging in the Finite Element Method

J. H. BRAMBLE AND A. H. SCHATZ

1. Introduction.

Let Ω be a bounded domain in Euclidean N-space $\mathbb{R}^N$ with smooth boundary $\partial\Omega$. For u a real valued function defined on Ω we shall consider the uniformly elliptic second order differential operator

$$Lu = -\sum_{i,j=1}^{N} \frac{\partial}{\partial x_j}\left(a_{ij}\frac{\partial u}{\partial x_i}\right) + cu$$

where a_{ij} and c are assumed smooth. The associated bilinear form is given by

$$B(v,w) = \sum_{i,j=1}^{N} \int_\Omega a_{ij} \frac{\partial v}{\partial x_i}\frac{\partial w}{\partial x_j}\, dx + \int_\Omega c\, v\, w\, dx\, .$$

For $f \in \mathbb{L}_2(\Omega)$, a weak solution u of

$$(1.1) \qquad Lu = f \quad \text{in} \quad \Omega$$

satisfies

$$(1.2) \qquad B(u,\varphi) = \int_\Omega f\, \varphi\, dx \equiv (f,\varphi)$$

for all functions φ which are continuous and piecewise continuously differentiable in Ω and which vanish near $\partial\Omega$. We can associate with (1.1) various kinds of boundary conditions. Examples of these are

$$\text{a)} \qquad u = 0 \quad \text{on} \quad \partial\Omega$$

or

$$\text{b)} \qquad \frac{\partial u}{\partial \nu} = 0 \quad \text{on} \quad \partial\Omega$$

or

$$\text{c)} \qquad \frac{\partial u}{\partial \nu} + u = 0 \quad \text{on} \quad \partial\Omega ,$$

where

$$\frac{\partial u}{\partial \nu} = \sum_{i,j=1}^{N} a_{ij} \frac{\partial u}{\partial x_j} n_i .$$

Here n_i is the component in the direction x_i of the outward normal to $\partial\Omega$ and $\frac{\partial}{\partial \nu}$ is called the co-normal derivative.

Let S_h be a linear space of "finite elements" and $u_h \in S_h$ an approximate solution to the boundary value problem (1.1) with a), b) or c) satisfied. For many different finite element methods proposed for these problems, the interior equations are as follows:

$$(1.3) \qquad B(u_h, \varphi) = (f, \varphi)$$

for all $\varphi \in S_h$ which vanish near $\partial\Omega$. In many different specific methods which have (1.3) in common, estimates for norms of the error $u = u_h$ in Sobolev spaces on all of Ω are well known. (For a summary of such results cf. [2].) Interior estimates in Sobolev norms for $u - u_h$ only satisfying (1.3) were given in [5] and in maximum-norm in [1].

Here we shall consider, instead of u_h as an approximation to u, certain "local averages of u_h". These, as will be seen subsequently, are formed by computing $K_h * u_h$, where K_h is a fixed function and $*$ denotes convolution. As we shall see, the function K_h has the following properties:

i) K_h has small support;

ii) K_h is "independent" of the specific choice of S_h or the operator L;

iii) $K_h * u_h$ is easily computable from u_h;

iv) $K_h * u_h$ approximates u to higher order than does u_h.

In the remainder of this paper, we shall describe the class of finite element subspaces which we consider, state our main result on the accuracy of $K_h * u_h$ and discuss the significance of the result. A local estimate in a problem with certain non-smooth data is presented as a consequence of the main results. Finally, we present a numerical example.

2. Notation, subspaces and the construction of K_h.

We shall denote by $C^s(\Omega)$, $s = 0, 1, 2, \ldots,$ the space of functions defined on Ω with uniformly continuous partial derivatives of order up to and including s on Ω. For $v \in C^s(\Omega)$ we set

$$|v|_{s,\Omega} = \sup_{\substack{|\alpha| \leq s \\ x \in \Omega}} |D^\alpha v(x)| ,$$

where α is a multi-index, $|\alpha| = \sum_{i=1}^{N} \alpha_i$ and $D^\alpha = (\partial/\partial x_1)^{\alpha_1} \ldots (\partial/\partial x_N)^{\alpha_N}$. For s real we define $H^s(\Omega)$, the Sobolev space with index s and for $v \in H^s(\Omega)$, $\|v\|_{s,\Omega}$ will denote its norm (cf. [4]). For example, for $s = 0, 1, 2, \ldots$, $\|\cdot\|_{s,\Omega}$ is given by

$$\|v\|_{s,\Omega} = \Big(\sum_{|\alpha| \leq s} \int_\Omega |D^\alpha v|^2 dx \Big)^{\frac{1}{2}} .$$

For s a positive non-integer, $H^s(\Omega)$ may be defined by interpolating between successive integers and for $s < 0$ by duality (cf. [4]).

The one-parameter family of spaces $\{S_h\}$, $0 < h \leq 1$ which we shall consider will be assumed to have the following properties:

1) For each h, $S_h \subset H^1(\Omega)$ and S_h is finite dimensional.

2) For $x \in \Omega_1 \subset\subset \Omega$ and $U \in S_h$ there are functions $\varphi_1, \ldots, \varphi_k$, which are piecewise polynomial with compact support such that
$$U(x) = \sum_{j=1}^{k} \sum_{\alpha \in \mathbb{Z}^N} a_\alpha^j \varphi_j(h^{-1}x - \alpha) .$$
Here $\Omega_1 \subset\subset \Omega$ means $\bar{\Omega}_1 \subset \Omega$, a_α^j are real coefficients and $\mathbb{Z}^N$ are the multi-integers. (This property may be described as an interior translation invariance property.)

3) For some positive integer r there is a constant C such that for $v \in H^s(\Omega)$, $1 \leq s \leq r$,
$$\inf_{\varphi \in S_h} \left(\|v-\varphi\|_{0,\varphi} + h\|v-\varphi\|_{1,\Omega} \right) \leq Ch^s \|v\|_{s,\Omega}.$$

4) Let $\Omega_1 \subset\subset \Omega$ and let w be an infinitely differentiable function with support in Ω_1. There is a constant C such that for $v \in S_h$
$$\inf_{\substack{\varphi \in S_h \\ \operatorname{supp}\varphi \subset \Omega}} \|wv - \varphi\|_{1,\Omega_1} \leq Ch\|v\|_{1,\Omega} .$$

It was shown in [1] that subspaces consisting of tensor products of one dimensional splines on a uniform mesh have all the requisite properties. Also, in [5], it was demonstrated that the triangular element subspaces in $\mathbb{R}^2$ defined in [3] are examples satisfying the above four conditions provided the triangulation is uniform. We emphasize that the uniformity is a condition which is only required locally. Thus we see that many of the finite element subspaces which are discussed in the literature satisfy the above conditions.

In order to define the function K_h we shall need to introduce the so-called smooth splines. In fact we shall choose K_h to be a particular smooth spline depending on the index r associated with the subspace S_h.

For t real define

$$\chi(t) = \begin{cases} 1, & |t| \leq \frac{1}{2} \\ 0, & |t| > \frac{1}{2} \end{cases}$$

and for $x \in \mathbb{R}^N$ set

$$\psi(x) = \prod_{j=1}^{N} \chi(x_j) .$$

For ℓ a positive integer set

$$\psi^{(\ell)}(x) = (\psi * \ldots * \psi)(x), \; (\ell - 1) \text{ times.}$$

The function $\psi^{(\ell)}$ is just the N-dimensional B-spline of Schoenberg [6]. The space of smooth splines of order ℓ on a mesh of width h consists of all functions of the form

$$U(x) = \sum_{\alpha \in \mathbb{Z}^N} a_\alpha \psi^{(\ell)}(h^{-1}x - \alpha) ,$$

for some coefficients a_α.

The proof of the following will be given in a forthcoming paper by the authors.

Proposition.

Let ℓ and t be two given positive integers. The smooth spline

$$K_h(x) = \sum_{\alpha \in \mathbb{Z}^N} k_\alpha \psi^{(\ell)}(h^{-1}x - \alpha)$$

may be chosen so that

a) $k_\alpha = 0$ when $|\alpha_j| > t-1$ for some j;

b) for $\Omega_0 \subset\subset \Omega_1$ and $v \in C^{2t}(\Omega_1)$ there is a constant C such that

$$|v - K_h * v|_{0,\Omega_0} \leq Ch^{2t} |v|_{2t,\Omega_1} \quad ;$$

and

c) for $v \in H^{2t}(\Omega_1)$ there is a constant C such that

$$\|v - K_h * v\|_{0,\Omega_0} \leq Ch^{2t} \|v\|_{2t,\Omega_1} \; .$$

This function K_h is the aforementioned function in terms of which our local averages will be defined.

Let us denote by $\mathring{S}_h(\Omega_1)$ the subspace of S_h whose elements consist of functions in S_h with support in Ω_1. Let us suppose now that $u_h \in S_h$ satisfies

$$B(u-u_h, \varphi) = 0$$

for all $\varphi \in \mathring{S}_h(\Omega_1)$. These equations are the same as (1.3) provided $Lu = f$. We now state our main result. The proof of this will be given in a forthcoming paper by the authors.

Theorem 1.

Let $\Omega_0 \subset\subset \Omega_1 \subset\subset \Omega$ and p be an arbitrary but fixed real number. Let $\ell = r-2$ and $t = r-1$. Then there is a constant C such that for $u \in H^{2r-2}(\Omega_1)$

$$\|u - K_h * u_h\|_{0,\Omega_0} \leq C\{h^{2r-2} \|u\|_{2r-2,\Omega_1} + \|u-u_h\|_{-p,\Omega_1}\}$$

and, for $u \in H^s(\Omega_1)$ with $s = 2r-2 + \lfloor N/2 \rfloor + 1$,

$$|u-K_h*u_h|_{0,\Omega_0} \leq C\{h^{2r-2}\|u\|_{s,\Omega_1} + \|u-u_h\|_{-p,\Omega_1}\}.$$

Let us consider some examples in order to illustrate the meaning of this result. Let $S_h^{(r)}$ consist of the smooth splines of order r (restricted to Ω). Then for $\varphi \in S_h^{(r)}$ we see that $K_h*\varphi \in S_h^{(2r-2)}$. It is known that for some $U_h \in S_h^{(2r-2)}$

$$u-U_h = O(h^{2r-2}) \quad \text{as} \quad h \to 0$$

for smooth u. The theorem says that in fact the special smooth spline K_h*u_h is such that

$$u - K_h*u_h = O(h^{2r-2}) \text{ as } h \to 0$$

in the interior of Ω provided that for some p

$$\|u-u_h\|_{-p,\Omega_1} = O(h^{2r-2}). \tag{2.1}$$

Let us consider a case where (2.1) is known to be true for $p = r-2$. Let $c > 0$ in the operator L. Then the solution of the Neumann problem

$$\begin{aligned} Lu &= f \quad \text{in} \quad \Omega \\ \frac{\partial u}{\partial \nu} &= 0 \quad \text{on} \quad \partial\Omega \end{aligned} \tag{2.2}$$

satisfies

$$B(u,\varphi) = (f,\varphi)$$

for all $\varphi \in H^1(\Omega)$. The solution $u_h \in S_h^{(r)}$ of

$$B(u_h,\varphi) = (f,\varphi)$$

for all $\varphi \in S_h^{(r)}$ exists and is unique and satisfies the estimate

$$\|u-u_h\|_{2-r,\Omega_1} \leq \|u-u_h\|_{2-r,\Omega} \leq Ch^{2r-2}\|u\|_{r,\Omega}$$

as may be found in [2]. More particularly, if we choose $r = 4$ (cubic splines), we obtain for $N = 2$

$$|u - K_h * u_h|_{0,\Omega_0} \leq Ch^6\{\|u\|_{8,\Omega_1} + \|u\|_{4,\Omega}\}.$$

Hence if u is locally smooth and globally less smooth ($u \in H^8(\Omega_1) \cap H^4(\Omega)$) we see that $K_h * u_h$ is a local 6th order approximation to u while u_h itself is in general only a 4th order approximation to u.

We emphasize that S_h need not be chosen to be the smooth splines (locally) but may be chosen from a much larger class of approximating subspaces of $H^1(\Omega)$.

3. The calculation of $K_h * u_h$ at mesh points.

Although $K_h * u_h$ can be calculated at arbitrary points, we shall write down explicitly the expression for it at mesh points.

Generally if u_h has the form

$$u_h(x) = \sum_{j=1}^{k} \sum_{\alpha \in \mathbb{Z}^N} a_\alpha^j \varphi_j(h^{-1}x-\alpha)$$

in Ω_1 then, for h sufficiently small and $\gamma \in \mathbb{Z}^N$,

$$(K_h * u_h)(h\gamma) = h^N \sum_{j=1}^{k} \sum_{\alpha \in \mathbb{Z}^N} a_{\gamma-\alpha}^j d_\alpha^j$$

where

$$d_\alpha^j = \sum_{\beta \in \mathbb{Z}^N} k_\beta (\psi^{(r-2)} * \varphi_j)(\alpha-\beta).$$

The values of d_α^j may be computed a priori, and once the values of a_α^j are known it is an easy matter to form $(K_h * u_h)(h\gamma)$.

In the case of smooth splines, $k = 1$ and $\varphi_1 = \psi^{(r)}$. Then $d_\alpha \equiv d_\alpha^1$, hence

$$(K_h * u_h)(h\gamma) = h^N \sum_{\alpha \in \mathbb{Z}^N} a_{\gamma-\alpha} d_\alpha$$

with

$$d_\alpha = \sum_{\beta \in \mathbb{Z}^N} k_\beta \psi^{(2r-2}(\alpha-\beta) .$$

Note that $d_\alpha = 0$ whenever $|\alpha_j|$ is large, for some j. Hence the above sum is finite.

The quantities k_α may be computed as follows. Let k_j' be the coefficients in the case $N=1$ and we take $k_j' = k_{-j}'$. Then

$$k_\alpha = \prod_{j=1}^{N} k'_{\alpha_j}$$

so that

$$K_h(x) = \prod_{j=1}^{N} \left[\sum_{n=-(t-1)}^{t-1} k_n' \psi_1^{(\ell)}(h^{-1}x_j - n) \right]$$

where $\psi_1^{(\ell)}$ corresponds to the case $N = 1$. We shall list below some tables of values of k_n' corresponding to various values of ℓ and t.

Similarly, let d_n' correspond to the case $N = 1$. Then

$$d_\alpha = \prod_{j=1}^{N} d'_{\alpha_j} = \prod_{j=1}^{N} \left[\sum_{n=-(t-1)}^{t-1} k_n' \psi_1^{(2r-2)}(\alpha_j - n) \right] .$$

We shall similarly list some values of d_j' for various indices ℓ and t.

Table 1: k_j', $\ell = r-2$, $t = r-1$

j\r	3	4	5	6
0	13/12	37/30	346517/241920	76691/45360
1	-1/24	-23/180	-81329/322560	-48061/113400
2		1/90	6337/161280	20701/226800
3			-3229/967680	-1573/113400
4				479/453600

Table 2: k_j', $\ell = t = r$

j\r	1	2	3	4
0	1	7/6	437/320	12223/7560
1		-1/12	-97/480	-919/2520
2			37/1920	311/5040
3				-41/7560

Table 3: d'_j

j\r	3	4	5	6
0	51/72	673/1080	$\frac{33055739}{58060800}$	$\frac{967,356,037}{1,828,915,200}$
1	11/72	4283/21600	$\frac{3589969}{16257024}$	$\frac{3,841,481,473}{16,460,236,800}$
2	-1/144	-61/5400	$\frac{-12162977}{1625702400}$	$\frac{31,253,191}{82,301,184,000}$
3		29/21600	$\frac{4795283}{2438553600}$	$\frac{48,179,483}{27,433,728,000}$
4		1/10800	$\frac{26273}{270950400}$	$\frac{89711}{514,382,400}$
5			$\frac{-58243}{812851200}$	$\frac{-2905789}{16,460,236,800}$
6			$\frac{-3229}{4877107200}$	$\frac{25867}{1,097,349,120}$
7				$\frac{117083}{82,301,184,000}$
8				479/164,602,368,000

4. An application to non-smooth f.

Let us suppose that u is the weak solution of the Neumann problem with $f \in \mathbb{L}_1(\Omega)$. The function u is well defined but is not in general in $H^1(\Omega)$. However $u_h \in S_h$ given by

$$B(u_h, \varphi) = (f, \varphi)$$

for all $\varphi \in S_h$ is well defined. We shall compute first a modified u_h as an approximation to u. Let us suppose that $\Omega_0' \subset\subset \Omega_0$ and that $\bar{\Omega}_0 \cap \bar{\Omega}_0'$ is empty. Assume further that $\mathrm{supp}(f) \subset \Omega_0'$.

Define $\tilde{u}_h \in S_h$ by

$$B(\tilde{u}_h, \varphi) = (K_h * f, \varphi)$$

for all $\varphi \in S_h$. We are now able to conclude from our general theorem that

$$|u - K_h * \tilde{u}_h|_{0,\Omega_0} \leq C h^{2r-2} \|f\|_{\mathbb{L}_1(\Omega)} .$$

Hence if f is "rough" in the interior of Ω and we want to approximate u away from the singularities of f then we may first "smooth" f and then smooth the corresponding approximation $\tilde{u}_h$. This is an example of high order local convergence of approximations to a globally non-smooth solution of a boundary problem.

5. The $\mathbb{L}_2$-projection.

The main result of the foregoing analysis is easily applied to the $\mathbb{L}_2$-projection. Although not of interest with regard to approximating solutions to differential equations, this is of independent interest in approximation theory. Let $u_h \in S_h$ satisfy

$$(u - u_h, \varphi) = 0$$

for all $\varphi \in S_h$. We may prove the following:

Theorem 2.

Let $\Omega_0 \subset\subset \Omega_1 \subset\subset \Omega$, $\ell = t = r$ and set $s = 2r + \lfloor N/2 \rfloor + 1$. Then there is a constant C such that, for $u \in H^s(\Omega_1) \cap \mathbb{L}_2(\Omega)$,

$$|u - K_h * u_h|_{0,\Omega_0} \leq Ch^{2r}\{\|u\|_{s,\Omega_1} + \|u\|_{0,\Omega}\}.$$

In this case we have carried out the following numerical example:

$$N = 1, \quad \Omega = (0,1), \quad \text{and} \quad u(x) = e^{-x}.$$

S_h consists of piecewise linear functions on a uniform subdivision of $(0,1)$ of length h. Thus $r = 2$. All the following numbers are correct to the figures given.

$u(\frac{1}{2}) = .6065306597$

h	$u_h(\frac{1}{2})$	$(K_h * u_h)(\frac{1}{2})$	$(u - K_h * u_h)(\frac{1}{2})$
1/16	.60633	.60653055	1×10^{-7}
1/32	.60648	.606530653	6.4×10^{-9}
1/64	.606518	.6065306593	4×10^{-10}

Theoretical power of h: 4.0

Computed power of h: 4.00038 .

References.

1. J. H. Bramble, J. A. Nitsche and A. H. Schatz, Interior maximum-norm estimates for Ritz-Galerkin methods, (in print).

2. J. H. Bramble and J. E. Osborn, Rate of convergence estimates for nonselfadjoint eigenvalue approximations, Math. Comp. 27 (1973), pp. 525-549.

3. J. H. Bramble and M. Zlámal, Triangular elements in the finite element method, Math. Comp. 24 (1970), pp. 809-820.

4. J. L. Lions and E. Magenes, Problèmes aux Limites non Homogènes et Applications, Vol. I, Travaux et Recherches Mathematiques no. 17, Dunod, Paris, 1968.

5. J. A. Nitsche and A. H. Schatz, Interior estimates for Ritz-Galerkin methods, Math. Comp. (in print).

6. I. J. Schoenberg, Contributions to the problem of approximation of equidistant data by analytic functions, part A, Quart. Appl. Math., 4 (1946), pp. 45-99, 112-141.

Department of Mathematics
Cornell University
Ithaca, New York 14850

Convergence of Nonconforming Methods

J.A. NITSCHE

0. Introduction

In connection with the numerical treatment of problems in statics or elastomechanics, i.e. of certain boundary value problems, non-conforming elements occurred originally as elements of approximating function spaces which do not have the required regularity. For instance, in the theory of plates, the principle of minimal energy gives rise to a variational problem quadratic in the second derivatives of the displacement. Then functions used as approximations which are not twice differentiable are non-conforming. If one treats a membrane problem, i.e. a variational problem which involves only first order derivatives, these trial functions possibly could be 'conforming' or have the then needed regularity. This means that conformity or non-conformity of approximating subspaces is intimately connected with a variational problem: we will speak of a non-conforming method if the approximation space is not contained in the domain of definition of the underlying variational principle. The importance of such methods is due to the fact that it is often difficult to construct subspaces of conforming elements in a numerically stable way.

In Section I, we consider as models the Dirichlet problem for the potential and bipotential operator and discuss, illustrated by several examples, the question of non-conforming elements using the classical Ritz- and Least-squares-method.

In the non-conforming case, the variational principle has to be extended to the given approximation spaces. The mathematical formulation and the corresponding error analysis is carried out in Section II. After some remarks concerning the finite element subspaces we will work with in Section III, we give in IV a first application: the non-conformity is caused by the violation of the boundary conditions. Typical examples in this respect are the penalty method of Babuška [4] and the modified least-squares-method of Bramble-Schatz [1] which are treated with the tools provided in Section II. The next application in Section V treats the case of 'nearly' Dirichlet boundary conditions, i.e. approximating subspaces are used which fulfill the boundary conditions only approximately in a certain sense.

The typical non-conforming case in which the approximation spaces do not have the required differentiability is discussed in Section VI. We consider in detail the use of the 'Adini-Clough-Melosh-spaces' for the numerical solution of the bipotential equation.

With respect to this section, we refer to the work of Ciarlet [1]. The concept of Section II provides an improvement of his error estimates in the generalized 'energy'-norm as well as the implementation of duality arguments in order to get estimates of the error in lower norms, for instance of the function itself.

Finally, Section VII is devoted to fields of approximation spaces: different subspaces are used to serve as approximation spaces for the functions, the first derivatives etc. In this context we refer to Oden [1]. We discuss in detail a modified Ritz-method for the bipotential equation using only approximation spaces for the functions and its first derivatives, which may be only one time differentiable.

For more practically oriented discussions of non-conforming elements, we mention the book of Zienkiewicz [1]. Also, for the sake of completeness, we refer to the work of Irons et al [1]. A thorough analysis of the 'patch-test' introduced by him would be beyond the scope of the present paper. We refer to the comments in the paper of Ciarlet [1].

I. Variational Principles for Linear Equations

In a (real) Hilbert space H let there be given the linear equation

(1) $$Au = f$$

with the following conditions on A:

(i) A positive definite, self-adjoint,

(ii) A^{-1} compact.

In order to characterize the solution of (1) by a variational principle we will consider a linear subspace H_1 of H and in H_1

(i) a symmetric and positive bilinear form $a(\cdot,\cdot)$,

(ii) a linear form $L(\cdot)$.

Then the solution u of (1) will be the solution of

(2) $$F(v) := a(v,v) - 2L(v) \Rightarrow \text{Min.} \qquad \text{for } v \in H_1$$

or of

(3) $$a(u,v) = L(v) \qquad \text{for } v \in H_1$$

if the following conditions are fulfilled:

(i) $u \in H_1$,

(ii) $F(u) \leq F(v)$ for $v \in H_1$.

We remark that the conditions imposed on a and L do not guarantee the existence of a solution of (2) respectively (3).

Examples

E.1 $$H = L_2(\Omega)$$

$$-\Delta u = f \quad \text{in } \Omega$$

$$u = 0 \quad \text{on } \partial\Omega .$$

E. 2 $\qquad H = L_2(\Omega)$

$$\Delta^2 u = f \text{ in } \Omega$$

$$u = u_n = 0 \text{ on } \partial\Omega .$$

The classical Ritz- and Least-Squares-Method can be described by:

E. 1. 1 $\qquad H_1 = \overset{\circ}{W}{}^1_2(\Omega)$

$$a(v, w) = D(v, w) = \sum \int_\Omega \int v,_i \, w,_i \, dx$$

$$L(v) = (f, v) = \int_\Omega \int f \, v \, dx .$$

E. 1. 2 $\qquad H_1 = \overset{\circ}{W}{}^1_2(\Omega) \cap W^2_2(\Omega)$

$$a(v, w) = (\Delta v, \Delta w)$$

$$L(v) = (f, \Delta v) .$$

E. 2. 1. $\qquad H_1 = \overset{\circ}{W}{}^2_2(\Omega)$

$$a(v, w) = (\Delta v, \Delta w)$$

$$L(v) = (f, v) .$$

The variational principle (2) leads quite naturally to an approximation method: For a given family $\mathfrak{S} = \{S_h \mid 0 < h \leq 1\}$ of (finite-dimensional) subspaces, we use that $u_h \in S_h$ as approximation of u which minimizes the functional F in S_h, i.e. which satisfies

(4) $$a(u_h, \chi) = L(\chi) \qquad \text{for } \chi \in S_h .$$

For this to make sense, the condition

(5) $$S_h \subseteq H_1$$

must be satisfied. This may in practice cause difficulties since (5) implies for instance:

a) in example E. 1. 1 and E. 1. 2: the functions of S_h must vanish on the boundary $\partial\Omega$,

b) in example E. 1. 2 and E. 2. 1: the functions of S_h must have partial derivatives of the second order, in particular the first derivatives have to be continuous.

These difficulties will be overcome by the following generalization: Let H' with $H_1 \subset H' \subset H$ be a linear space such that the inclusion $S_h \subseteq H'$ for the family $\mathfrak{S}$ used is easily achieved. Further let the symmetric bilinear form a' be an extension of a to H'. Then the generalized problem

$$\text{(6)} \qquad a'(u_h, \chi) = L(\chi) \qquad \text{for } \chi \in S_h$$

will - under certain mild conditions on a' - give a solution $u_h \in S_h$ which may be considered as an approximation to u. (Remark: because of $H' \subset H$, the functional L is defined in S_h).

The convergence of u_h to u will depend on the properties of $\{S_h\}$ as well as on the extension $H_1 \rightarrow H'$ on the one hand and $a \rightarrow a'$ on the other. In the case of boundary value problems, the underlying ideas can be categorized as follows:

I. Boundary Conditions

a. H' is chosen without regard for any boundary condition. Then the extension a' has to contain 'boundary'-terms in order to force the solution of (6) to have 'small' boundary values. We mention the methods

E. 1. 1. a (Babuška [4]) $H' = W_2^1(\Omega)$

$$a'(v, w) = D(v, w) + \kappa_h \langle v, w \rangle \; .$$

E. 1. 1. b (Nitsche [3]) $H' = W_2^2(\Omega)$

$$a'(v,w) = D(v,w) + \kappa_h \langle v,w\rangle - \langle v,w_n\rangle - \langle v_n,w\rangle .$$

E. 1. 2. a (Bramble-Schatz [1]) $H' = W_2^2(\Omega)$

$$a'(v,w) = (\Delta v, \Delta w) + \kappa_h \langle v,w\rangle .$$

In these examples, κ_h are properly chosen weight factors with $\kappa_h \to \infty$ for $h \to 0$.

b. H' is chosen to fulfill a weakened boundary condition, which for actual subspaces S_h is easy to fulfill: For instance, the values of the functions, of their tangential derivatives etc. have to vanish at certain points of the boundary depending on h. In this case the bilinear form a may be unchanged. We mention Nitsche [4] and Berger-Scott-Strang [1].

II. Differentiability

a. H' is chosen without regard for the regularity requirements of H_1. There are the possibilities:

1. There is - depending on the subspaces S_h finally used - a subdivision $\{\Omega_i \mid i = 1, 2, \ldots, I(h)\}$ of Ω such that the restriction of any $v' \in H'$ to Ω' coincides with the restriction to Ω_i of a $v \in H_1$.

If now $a(\cdot,\cdot)$ has the representation

$$a(v,w) = \sum a_i(v|_{\Omega_i}, w|_{\Omega_i}) \tag{7}$$

for $v, w \in H_1$ the right hand side is also defined for $v, w \in H'$. The bilinear form $a'(\cdot,\cdot)$ then has to include additional 'interface' terms in order to force the jumps of a solution of (6) or of its derivatives across the interfaces $\partial\Omega_i \cap \partial\Omega_j$ to become 'small'. We mention

E. 2. 1. a (Babuška-Zlámal [1])

$$\bar{\Omega} = \overline{\bigcup \Omega_i} \quad \text{with} \quad \Omega_i \cap \Omega_k = \phi ,$$

$$H' = \overset{\circ}{W}{}^1_2 \cap W^2_2(\Omega_1) \cap \ldots \cap W^2_2(\Omega_I)$$

$$a'(v,w) = \sum \iint_{\Omega_i} \Delta v \Delta w + \kappa_h \sum \int_{\partial\Omega_i \cap \partial\Omega_j} [v,_n][w,_n] .$$

Here $[\Phi,_n]$ denotes the jump of the normal derivative of the function Φ .

2. In addition to a space H'_0, further spaces H'_i, H'_{ij} etc. are chosen as auxiliary spaces for the first, second derivatives etc. The space H' is then the product of H'_0 , H'_i etc. Any derivative Dv occuring in $a(\cdot, \cdot)$ is then replaced by an element of the corresponding space H'_D , thus forming one part of the extended form $a'(\cdot, \cdot)$. The other part consists of terms connecting the elements of H'_0 with those of H'_i etc. such that if $u' = (u'_0, u'_i, \ldots) \in H'$ is the solution of (6), the components u'_i will be 'nearly' the derivative $u'_{0,i}$ etc.

E. 1. 1. c

$$H' = \overset{\circ}{W}{}^1_2(\Omega) \times \prod_0^n L_2(\Omega) ,$$

$$a'(v,w) = \sum (v_i, w_i) + \kappa_h \sum (v_{0,i} - v_i, w_{0,i} - w_i) .$$

E. 1. 2. b

$$H' = \overset{\circ}{W}{}^1_2(\Omega) \times \prod_0^n W^1_2(\Omega) ,$$

$$a'(v,w) = (\sum v_{i,i}, \sum w_{j,j}) + \kappa_h \sum (v_{0,i} - v_i, w_{0,i} - w_i) +$$

$$+ \tilde{\kappa}_h \sum (v_{i,j} - v_{j,i}, w_{i,j} - w_{j,i}),$$

$$L'(v) = (f, \sum v_{i,i}) .$$

E. 2. 1. b

$$H' = \prod_0^n \overset{\circ}{W}{}_2^1(\Omega)$$

$a'(\cdot,\cdot)$ is the same as in E. 1. 2. b but the linear form now is

$$L'(v) = L(v) = (f, v_0)\,.$$

In view of the identity

$$\iint u,_{ij}\; v,_{ij} = \iint u,_{ii}\; v,_{jj}$$

for $u \in \overset{\circ}{W}{}_2^2(\Omega)$ and $v \in W_2^3(\Omega)$, we have for $u, v \in \overset{\circ}{W}{}_2^2(\Omega) \cap W_2^3(\Omega)$

$$\iint \Delta u\; \Delta v = \sum \iint u,_{ij}\; v,_{ij}\,.$$

Therefore the principle described in E. 2. 1. b may be changed to

E. 2. 1. c

$$H' = \prod_0^n \overset{\circ}{W}{}_2^1(\Omega)$$

$$a'(v, w) = \sum (v_{i,j}, w_{i,j}) + \kappa_h \sum (v_{0,i} - v_i, w_{0,i} - w_i)$$

$$L(v) = (f, v_0)\,.$$

b. H' is chosen to fulfill the required regularity in a weakened form. In practice, similar to the situation discussed in II. a, the elements of H' will be sufficiently smooth in subdomains Ω_i and the jumps of certain derivatives across $\partial\Omega_i \cap \partial\Omega_j$ are 'small'. In this case the form $a'(\cdot,\cdot)$ may be taken from (7).

II. Non-Conforming Methods, Error Analysis

In this section we will discuss the behavior of the difference $u - u_h$ of the solutions of the two problems

$$a(u, v) = L(v) \qquad \text{for } v \in H_1 \tag{1}$$

$$a'(u_h, \chi) = L(\chi) \qquad \text{for } \chi \in S_h \tag{2}$$

under the following assumptions:

(i) H_1, H' are pre-Hilbert-spaces with norms or semi-norms $\|\cdot\|_1$, $\|\cdot\|'$, the second norm possibly depending on a parameter h with $0 < h \leq 1$. The inclusions $H_1 \subset H' \subset H$ are continuous.

(ii) $a(\cdot,\cdot)$, $a'(\cdot,\cdot)$ are symmetric, positive and bounded bilinear forms on H_1 respectively H' with $a \leq a'$, i.e., $a'(v, w) = a(v, w)$ for $v, w \in H_1$.

(iii) $L(\cdot)$ is a bounded linear functional on H, or at least on H', hence also on H_1.

(iv) $\mathfrak{S} = \{S_h \mid 0 < h \leq 1\}$ is a family of finite dimensional subspaces of H'.

(v) The problems (1) and (2) have unique solutions.

The equation (2) is a finite system of linear equations for $u_h \in S_h$. A necessary and sufficient condition for uniqueness is: for any $\varphi \in S_h$ there is a $\psi \in S_h$ with $a'(\varphi, \psi) \neq 0$. Therefore,

$$|||\varphi||| = \sup \{a'(\varphi, \psi) \mid \psi \in S_h \wedge \|\psi\|' \leq 1\} \tag{3}$$

defines a second norm in S_h. Since S_h is finite-dimensional, $|||\cdot|||$ and $\|\cdot\|'$ are equivalent on S_h. With an

appropriate constant c_h, we have therefore

(4) $\qquad \|\varphi\|' \leq c_h \sup \{a'(\varphi,\psi) \mid \psi \in S_h \wedge \|\psi\|' \leq 1\}$

for $\varphi \in S_h$.

Assumption (ii) guarantees the existence of a constant C_h so that

(5) $\qquad |a'(v,w)| \leq C_h \|v\|' \|w\|' \qquad$ for $v, w \in H'$.

In the following,

(6) $$e = e_h = u - u_h$$

will denote the error of the approximation method (2),

(7) $$\varepsilon = \varepsilon_h = u - U_h$$

will denote the error in some approximation $U_h \in S_h$ to u (to be chosen later). Then

(8) $$\Phi = e - \varepsilon = U_h - u_h \in S_h .$$

Subtraction of (2) from (1) leads to

(9) $\qquad a'(e,\chi) = N(u,\chi) = a'(u,\chi) - L(\chi) \qquad$ for $\chi \in S_h$,

where $N(v,w) := a'(v,w) - L(w)$ for $v \in H_1$ and $w \in H'$. The condition

(10) $\qquad N(u,\chi) = 0 \qquad$ for $u \in H_1 \wedge \chi \in S_h$

can be interpreted as a 'projection'-condition: We consider the mapping $P_h : u \mapsto u_h = P_h u$. In case $u = P_h U \in H_1 \cap S_h$ we have $e = (P_h - P_h^2)U \in S_h$ and $a'(e,\chi) = 0$. (4) then gives $e = 0$.

With (8) we get from (9)

$$a'(\Phi, \chi) = -a'(\varepsilon, \chi) + N(u, \chi) . \tag{11}$$

Now we define

$$\gamma_h(u) = \inf\{ \|u - \chi\|' \mid \chi \in S_h \} , \tag{12}$$

$$\Gamma_h(u) = \sup\{ |N(u, \chi)| \mid \chi \in S_h \wedge \|\chi\|' \leq 1\} . \tag{13}$$

U_h respectively ε then may be chosen so that $\|\varepsilon\|' \leq \gamma_h(u)$ and we get

$$|a'(\varepsilon, \chi)| \leq C_h \gamma_h(u) \|\chi\|' . \tag{14}$$

Similarly we have

$$|N(u, \chi)| \leq \Gamma_h(u) \|\chi\|' . \tag{15}$$

Therefore (4) combined with (11) gives

$$\|\Phi\|' \leq c_h \{C_h \gamma_h(u) + \Gamma_h(u)\} \tag{16}$$

or

$$\|e\|' \leq (1 + c_h C_h)\gamma_h(u) + c_h \Gamma_h(u) . \tag{17}$$

With the help of a duality argument we are able to derive error estimates in H . We choose an invertible operator $\tilde{A}'$ with

$$D(\tilde{A}') \subseteq H' , \quad R(\tilde{A}') = H \tag{18}$$

and define

$$\tilde{N}(v, w) = a'(v, w) - (v, \tilde{A}'w) \quad \text{for} \quad v \in H' , \ w \in D(\tilde{A}'). \tag{19}$$

The choice $v = e$, $\tilde{A}'w = e$ gives

(20) $$\|e\|^2 = a'(e,w) - \tilde{N}(e,w)$$

and because of (9) with an arbitrary $\chi \in S_h$

(21) $$\|e\|^2 = a'(e,w-\chi) + N(u,\chi) - \tilde{N}(e,w) .$$

In the applications, (21) will be an improved error estimate compared with (17) using the continuous imbedding $H' \to H$. In order to discuss this in a general framework, we introduce the linear space

(22) $$H'' = \{v \mid v \in D(\tilde{A}') \wedge \tilde{A}'v \in H\} = \tilde{A}'^{-1}(H)$$

with the norm

(23) $$\|v\|'' = \|\tilde{A}'v\| .$$

The quantity Δ_h defined by the inequality

(24) $$|\tilde{N}(v,w)| \leq \Delta_h \|v\|' \|w\|'' \quad \text{for } v \in H' \wedge w \in H''$$

measures the 'distance' of the chosen operator $\tilde{A}'$ from the operator A' connected with the bilinear form a', i.e. of A' defined by

(25) $$a'(v,w) = (v, A'w) \quad \text{for} \quad v \in H' \wedge w \in D(A') .$$

Accepting the perturbation term $\tilde{N}$ and using only the partial duality (19) instead of (25) provides a greater flexibility in the applications.

With (24) we find the estimate

(26) $$|\tilde{N}(e,w)| \leq \Delta_h \|e\|' \|w\|'' = \Delta_h \|e\| \|e\|'$$

for the third term of (21).

Next we introduce parallel to (13)

(27) $$\Gamma'(u) = \sup \{ |N(u,v)| \mid v \in H' \wedge \|v\|' \leq 1\} .$$

If now the assumption

$$(28) \qquad H'' \subseteq H_1$$

holds true, then $w = \tilde{A}^{-1}e \in H_1$, hence $N(u,w) = 0$ and we can estimate the second term on the right hand side of (21) using assumption (ii) to get

$$(29) \qquad |N(u,\chi)| = |N(u,w-\chi)| \leq \Gamma'(u) \|w-\chi\|' .$$

Finally, we introduce

$$(30) \qquad \delta_h = \sup \{\gamma_h(v) \,|\, v \in H'' \wedge \|v\|'' \leq 1\} .$$

Then the function χ in (21) may be chosen such that

$$(31) \qquad \|w - \chi\|' \leq \delta_h \|w\|'' = \delta_h \|e\| .$$

Combining the inequalities (26) and (29) - (31) gives

$$(32) \qquad \|e\|^2 \leq C_h \|e\|' \cdot \delta_h \|e\| + \Gamma'(u)\delta_h \|e\| + \Delta_h \|e\| \, \|e\|'$$

or

$$(33) \qquad \|e\| \leq (C_h\delta_h + \Delta_h) \|e\|' + \delta_h \, \Gamma'(u)$$

There is another way of estimating the first two terms in (21) which will be discussed in Sections V and VI.

III. Finite Element Subspaces

Before analyzing approximation methods, we will make here a few remarks about the finite dimensional subspaces we will work with. We consider a family

$$\mathfrak{S} = \{S_h\}$$

of subspaces of $H = L_2(\Omega)$, with $h \in (0,1)$ or $h = h_1, h_2, \ldots$ with $h_i \to 0$.

Definition: $\mathfrak{S}$ is of approximation class $(\bar{k},\bar{r})_{\Omega}$, respectively $(\bar{k},\bar{r})_{\Omega,\partial\Omega}$, if

(i) $S_h \subseteq W_2^{\bar{k}}(\Omega)$ for $S_h \in \mathfrak{S}$, respectively also

$S_h \subseteq W_2^{\bar{k}}(\partial\Omega)$.

(ii) For any $u \in W_2^r(\Omega)$ with $r \leq \bar{r}$ there is a $\chi \in S_h$ with

(1) $$\|u - \chi\|_k \leq c\, h^{r-k} \|u\|_r \quad \text{for} \quad k \leq \text{Min}(\bar{k}, r)$$

respectively for $u \in W_2^r(\Omega) \cap W_2^r(\partial\Omega)$

(2) $$\|u - \chi\|_k + |u - \chi|_k \leq c\, h^{r-k} \{\|u\|_r + |u|_r\}.$$

Definition: $\mathfrak{S}$ is of inverse class $(\bar{s},\bar{k})_{\Omega}$, respectively $(\bar{s},\bar{k})_{\Omega,\partial\Omega}$, if

(i) $S_h \subseteq W_2^{\bar{k}}(\Omega)$ respectively also

$S_h \subseteq W_2^{\bar{k}}(\partial\Omega)$ for $S_h \in \mathfrak{S}$.

(ii) For $\chi \in S_h$ and $0 \leq k \leq \bar{k}$.

(3) $$\|\chi\|_k \leq c\, h^{s-k} \|\chi\|_s \quad \text{for} \quad 0 \leq s \leq \text{Min}(k, \bar{s})$$

respectively

(4) $$\|\chi\|_k + |\chi|_k \leq c\, h^{s-k} \{\|\chi\|_s + |\chi|_s\}.$$

Definition: $\mathfrak{S}$ has nearly-zero-boundary values of order $\bar{k}$ if

(i) $S_h \subseteq W_2^{\bar{k}}(\Omega) \cap W_2^{\bar{k}}(\partial\Omega)$ for $S_h \in \mathfrak{S}$.

(ii) for $\chi \in S_h$ and $0 \leq k \leq \bar{k}$

(5) $$|D^l\chi| \leq c\, h^{k-l+\frac{1}{2}} \|\chi\|_k \quad \text{for} \quad l < k.$$

Subspaces of functions which contain all piecewise polynomials of degree $\leq \bar{r}-1$ on a subdivision of R^n and which are continuously differentiable of order $\bar{k}-1$ are in general of approximation class $(\bar{k}, \bar{r})$.

References: Babuška [1], [2], Bramble-Zlámal [1].

In case the subdivision is regular, the above mentioned subspaces will also be of inverse class $(\bar{k}-1, \bar{k}-1)$.

References: Babuška [3], Nitsche [1], [2].

Remark: Normally only conditions (1) and (3) are considered. In view of the trace theorem, (1) implies

$$|u - \chi|_k \leq c\, h^{r-k-\frac{1}{2}} \|u\|_r \quad .$$

The relation (2) was discussed in Nitsche [4].

From the general finite element subspaces mentioned above, one can get subspaces with nearly-zero-boundary conditions in the following way:

The values of the functions, tangential and normal derivatives etc. are set to zero at a number of points of $\partial\Omega$.

References: Berger-Scott-Strang [1], Nitsche [4].

Though we will not make use of it, we want to mention one special direction in order to overcome difficulties with boundary conditions, viz. the introduction of isoparametric elements.

References: Ciarlet-Raviart [1], Zlámal [1].

IV. Application 1: Subspaces Independent of Boundary Conditions

First we discuss the method of Babuška, i. e. E. 1. 1. a, with the tools provided in Section II. In $H' = W_2^1(\Omega)$ we use the norm induced by the bilinear form a'

$$(1) \qquad \|v'\| = a'(v,v)^{\frac{1}{2}} = \{D(v,v) + \kappa_h |v|^2\}^{\frac{1}{2}}$$

which gives - see (II. 4), (II. 5) -

$$(2) \qquad c_h = C_h = 1 .$$

Further we have

$$(3) \qquad L(v) = (f,v) = (-\Delta u, v) = D(u,v) - \langle u_n, v\rangle$$

and therefore with $u \in \overset{\circ}{W}{}_2^1(\Omega) \cap W_2^2(\Omega)$ - see (II. 9) -

$$(4) \qquad N(u,v) = -\langle u_n, v\rangle \quad \text{for } v \in H'$$

and so - see (II. 13), (II. 27) -

$$(5) \qquad \Gamma_h(u) \le \Gamma'(u) \le \kappa_h^{-\frac{1}{2}} |u_n| \le c\, \kappa_h^{-\frac{1}{2}} \|u\|_2 .$$

We may define $\tilde{A}'$ by

$$(6) \qquad \tilde{A}'w = v \Leftrightarrow \begin{cases} -\Delta w = v & \text{in } \Omega \\ w = 0 & \text{in } \partial\Omega \end{cases} .$$

Then $H'' = \overset{\circ}{W}{}_2^1(\Omega) \cap W_2^2(\Omega)$ and $\|\cdot\|''$ is equivalent to the norm in $W_2^2(\Omega)$, condition (II. 28) is met. We get by integration by parts

$$(7) \qquad \tilde{N}(v,w) = \langle v, w_n\rangle .$$

A bound can be obtained by

$$(8)\qquad |\tilde{N}(v,w)| \le |v|\,|w_n|$$

$$\le c\,\kappa_h^{-\frac{1}{2}}\,\|v\|'\,\|w\|_2 \le c\,\kappa_h^{-\frac{1}{2}}\,\|v\|\,\|w\|''$$

which implies - see (II. 24) -

$$(9)\qquad \Delta_h \le c\,\kappa_h^{-\frac{1}{2}}\,.$$

Now assume the family $\mathfrak{S}$ of approximating subspaces to be of approximation class $(1,\bar{r})_{\Omega,\partial\Omega}$ with at least $\bar{r} \ge 2$. Then - see (II. 12), (II. 30) -

$$(10)\qquad \begin{aligned} \gamma_h(u) &\le c(h^{r-1} + h^r\,\kappa_h^{\frac{1}{2}})\,\|u\|_r\,, \qquad (r \le \bar{r}) \\ \delta_h(u) &\le c(h + h^2\,\kappa_h^{\frac{1}{2}})\,. \end{aligned}$$

Choosing $\kappa_h = c\,h^{-2}$ gives for instance in the case $r = 2$ the error estimates - see (II. 17), (II. 33) -

$$(11)\qquad \begin{aligned} \|e\|' &\le c\,h\,\|u\|_2 \\ \|e\| &\le c\,h^2\,\|u\|_2 \end{aligned}\quad .$$

Since

$$(12)\qquad \{D(v,v) + |v|^2\}^{\frac{1}{2}}$$

is equivalent to the norm in $W_2^1(\Omega)$ we may split the estimate (11_1) into

$$(13)\qquad \begin{aligned} \|e\|_1 &\le c\,h\,\|u\|_2\,, \\ |e| &\le c\,h^2\,\|u\|_2\,. \end{aligned}$$

The results of Babuška [4], [5] are somewhat different since only the approximation class $(1,\bar{r})_\Omega$ is assumed.

As further application we consider the method of Bramble-Schatz, i.e. example E.1.2.a. Similar to (1), we use also here

$$\|v\|' = a'(v,v)^{\frac{1}{2}} = \{\|\Delta v\|^2 + \kappa_h |v|^2\}^{\frac{1}{2}} \tag{14}$$

as norm in $H' = W_2^2(\Omega)$, with respect to which H' will not be complete. We have

$$c_h = C_h = 1 . \tag{15}$$

Simple calculation gives

$$N(u,v) = 0 \quad \text{for} \quad u \in H_1 \wedge v \in H' . \tag{16}$$

In this example we use $\tilde{A}'$ defined by

$$\tilde{A}'w = v \Leftrightarrow \begin{cases} \Delta^2 w = v & \text{in} \quad \Omega \\ w = \Delta w = 0 & \text{on} \quad \partial\Omega . \end{cases} \tag{17}$$

Then H'' is a closed subspace of $\mathring{W}_2^1(\Omega) \cap W_2^4(\Omega)$ and $\|\cdot\|''$ is equivalent to $\|\cdot\|_4$. We have

$$\tilde{N}(v,w) = -\langle v, \Delta w_n \rangle . \tag{18}$$

Let us assume $\mathfrak{S}$ to be of approximation class $(2,\bar{r})_{\Omega,\partial\Omega}$ with $\bar{r} \geq 4$. Then

$$\begin{aligned} \gamma_h(u) &\leq c\,\{h^{r-2} + h^r \kappa_h^{+\frac{1}{2}}\}\,\|u\|_r \qquad (r \leq \bar{r}) \\ \delta_h(u) &\leq c\,\{h^2 + h^4 \kappa_h^{+\frac{1}{2}}\} . \end{aligned} \tag{19}$$

Because of (16) we have

$$\Gamma_h(u) = \Gamma'(u) = 0 . \tag{20}$$

On the other hand, we get with $|\Delta w_n| \leq c\,\|w\|_4 \leq c\,\|w\|''$

and $|v| \leq \kappa_h^{-\frac{1}{2}} \|v\|'$ the bound - see (II. 24) -

(21) $$\Delta_h \leq c\, \kappa_h^{-\frac{1}{2}} \quad .$$

The choice $\kappa_h = c\, h^{-4}$ leads to the error estimates

(22) $$\begin{aligned} \|\Delta e\| &\leq c\, h^{r-2}\, \|u\|_r\,, \\ |e| &\leq c\, h^r\, \|u\|_r\,, \qquad (r \leq \bar{r}) \\ \|e\| &\leq c\, h^r\, \|u\|_r\,. \end{aligned}$$

We should mention here that the assumptions of Bramble-Schatz [1] differ. They only use approximation class $(2, \bar{r})_\Omega$ and get still the same error estimates (22).

V. Application 2: Subspaces with 'Nearly' Dirichlet Conditions

As model we consider the Dirichlet problem for Poisson's equation E. 1. In addition to certain approximation properties stated below, the family $\mathfrak{S}$ is assumed to have nearly zero boundary values of order 1 (see (III. 5)). Then we may use

(1) $$a'(v, w) = D(v, w)\,.$$

We have, similar to (IV. 4),

(2) $$N(u, \chi) = -\langle u_n, \chi\rangle \quad ,$$

which in the present case can be estimated by

(3) $$|N(u, \chi)| \leq c\, h^{3/2}\, \|u\|_2\, \|\chi\|_1 \quad .$$

Because of - see (IV. 12) -

$$(4)\qquad \begin{aligned}\|\chi\|_1 &\le c\,\{D(\chi,\chi) + |\chi|^2\}^{\frac{1}{2}} \\ &\le c\,\{D(\chi,\chi)^{\frac{1}{2}} + h^{3/2}\|\chi\|_1\}\end{aligned}$$

there is a $h_0 > 0$ such that for $h \le h_0$

$$(5)\qquad \|\chi\|_1 \le c\, D(\chi,\chi)^{\frac{1}{2}} = c\, a'(\chi,\chi)^{\frac{1}{2}} \quad \text{for } \chi \in S_h .$$

We may also choose in this example $H' = W_2^1(\Omega)$ with the semi-norm

$$(6)\qquad \|v\|' = a'(v,v)^{\frac{1}{2}}$$

and get from (3)

$$(7)\qquad \Gamma_h(u) = c\, h^{3/2}\, \|u\|_2 \ .$$

The choice (6) of the norm in H' gives - see (II. 4), (II. 5)-

$$(8)\qquad c_h = C_h = 1 \ .$$

The disadvantage is that now we have only - see (II. 27) -

$$(9)\qquad \Gamma'(u) \le |u_n|$$

and so the way of estimating the error in the norm of H given in Section II has to be modified.

If the family $\mathfrak{S}$ is of approximation class $(1,\bar{r})_{\Omega,\partial\Omega}$, we have

$$(10)\qquad \gamma_h(u) \le c\, h^{r-1}\|u\|_r \qquad (r \le \bar{r})$$

and therefore - see (II. 17) -

$$(11)\qquad \|e\|' \le c\,\{h^{r-1}\,\|u\|_r + h^{3/2}\,\|u\|_2\} \ .$$

An estimate of the error in the norm of $L_2(\partial\Omega)$ can be derived in the following way: using the notation (II. 6)-(II.8) we have because of (5)

$$
\begin{aligned}
|e| &\leq |\varepsilon| + |\Phi| \\
&\leq c\,h^r \|u\|_r + c\,h^{3/2}\, D(\Phi,\Phi)^{\frac{1}{2}} \\
&\leq c\,h^r \|u\|_r + c\,h^{3/2}\{\|\varepsilon\|_1 + \|e\|'\} \\
&\leq c\,\{h^r \|u\|_r + h^3 \|u\|_2\}\,.
\end{aligned} \tag{12}
$$

Combined with this, the estimate (11) gives

$$\|e\|_1 \leq c\,\{h^{r-1}\|u\|_r + h^{3/2}\|u\|_2\}\,. \tag{13}$$

Using the operator $\tilde{A}'$ defined by (IV.6) gives

$$\tilde{N}(v,w) = \langle v, w_n \rangle \tag{14}$$

and so with $v = e$, $w = \tilde{A}'^{-1} e$,

$$|\tilde{N}(v,w)| \leq \|w\|_2\, |e| \leq c\,\|e\|\{h^r\|u\|_r + h^3\|u\|_2\}\,. \tag{15}$$

This will be the estimate of the third term in (II.21). The essential point here is the estimate of the second term of the same formula. We have, with $\chi \in S_h$ a suitable function approximating $w \in \mathring{W}_2^1(\Omega) \cap W_2^2(\Omega)$ defined by (IV.6),

$$
\begin{aligned}
|N(u,\chi)| &= |\langle u_n, \chi \rangle| \leq |u_n|\,|w-\chi| \\
&\leq c\,h^2\|u\|_2\,\|w\|_2 \leq c\,h^2\|u\|_2\,\|e\|\,.
\end{aligned} \tag{16}
$$

On the other hand we have

$$|a'(e, w-\chi)| \leq \|e\|'\,\|w-\chi\|_1 \leq c\,h\,\|e\|\,\|e\|'\,.$$

The above inequalities combined give -see (II.21) - in case $\bar{r} \geq 2$

$$\|e\| \leq c\,h^2\|u\|_2\,.$$

This of course is the roughest estimate we can get, for more details see: Nitsche [4].

VI. Application 3: Subspaces with Reduced Differentiability

For the sake of simplicity we restrict ourselves here to an example already treated by Myoshi [1] and Ciarlet [1]:

(i) as model, Dirichlet's problem for the bipotential operator in two dimensions in the unit square $\Omega = Q$ is considered.

(ii) as typical example of non-conformity, the so-called 'Adini-Clough-Melosh' functions are taken in connection with the Ritz method.

The subspaces mentioned can be described in the following way: Let $P_{ik} = (x_i, y_k) = (hi, hk)$ with $h = 1/(N+1)$ and $0 \leq i$, $k \leq N+1$ be the grid points of a uniform mesh covering Q. With any (i,k) we associate the square

$$Q_{ik} = \{(x,y) \mid 0 < x - x_i,\ y - y_k < h\} \ . \tag{1}$$

The Adini-Clough-Melosh space S_h is then characterized by

(i) restricted to any Q_{ik}, the functions in S_h are linear combinations of a polynomial of third degree in x, y and of the two terms x^3y, xy^3,

(ii) the values of the function and its first derivatives in four adjacent squares coincide at the common grid point.

The space $\mathring{S}_h$ consists of those $\chi \in S_h$ with $\chi = \chi_x = \chi_y = 0$ at the points $P_{ik} \in \partial Q$. $\mathring{S}_h$ is of dimension $3\,N^2$, a basis may be constructed by means of the functions (see Myoshi [1])

$$\Phi^0(x,y) = 1-3x^2-xy-3y^2+2x^3+3x^2y+3xy^2+2y^3-2x^3y-2xy^3,$$

$$(2)\quad \Phi^1(x,y) = x-2x^2-xy+x^3+2x^2y-x^3y\ ,$$

$$\Phi^2(x,y) = \Phi^1(y,x)$$

for $0 \leq x, y \leq 1$ and $\Phi^\nu = 0$ for $x > 1$ or $y > 1$ combined with

$$(3)\quad \begin{aligned} \Phi^0(x,y) &= \Phi^0(-x,y) = \Phi^0(-x,-y) = \Phi^0(x,-y)\ , \\ \Phi^1(x,y) &= -\Phi^1(-x,y) = -\Phi^1(-x,-y) = \Phi^1(x,-y)\ . \end{aligned}$$

Then

$$(4)\quad \varphi^\nu_{ik}(x,y) = \Phi^\nu(h^{-1}(x-x_i),\ h^{-1}(y-y_k))$$

form a local basis in $\mathring{S}_h$, any $\chi \in \mathring{S}_h$ has the representation

$$(5)\quad \chi = \sum_{i,k=1}^{N} \xi_{ik}\varphi^0_{ik} + \eta^1_{ik}\varphi^1_{ik} + \eta^2_{ik}\varphi^2_{ik}\ .$$

$\mathring{S}_h$ is not contained in the domain of definition $H_1 = \mathring{W}^2_2(Q)$ of the Ritz-functional for the bipotential operator. But we have $\mathring{S}_h \subseteq H'$ with

$$(6)\quad H' = \mathring{W}^1_2(Q) \cap \{\bigcap_{i,k} W^2_2(Q_{ik})\}\ .$$

The simplest way of extending (see examples E. 2. 1 and E. 2. 1. c) the Ritz functional

$$(7)\quad a(v,w) = \iint \Delta v\,\Delta w = \sum_{p,q=1}^{2} \iint v,_{pq}w,_{pq} \quad \text{for } v,w \in H_1$$

from H_1 to H' is to use

$$(8)\quad a'(v,w) = \sum_{i,k} \iint_{Q_{ik}} v_{xx}w_{xx} + 2v_{xy}w_{xy} + v_{yy}w_{yy}\ .$$

On H' ,

$$\|v\|' = a'(v,v)^{\frac{1}{2}} \tag{9}$$

defines only a semi-norm, any piecewise bilinear function annihilates the '-norm. Since such a function (not identical zero) has jumps in the first derivatives at the grid points, (9) does define a norm in S_h. Direct calculation, using the representation (5), shows that the '-norm (9) is equivalent (independently of h) to

$$\|\chi\|^{\wedge 2} = \sum \{(\nabla_1 \xi_{ik} - \eta^1_{ik})^2 + (\nabla_2 \xi_{ik} - \eta^2_{ik})^2 +$$
$$+ h^2[(\nabla_1\nabla_2 \xi_{ik})^2 + (\nabla \eta^1_{ik})^2 + (\nabla \eta^2_{ik})^2]\} .$$

Here ∇_1 and ∇_2 are the first divided differences with respect to the first and second index and $(\nabla\eta)^2 = (\nabla_1\eta)^2 + (\nabla_2\eta)^2$.

For completeness we mention: in general, $\chi \in \overset{\circ}{S}_h$ will not satisfy the boundary condition $\chi_n = 0$ on ∂Q, still, the inequality

$$|\chi_n| \leq c\, h^{\frac{1}{2}} \|\chi\|' \tag{11}$$

is valid. Since the first derivatives of $\chi \in \overset{\circ}{S}_h$ or of $v \in H'$ have jumps across the grid lines, we have by integration by parts

$$\begin{aligned} N(u,v) &= a'(u,v) - (\Delta^2 u, v) \\ &= \sum \oint_{\partial Q_{ik} \cap \partial Q_{lm}} u_{,nn}[v_{,n}]ds \end{aligned} \tag{12}$$

with $[\cdot]$ denoting the jump. Using the representation (5), we get by direct calculation

$$N(u,\chi) = \sum(\nabla_2\xi_{ik} - \eta_{ik}^2)h^2 \int_0^h \alpha(t/h)\nabla_1^2 u_{xx}(x_i, y_k+t)dt + \ldots \tag{13}$$

$$= \sum(\nabla_2\xi_{ik} - \eta_{ik}^2)h^2 I_{ik} + \ldots$$

with analogous terms indicated by the dots. $\alpha(s)$ is a polynomial of degree 3 with the property

$$\int_0^1 \alpha(s)ds = 0 . \tag{14}$$

Because of the identity (in one variable)

$$\nabla_1^2 f(x_i) = h^{-2} \int_{-h}^h (h-|t|) f''(x_i+t)dt, \tag{15}$$

we have

$$I_{ik} = h^{-2}\int_{x_{i-1}}^{x_{i+1}} dx \int_{y_k}^{y_{k+1}} dy\, \alpha(h^{-1}(y-y_k))(h-|x-x_i|)\partial_x^4 u(x,y). \tag{16}$$

By Schwarz's inequality we get

$$|I_{ik}| \leq c \|u\|_{W_2^4(Q_{i-1,k} \cup Q_{ik})} \tag{17}$$

and therefore

$$\sum I_{ik}^2 \leq c \|u\|_4^2 \tag{18}$$

With the help of this, we derive from (13)

$$|N(u,\chi)| \leq c\, h^2 \|u\|_4 \|\chi\|' \tag{19}$$

or - see (II.13) -

$$\Gamma_h(u) \leq c\, h^2 \|u\|_4 . \tag{20}$$

Standard approximation arguments supply the estimate - see (II.12) -

(21) $$\gamma_h(u) \leq c\, h^2 \|u\|_4$$

which in turn gives - see (II. 17) -

(22) $$\|e\|' \leq c\, h^2 \|u\|^4 .$$

That error-estimate can be improved along the lines given in Section II. We use the operator $\tilde{A}'$ defined by

(23) $$\tilde{A}'w = v \Leftrightarrow \begin{cases} \Delta^2 w = v & \text{in } Q \\ w = w_n = 0 & \text{on } \partial Q \end{cases} .$$

Then we have - see (II. 19) - because of $H'' \subseteq H_1$ - see (II. 22), (II. 28)-

(24) $$\tilde{N}(v, w) = N(w, v) \quad \text{for } v \in H'.$$

Using (II. 28) and assumption (ii) of Section II once more, we have

(25) $$\tilde{N}(e, w) = N(w, u - u_h) = -N(w, u_h) .$$

With w defined by $\tilde{A}'w = e$, we therefore get -see (II. 21)-

(26) $$\|e\|^2 = a'(e, w - \chi) + N(u, \chi) + N(w, u_h) .$$

In Section V, we already gave a modification in deriving an error estimate in the norm of H, based on relation (26). Because of (V. 2) and (V. 14) we could have used (26) also in V. We now make use of the fact that the solution u of the original problem is contained in H'' (see (II. 22)). Let us introduce quantities δ''_h, Δ''_h in the following way:

(i) for any $v \in H''$

(27) $$\Gamma_h(v) \leq \delta''_h \|v\|'' ;$$

(ii) for any pair $v, w \in H''$ there is a $\chi \in S_h$ such that

$$\|w - \chi\|' \leq \delta_h'' \|w\|'' \tag{28}$$

$$|N(v, \chi)| \leq \Delta_h'' \|v\|'' \|w\|'' .$$

Then (II. 33) can be replaced by

$$\|e\| \leq \delta_h''(\gamma_h(u) + \Gamma_h(u)) + \Delta_h'' \|u\|'' . \tag{29}$$

We have assumed here for simplicity that $c_h = C_h = 1$. We derive (29) from (26), estimating the three terms in (26) separately. First we have, using (II. 17),

$$|a'(e, w - \chi)| \leq c \|e\|' \|w - \chi\|' \tag{30}$$

$$\leq c\, \delta_h''(\gamma_h(u) + \Gamma_h(u)) \|e\| .$$

Next, almost from the definition,

$$|N(u, \chi)| \leq \Delta_h'' \|u\|'' \|e\| . \tag{31}$$

Finally, using an appropriate $U_h \in S_h$, we can write

$$|N(w, u_h)| \leq |N(w, U_h)| + |N(w, u_h - U_h)|$$

$$\leq \Delta_h'' \|u\|'' \|e\| + \Gamma_h(w) \|u_h - U_h\|' \tag{32}$$

$$\leq \Delta_h'' \|u\|'' \|e\| + \delta_h'' \|e\| \{\|u - u_h\|' + \|u - U_h\|'\} .$$

Since the last term on the right hand side of (32) is bounded by the right hand side of (30), we have in fact (29).

Now we go back to the special method of this section. First we remember $H'' = \mathring{W}_2^2(Q) \cap W_2^4(Q)$, the ''-norm is just the norm in $W_2^4(Q)$. Next, choosing χ in (28) to be the interpolant of w, we get

$$(33)\qquad \begin{aligned} \delta''_h &\le c\,h^2 \;, \\ \Delta''_h &\le c\,h^3 \;. \end{aligned}$$

(33_1) is merely (21), (33_2) improves the estimate (19) but of course the '-norm of χ is replaced by the "-norm of w. We will not give the details, the relation (14) has to be used and a careful but purely technical analysis of the terms in (13) is necessary.

With (33) and (20), (21) we get finally from (29)

$$\|e\| \le c\,h^3\;\|u\|_4 \;.$$

VII. Application 4: The Use of Fields

In Section I we gave some examples of how one might modify respectively generalize a given variational principle by introducing a field of approximating subspaces, see E. 1. 1. c, E. 1. 2. b and E. 2. 1. b as well as E. 2. 1. c. The last two examples seem to be of most interest. In this section, we will discuss the last mentioned method.

We construct an approximation scheme for the boundary value problem

$$(1_1)\qquad \Delta^2 z = f \qquad \text{in } \Omega \,,$$

$$(1_2)\qquad z = z_n = 0 \qquad \text{on } \partial\Omega$$

in the following way: Let there be given a family of $n+1$ finite-dimensional subspaces

$$(2)\qquad S^h = \{S^h_0,\, S^h_1, \ldots, S^h_n\} \,.$$

The vector-valued function $u^h = \{u^h_i \mid i = 0, \ldots, n\}$ is chosen according to

$$(3)\qquad a'(u^h, \chi) = L(\chi) \qquad \text{for} \quad \chi \in S^h$$

with

$$a'(v,w) = \sum(v_{0,i}-v_i, w_{0,i}-w_i) + \lambda^2 \sum(v_{i,j}, w_{i,j})$$

(4)

$$L(v) = \lambda^2(f, v_0) \; .$$

Here λ will depend on h. The solution u^h of (3) is considered to be an approximation to the vector $u = \{u_i | u_0 = z \wedge u_i = z,_i\}$.

The boundary conditions (1_2) have to be imposed in a proper way. In view of the previous chapters there are the possibilities of

(i) choosing S^h without regard for boundary conditions and adding proper terms to a' - see Section IV;

(ii) imposing nearly zero boundary values of order 1 on the spaces S_i^h - see Section V.

For the sake of simplicity, we will restrict ourselves in the presentation given here to the case

(5) $$S_i^h \subseteq \overset{\circ}{W}{}_2^1(\Omega) \; .$$

By 'superposition' of the ideas of Section IV and V, the corresponding results for the two above mentioned generalizations are obtained without difficulties.

We choose in the case of (5) $H' = \prod \overset{\circ}{W}{}_2^1(\Omega)$ with the norm (easily verified to be a norm)

(6) $$\|v\|' = a'(v,v)^{\frac{1}{2}} \; .$$

So we have

(7) $$c_h = C_h = 1 \; .$$

The definition of u implies $u_{0,1} - u_i = u_{i,j} - u_{j,i} = 0$ and therefore

(8) $$N(u,\chi) = \lambda^2\{\sum(z,_{ij},\chi_{i,j}) - (\Delta^2 z,\chi_0)\}$$
$$= \lambda^2 \sum(\Delta z,_i,\chi_{0,i} - \chi_i) .$$

This gives

(9) $$|N(u,\chi)| \leq \lambda^2 \|z\|_3 \|\chi\|' .$$

Since no special assumptions on χ are used, we get

(10) $$\Gamma_h(u) \leq \Gamma'(u) \leq \lambda^2 \|z\|_3 .$$

The norm (6) in H' can be bounded by

(11) $$\|v\|' \leq c\{\|v_0\|_1 + \sum\|v_i\| + \lambda \sum \|v_i\|_1\} .$$

Now we impose approximability assumptions on S^h:

(12) $$\left.\begin{array}{l} S_0^h \text{ is of approximation class } (1,\bar{r})_\Omega \\ S_i^h \text{ is of approximation class } (1,\bar{r}-1)_\Omega \end{array}\right\} \text{with } \bar{r} \geq 3 .$$

The essential point is that the first index in the approximability assumptions is a 1, i.e., we use only subspaces of $W_2^1(\Omega)$ without a higher differentiability. Because of (11), these assumptions give

(13) $$\gamma_h(u) = \inf\|u-\chi\|' \leq c\{h^{r-1} + \lambda h^{r-2}\}\|z\|_r \quad (r \leq \bar{r}) .$$

In case of

(14) $$\lambda \leq c\, h$$

we have, in particular,

(15) $$\gamma_h(u) \leq c\, h^{r-1}\|z\|_r .$$

So we get the error estimate

$$\|e\|' \leq c\{h^{r-1}\|z\|_r + \lambda^2\|z\|_3\} . \tag{16}$$

The error $e_{i,j} = z,_{ij} - u^h_{i,j}$ of the second derivatives of the solution of the boundary value problem (1) is subsumed in $\|e\|'$ since a consequence of (16) is

$$\|z,_{ij} - u^h_{i,j}\| \leq c\{h^{r-1}\lambda^{-1}\|z\|_r + \lambda\|z\|_3\} \quad (i,j=1,\ldots,n). \tag{17}$$

In order to find an error estimate for the solution z itself, we introduce the function V defined by

$$\begin{aligned} \Delta^2 V &= e_0 &&\text{in} \quad \Omega \\ V &= V_n = 0 &&\text{on} \quad \partial\Omega \end{aligned} \tag{18}$$

and put

$$v_0 = V , \quad v_i = V,_i . \tag{19}$$

By standard a priori estimates we have $V \in \mathring{W}^2_2(\Omega) \cap W^4_2(\Omega)$ in case $e_0 \in L_2(\Omega)$ with $\|V\|_4 \leq c\ \|e_0\|$. By integration by parts we get

$$\begin{aligned} a'(e,v) &= \lambda^2 \sum (e_{i,j}, V,_{ij}) = -\lambda^2 \sum (e_i, \Delta V,_i) \\ &= \lambda^2 \sum (e_{0,1} - e_i, \Delta V,_i) + \lambda^2 (e_0, \Delta^2 V) \end{aligned} \tag{20}$$

or

$$\|e_0\|^2 = -\sum (e_{0,i} - e_i, \Delta V,_i) + \lambda^{-2}\, a'(e,v) . \tag{21}$$

The first term of the right hand side in (21) is bounded by

$$\begin{aligned} |\sum (e_{0,i} - e_i, \Delta V,_i)| &\leq c\|e\|'\ \|V\|_3 \\ &\leq c\|e_0\|\ \|e\|' . \end{aligned} \tag{22}$$

In order to get a bound for the second term, we use (II.9) and get with an arbitrary $\chi \in S^h$

$$a'(e,v) = a'(e,v-\chi) + N(u,\chi) . \tag{23}$$

In the representation (8) of the functional N, we may add and subtract $v_{0,i} = v_i = V,_i$ in the second factor in the scalar-product and then estimate

$$|N(u,\chi)| \leq \lambda^2 \|z\|_3 \{ \|V-\chi_0\|_1 + \sum \|V,_i - \chi_i\| \} . \tag{24}$$

Because of the assumptions (12) on the subspace S^h and of the regularity of V (see (18)) there is a $\chi \in S^h$ with

$$\|V-\chi_0\|_1, \|V,_i - \chi_i\| \leq c\, h^{r-1} \|V\|_r \leq c\, h^{r-1} \|e_0\| \tag{25}$$

$$(r \leq \text{Min}\,(4,\bar{r})) .$$

With the choice of χ in this manner, we get

$$|N(u,\chi)| \leq c\,\lambda^2\, h^{r-1} \|e_0\| \|z\|_3 \qquad (r \leq \text{Min}\,(4,\bar{r})). \tag{26}$$

Similarly, we get

$$\begin{aligned} |a'(e,v-\chi)| &\leq \|e\|' \|v-\chi\|' \\ &\leq c\|e\|' \{ \|V-\chi_0\|_1 + \sum \|V,_i - \chi_i\| + \lambda \sum \|V,_i - \chi_i\|_1 \} \\ &\leq c(h^{r-1} + \lambda\, h^{r-2}) \|e_0\| \|e\|' \end{aligned} \tag{27}$$

We impose here the condition (14) on λ, too. Then (21) together with the estimates shown gives

$$\|e_0\| \leq c(1 + h^{r-1}\lambda^{-2}) \|e\|' + c\, h^{r-1} \|z\|_3 \tag{28}$$

and because of (16)

$$\|e_0\| \leq c(h^{r-1} + \lambda^2) \{ h^{r-1} \lambda^{-2} \|z\|_r + \|z\|_3 \} \tag{29}$$

$$(r \leq \text{Min}\,(4,\bar{r})) .$$

Finally we turn to the error of the first derivatives. It will be simplest to get first an estimate of e_0 in the norm of $W_2^1(\Omega)$. In the present situation we use an auxiliary function V defined by (compare with (18))

$$\Delta^2 V = \Delta e_0 \quad \text{in} \quad \Omega \tag{30}$$
$$V = V_n = 0 \quad \text{on} \quad \partial\Omega .$$

Here Δe_0 has to be taken in the distributional sense. For $e_0 \in \overset{\circ}{W}{}_2^1(\Omega)$, we have $V \in \overset{\circ}{W}{}_2^2(\Omega) \cap W_2^3(\Omega)$ with $\|V\|_3 \leq c\|e_0\|_1$. Since e_0 has zero boundary values we have with $\underline{c} > 0$

$$\begin{aligned} \underline{c}\|e_0\|_1^2 &\leq D(e_0, e_0) = -(e_0, \Delta e_0) \\ &= -(e_0, \Delta^2 V) = \sum (e_{0,i}, \Delta V_{,i}) \\ &= \sum (e_{0,i} - e_i, \Delta V_{,i}) + \sum (e_i, \Delta V_{,i}) . \end{aligned} \tag{31}$$

By integration by parts we get (see (20))

$$\begin{aligned} \sum (e_i, \Delta V_{,i}) &= -\sum (e_{i,j}, V_{,ij}) \\ &= -\lambda^{-2} a'(e, v) . \end{aligned} \tag{32}$$

Therefore we may combine (compare with (24), (27)) the above estimates:

$$\begin{aligned} \underline{c}\|e_0\|_1^2 &\leq \sum (e_{0,i} - e_i, \Delta V_{,i}) - \lambda^{-2} a'(e, v-\chi) - \lambda^{-2} N(u,\chi) \\ &\leq c\|e\|' \|V\|_3 + c\lambda^{-2}\|e\|' \{\|V-\chi_0\|_1 + \\ &\quad + \sum \|V_{,i} - \chi_i\| + \lambda \sum \|V_{,i} - \chi_i\|_1\} + \\ &\quad + c\|z\|_3 \{\|V-\chi_0\|_1 + \sum \|V_{,i} - \chi_i\|_0\} \\ &\leq c\|e_0\|_1 \|e\|' \{1 + h^2\lambda^{-2}\} + c h^2 \|e_0\|_1 \|z\|_3 . \end{aligned} \tag{33}$$

With the estimate (16) of $\|e\|'$, we have

$$\|e_0\|_1 \leq c(h^2 + \lambda^2)\{(h^{r-1}\lambda^{-2})\|z\|_r + 2\|z\|_3\}. \tag{34}$$

The derivatives $u^h_{0,i}$ may be used to serve as approximations to the first derivatives of the solution z of (1). On the other hand, u^h_i may be thought of as being approximations of $z,_i$. In view of

$$\|e_i\| \leq \|e_0\|_1 + \|e\|' \tag{35}$$

and (16), the right hand side of (34) is also a bound for $\|e_i\|$. In order to illustrate the error estimates (17), (29), and (34) (35), we specialize:

a. $\bar{r} = 3$. This means the approximation chosen for the function z and the first derivatives are quadratic, respectively linear, finite elements. Then the choice $\lambda = c\,h$ gives

$$\begin{aligned} \|z-u^h_0\| &\leq c\,h^2\|z\|_3, \\ \|z,_i-u^h_{0,i}\|,\ \|z,_i-u^h_i\| &\leq c\,h^2\|z\|_3, \\ \|z,_{ij}-u^h_{i,j}\| &\leq c\,h\,\|z\|_3. \end{aligned} \tag{36}$$

b. $\bar{r} = 4$. The approximation spaces are now cubic, respectively quadratic, finite elements. Then replacing the 3-norm of z on the right hand sides of (36) by the 4-norm, we have the powers of h

1. for the choice $\lambda = c\,h$: 2, 2, 1 ,
2. for the choice $\lambda = c\,h^{3/2}$: 3, 2, 3/2 .

Appendix: Notation Used

c	:	a generic numerical constant which may be different at different places
$\mathbb{R}^n$	:	n-dimensional euclidean space
Ω	:	an open and bounded domain in $\mathbb{R}^n$
$\partial\Omega$	:	boundary of Ω, sufficiently smooth
$C^\infty(\Omega)$	:	space of functions with derivatives continuous in $\bar{\Omega}$ of arbitrary order
$C_0^\infty(\Omega)$	:	subspace of $C^\infty(\bar{\Omega})$ of functions with compact support in Ω
α	:	$= (\alpha_1, \ldots, \alpha_n)$ multi-index, $\alpha_i \geq 0$ and integer
$\lvert\alpha\rvert$	:	$= \sum \alpha_i$
D^α	:	$= \dfrac{\partial^{\lvert\alpha\rvert}}{\partial x_1^{\alpha_1} \ldots \partial x_n^{\alpha_n}}$
D^k	:	= general k-th derivative, i.e. $= D^\alpha$ for any $\lvert\alpha\rvert = k$
$\iint(\ )$	:	$\iint_\Omega (\ )dx$
$\oint(\ )$	:	$\int_{\partial\Omega} (\ )do$
(u, v)	:	$\iint uv$
$\lVert u \rVert$	:	$(u,u)^{\frac{1}{2}}$ norm in $L_2(\Omega)$

$\|\nabla^k u\|$	:	$\{\sum_{|\alpha|=k} \|D^\alpha u\|^2\}^{\frac{1}{2}}$
$\|u\|_r$	:	$\{\sum_{k \le r} \|\nabla^k u\|^2\}^{\frac{1}{2}}$
$\langle u, v\rangle$	:	$= \oint uv$
$|u|$	:	$= \langle u, u\rangle^{\frac{1}{2}}$ norm in $L_2(\partial\Omega)$
$|\nabla^k u|$ $|u|_k$	:	defined similar to $\|\nabla^k u\|$, see Bramble-Nitsche [1]
$W_2^k(\Omega)$	:	completion of $C^\infty(\bar{\Omega})$ with respect to $\|\cdot\|_k$
$\mathring{W}_2^k(\Omega)$	:	completion of $C_0^\infty(\Omega)$ with respect to $\|\cdot\|_k$
$W_2^k(\partial\Omega)$	:	completion of $C^\infty(\partial\Omega)$ with respect to $\|\cdot\|_k$

References

Aubin, J. P.

1. Approximation des espaces de distributions et des opérateurs differentiels
   Bull. Soc. Math. France Mem. 12 (1967), 1-139

Babuška, I.

1. Approximation by hill functions
   Comm. Math. Univ. Carolinea 11 (1970), 787-811
2. Approximation by hill functions II
   Techn. Note BN-708, Univ. of Maryland (1971)
3. Error bounds for the finite element method
   Numer. Math. 16 (1971), 322-333.
4. Numerical solution of boundary value problems by the perturbated variational principle
   Techn. Note BN-624, Univ. of Maryland (1969)

Babuška, I.
5. Finite element method with penalty
Techn. Note BN-710, Univ. of Maryland (1971)

Babuška, I. and M. Zlámal
1. Nonconforming elements in the finite element method
Techn. Note BN-729, Univ. of Maryland (1972)

Berger, A. E. and R. Scott and G. Strang
1. Approximate boundary conditions in the finite element method
Symposia Mathematica, Acad. Press (1972)

Bramble, J. H. and J. A. Nitsche
1. A generalized Ritz-least squares method for Dirichlet problems
SIAM J. Numer. Anal. 10 (1973), 81-93

Bramble, J. H. and A. Schatz
1. Rayleigh-Ritz-Galerkin methods for Dirichlet's problem using subspaces without boundary conditions
Comm. p. appl. Math. 23 (1970), 653-675
2. Least squares method for 2m-th order elliptic boundary value problems
Math. of Comp. 25 (1971), 1-32

Bramble, J. H. and M. Zlámal
1. Triangular elements in the finite element method
Math. of Comp. 24 (1970), 809-820

Ciarlet, P. G.
1. Conforming and non-conforming finite element methods for solving the plate problem
(to appear)

Ciarlet, P. G. and P. A. Raviart

1. Interpolation theory over curved elements, with applications to finite element methods
Comp. Methods in Appl. Mech. and Eng. 1 (1972), 217-249

Irons, B. M. and A. Razzaque

1. Experiments with the patch-test for convergence of finite elements
in "The mathematical foundations of the finite element method with application to partial differential equations", K. Aziz and I. Babuška eds., Acad. Press, New York and London, 1972, 557-587.

Miyoshi, T.

1. Convergence of finite element solutions represented by a non-conforming basis
Kumamoto J. Sci. Math. 9 (1972), 11-20

Nitsche, J. A.

1. Umkehrsätze für Spline-Approximationen
Comp. Math. 21 (1969), 400-416
2. Lineare Spline-Functionen und die Methode von Ritz für elliptische Randwertprobleme
Arch. rat. Mech. Anal. 36 (1970), 348-355
3. Über ein Variationsprinzip zur Lösung von Dirichlet-Problemen bei Verwendung von Teilräumen, die keinen Randbedingungen unterworfen sind
Abh. d. Hamb. Math. Sem. 36 (1971), 9-15
4. A projection method for Dirichlet-Problems using subspaces with nearly zero boundary conditions
in "The mathematical foundations of the finite element method with application to partial differential equations", K. Aziz and I. Babuška eds., Acad. Press, New York and London, 1972, 603-627.

Oden, J. T.
1. Some contributions to the mathematical theory of mixed finite element approximations (to appear)

Zienkiewicz, O. C.
1. The finite element method in engineering science (2nd ed.) McGraw-Hill, New York and London, 1971.

Zlámal, M.
1. Curved elements in the finite element method (to appear).

Institut für Angewandte Mathematik
Universität Freiburg
78 Freiburg i. Br.
Hebelstr. 40
Germany

Some Convergence Results for Galerkin Methods for Parabolic Boundary Value Problems

VIDAR THOMÉE

1. Introduction.

We shall consider the parabolic initial-boundary value problem

$$\frac{\partial u}{\partial t} = -Lu \equiv \frac{d}{dx}\left(p\frac{du}{dx}\right) - qu\ , \quad 0 \leq x \leq 1\ , \qquad t > 0\ ,$$

$$(1.1) \qquad u(0,t) = u(1,t) = 0\ , \quad t > 0\ ,$$

$$u(x,0) = v(x), \quad 0 \leq x \leq 1\ ,$$

where p and q are smooth functions of x only, with p positive and q non-negative. As is well known, the exact solution of this problem may be represented in the form

$$(1.2) \qquad u(x,t) = \sum_{j=1}^{\infty} \beta_j e^{-\lambda_j t} \varphi_j(x) \ \text{ with } \ \beta_j = (v, \varphi_j) = \int_0^1 v(x)\varphi_j(x)dx,$$

where $\{\lambda_j\}_1^\infty$ and $\{\varphi_j\}_1^\infty$ are the eigenvalues (in increasing order) and (orthonormal) eigenfunctions of the corresponding two-point boundary value problem

$$(1.3) \qquad Lv = \lambda v\ , \quad v(0) = v(1) = 0\ .$$

The eigenvalues are simple, positive and tend to infinity with j.

One consequence of the growth of the eigenvalues is that the components $\beta_j \varphi_j$ of v with large j are damped for t positive so that $u(t)$ is then smooth even if v is not. This can be expressed in quantitative form as follows. For $s \geq 0$, let $\dot{H}^s$ be the space of functions in $L_2 = L_2(0,1)$ for which

$$(1.4) \qquad \|v\|_s = \left(\sum_{j=1}^{\infty} \lambda_j^s \beta_j^2\right)^{\frac{1}{2}} < \infty , \quad \text{where } \beta_j = (v, \varphi_j) .$$

It can be shown (cf. e.g. [2], [8]) that for s a non-negative integer, $\dot{H}^s$ consists of the functions in $H^s = W_2^s(0,1)$ which satisfy the boundary conditions $L^j v(0) = L^j v(1) = 0$ for $j < s/2$. In particular, $\dot{H}^\infty = \bigcap_{s \geq 0} \dot{H}^s$ consists of all $v \in C^\infty[0,1]$ for which $L^j v(0) = L^j v(1) = 0$ for all $j \geq 0$. It is easy to see that $\dot{H}^\infty$ is dense in $L_2 = \dot{H}^0$. The smoothing property of the solution operator of (1.1) can then be expressed in the following way: For any $v \in L_2$, the solution $u(t)$ belongs to $\dot{H}^\infty$ for $t > 0$ and, for each $s_1 \geq s_2$, there is a constant C such that

$$(1.5) \qquad \|u(t)\|_{s_1} \leq C t^{-(s_1 - s_2)/2} \|v\|_{s_2} .$$

We want to consider the approximate solution of (1.1) by Galerkin's method. For this purpose we shall consider a family $\mathcal{S}_\mu = \{S_h\}$ of finite dimensional subspaces S_h of $\dot{H}^1$ depending on the (small) parameter $h \leq 1$, for which (with $\|v\| = \|v\|_0 = (v,v)^{\frac{1}{2}}$)

$$(1.6) \qquad \inf_{\chi \in S_h} (\|v - \chi\| + h\|v - \chi\|_1) \leq C h^s \|v\|_s , \quad 1 \leq s \leq \mu .$$

Such an inequality will be satisfied in particular if for each h the subspace S_h consists of functions in $C^k[0,1]$ with $0 \leq k \leq \mu - 2$ which are piecewise polynomials of degree $\mu - 1$ on a partition $0 = x_0 < x_1 < \ldots < x_{N_h} = 1$ of $[0,1]$ with $x_j - x_{j-1} \leq h$.

Introducing now the bilinear form

$$L(v,w) = \int_0^1 (p \frac{dv}{dx}\frac{dw}{dx} + qvw)dx ,$$

the continuous time Galerkin problem is the following: Find $U = U(t) \in S_h$ for $t \geq 0$ such that

$$(1.7) \qquad (\frac{\partial U}{\partial t}, \chi) + L(U,\chi) = 0 \quad \text{for all } \chi \in S_h ,$$
$$U(0) = V ,$$

where V is some conveniently chosen approximation of v. This leads to an initial-value problem for a finite linear system of ordinary differential equations for the determination of $U(t)$ which is easily seen to have a unique solution. This initial-value problem shares some simple properties with the original parabolic problem; setting $\chi = U$ in (1.7) we find for instance easily the stability inequality

$$(1.8) \qquad \|U(t)\|^2 + 2\int_0^t L(U,U)ds \leq \|V\|^2.$$

We have collected in Section 2 of this paper some error estimates for the Galerkin problem (1.7) by such energy considerations. Estimates of this type have been obtained in varying degrees of generality in, e. g., Price and Varga [10], Douglas and Dupont [3], Fix and Nassif [6] and Dupont [5]. The main result in [5] in the one-dimensional case we consider takes the form (Theorem 2.1 below) that if $V = Jv$ where $J = J_h$ is a projection of $\dot{H}^\mu$ into S_h such that

$$(1.9) \qquad \|Jv - v\| \leq Ch^\mu \|v\|_\mu ,$$

then

$$(1.10) \qquad \|U(t) - u(t)\| \leq Ch^\mu \|v\|_\mu \quad \text{for} \quad t \geq 0 .$$

Examples of projections satisfying (1.9) are the projection P_0 with respect to the L_2 inner product, the elliptic projection P_1 to be defined presently, and various interpolation operators. Depending on the projection one can often obtain weaker error estimates for less smooth v; for $J = P_0$ we have (Theorem 2.2),

$$\|U(t) - u(t)\| \le Ch^s \|v\|_s \quad \text{for} \quad 0 \le s \le \mu, \ t > 0 .$$

The proof of (1.10) makes use of a projection P_1 of $\dot{H}^1$ into S_h defined by

$$L(P_1 v, \chi) = L(v, \chi) \quad \text{for all} \quad \chi \in S_h . \tag{1.11}$$

As a consequence of (1.6), this projection satisfies (cf. e.g. [1])

$$\|(I - P_1)v\|_j \le Ch^{s-j} \|v\|_s , \quad j = -1, 0, 1 , \tag{1.12}$$

where for $j = -1$, the norm is defined by

$$\|v\|_{-1} = \sup_{w \in \dot{H}^1} \frac{(v, w)}{\|w\|_1} = \sum_{j=1}^{\infty} \lambda_j^{-1} \beta_j^2 .$$

It was proved by Wheeler [15], Douglas, Dupont and Wahlbin [4] that if S_μ is a family of subspaces of $C^k[0,1]$ $(k \le \mu-2)$ of piecewise polynomials of degree $\mu-1$ of the type considered above, on a quasi-uniform partitioning (so that $ch \le x_j - x_{j-1} \le h$ with $c > 0$ independent of h), then

$$|(I-P_1)v| \le Ch^\mu |v^{(\mu)}| \quad \text{where} \quad |v| = \sup_{0 \le x \le 1} |v(x)| . \tag{1.13}$$

Using this, it was proved by Wheeler [14] that if $V = P_1 v$ then we also have the maximum-norm estimate

$$|U(t) - u(t)| \le Ch^\mu \|v\|_{\mu+1} .$$

The proof can be easily extended to yield (Theorem 2.3) the same estimate for projections satisfying a somewhat more restrictive condition than (1.9), namely

$$\|(J-P_1)v\|_1 \le Ch^\mu \|v\|_{\mu+1} . \tag{1.14}$$

Our first aim is now to prove that for t bounded away from zero the restriction (1.14) is unnecessary; in fact we shall see (Theorem 2.4 below; a similar result was obtained independently by Wahlbin [13]) that, assuming only (1.9),

$$|U(t) - u(t)| \leq Ct^{-\frac{1}{2}}h^{\mu}\|v\|_{\mu} .$$

The proof depends on a smoothing property of the solution operator of the Galerkin problem; we have

$$\|U(t)\|_1 \leq Ct^{-\frac{1}{2}}\|V\|_0 . \tag{1.15}$$

To prove (1.15), set $\chi = \partial U/\partial t$ in (1.7) and multiply by $2t$. We obtain, for $t \geq 0$,

$$t\frac{d}{dt}L(U, U) = \frac{d}{dt}(tL(U, U)) - L(U, U) \leq 0$$

so that, using also (1.8),

$$tL(U, U) \leq \int_0^t L(U, U)ds \leq \tfrac{1}{2}\|V\|^2 .$$

In Section 3 we shall analyze the error in the Galerkin method by comparing the eigenfunction expansion (1.2) with a corresponding discrete eigenfunction expansion of U. Thus, let $\{\Lambda_j\}_1^{N_h}$ and $\{\Phi_j\}_1^{N_h}$ be the eigenvalues (in increasing order) and (orthonormal) eigenfunctions of the discrete eigenvalue problem: Find $V \in S_h$ and Λ such that

$$L(V,\chi) = \Lambda(V,\chi) \quad \text{for all} \quad \chi \in S_h . \tag{1.16}$$

It is easy to see that the solution of the Galerkin problem (1.7) can then be represented as

$$U(t) = \sum_{j=1}^{N_h} B_j e^{-\Lambda_j t}\Phi_j \quad \text{where} \quad B_j = (V,\Phi_j) . \tag{1.17}$$

Employing estimates for $\Lambda_j - \lambda_j$ and $\Phi_j - \varphi_j$, we shall be able to prove for instance (Theorem 3.2) that for $t \geq t_0 > 0$, with $V = P_0 v$ and assuming (1.13), we have full $O(h^\mu)$ convergence in the maximum norm even if v is only in L_1,

$$(1.18) \qquad |U(t) - u(t)| \le C(t_0) h^\mu \|v\|_{L_1} \quad \text{for } t \ge t_0 > 0 .$$

Without (1.13), we may still prove the corresponding estimate in L_2 for v in L_2 (Theorem 3.1 below; a similar result has been obtained independently by Helfrich [7]). This time the choice of V is not as arbitrary as in Section 2; (1.18) uses the smoothing effect of the L_2-projection (cf. also Theorem 3.3 where $V = P_1 v$). Similar results for finite difference methods were described in Kreiss, Thomée and Widlund [8] and Thomée and Wahlbin [12]. The eigenfunction expansions (1.2) and (1.17) were used in a heuristic way for error estimation by Strang and Fix ([11], Chapter 7).

The above theory assumes that the ordinary differential equations with respect to t are solved exactly. In Section 4 we shall study the error introduced by discretizing also in t. For this purpose, let k be the time step and let $r(\tau)$ be a rational function with

$$(1.19) \qquad r(\tau) = e^{-\tau} + 0(\tau^{\nu+1}) \quad \text{as } \tau \to 0 .$$

If $r(\tau) = b(\tau)/a(\tau)$ with

$$a(\tau) = \sum_0^\alpha a_j \tau^j, \quad b(\tau) = \sum_0^\beta b_j \tau^j, \quad a_\alpha \ne 0, \; b_\beta \ne 0, \; a_0 = b_0 = 1,$$

we shall assume

$$(1.20) \qquad a(\tau) > 0 \quad \text{and} \quad |r(\tau)| < 1 \quad \text{for } \tau > 0 .$$

In particular, this implies $\beta \le \alpha$. Using the Padé approximations of $e^{-\tau}$, we can achieve $\nu = \alpha + \beta$ in (1.19); we shall assume below that $2\nu \ge \alpha$. We define for $v, w \in \dot{H}^\infty$, with $v_j = (v, \varphi_j)$, $w_j = (w, \varphi_j)$,

$$A_k(v, w) = (a(kL)v, w) = \sum_1^\infty a(k\lambda_j) v_j w_j ,$$

$$B_k(v,w) = (b(kL)v,w) = \sum_1^{\infty} b(k\lambda_j)v_j w_j .$$

Clearly, by the latter representation, A_k is defined on $\dot{H}^\alpha \times \dot{H}^\alpha$ which can also be made obvious by integrating by parts α times in the first representation, and similarly for $B_k(v,w)$. By the positivity of $a(\tau)$, $A_k(v,v)$ is positive definite on $\dot{H}^\alpha$. Assuming now that $S_h \subset \dot{H}^\alpha$, we may therefore define an approximation $U_n \in S_h$ of $u(nk)$ recursively for $n = 1, 2, \ldots$ by

$$A_k(U_{n+1}, \chi) = B_k(U_n, \chi) \quad \text{for all} \quad \chi \in S_h , \tag{1.21}$$

with a suitable choice of U_0.

A special case is the trapezoidal (Crank-Nicolson) rule

$$(U_{n+1}, \chi) + \frac{k}{2} L(U_{n+1}, \chi) = (U_n, \chi) - \frac{k}{2} L(U_n, \chi) \quad \text{for all} \quad \chi \in S_h, \tag{1.22}$$

which corresponds to $r(\tau) = (1-\tau/2)/(1+\tau/2)$, or more generally methods based on subdiagonal and diagonal Padé approximations of $e^{-\tau}$. Zlámal [16] considered the method corresponding to

$$r(\tau) = 1 - \frac{\tau}{1+b\tau} - \frac{\beta}{4}\left(\frac{\tau}{1+b\tau}\right)^2, \quad b = \tfrac{1}{2}(1+\tfrac{1}{3}\sqrt{3}), \quad \beta = \tfrac{2}{3}\sqrt{3} . \tag{1.23}$$

Error estimates for this type of procedures were given in special cases in e.g. Dupont [5], Zlámal [16]. In the present generality, such estimates were obtained in Bramble and Thomée [2] under the assumption that $\kappa = k/h^2$ = constant. The theory in [2] also covers situations in higher dimensions and with subspaces which are not necessarily contained in $\dot{H}^\alpha$. We shall present here only the case of the theory in [2] in the situation described above but assuming only that k/h^2 is bounded below. We shall prove (Theorem 4.1) that if in addition to (1.20) we assume $a_\alpha \neq b_\alpha$ then for $\sigma > \max(\mu, 2\nu)$, with $U_0 = v$,

$$\| U_n - u(nk) \| \le C(\kappa^{\alpha/2} h^\mu + k^\nu) \| v \|_\sigma . \tag{1.24}$$

The condition $a_\alpha \neq b_\alpha$ is satisfied in particular for subdiagonal and odd order diagonal Padé approximations. For the trapezoidal rule (1.22) ($\nu = 2$), Dupont [5] proved the corresponding estimate independent of κ and with $\sigma = \max(\mu, 4)$ and, for the case (1.23) ($\nu = 3$), Zlámal [16] showed an estimate independent of κ.

In the same way as we could represent the solution $U(t)$ of the semi-discrete problem (1.7) in terms of the eigenvalues and eigenfunctions in (1.17), we can now represent the solution of the completely discrete problem (1.21) in the form

$$U_n = \sum_{j=1}^{N_h} B_j r(k\Lambda_j)^n \Phi_j, \qquad B_j = (V, \Phi_j). \tag{1.25}$$

We shall end Section 4 by using this representation to prove some complements to estimates of the type (1.18) for the semi-discrete problem in that we shall show that under certain conditions on $r(\tau)$, if $U_0 = V$ is bounded in some weak sense, then for $t > 0$ the additional error introduced by discretizing also in time is $O(k^\nu)$ for small k. Altogether this will lead to a total $O(h^\mu + k^\nu)$ error estimate without any restriction on the relation between k and h.

The possibility of obtaining error estimates of high order for t positive when v is non-smooth, in the case of the semi-discrete problem, depends on the presence of exponential factors in (1.2) and (1.17). In order to obtain similar effects in (1.25), we have to make certain assumptions on the behavior of $r(\tau)$ for large τ. We shall consider three such assumptions, namely

(i) $\quad |r(\tau)| < 1 \quad \text{for} \quad \tau > 0,$

(ii) $\quad \sup_{\tau > \delta} |r(\tau)| < 1 \quad \text{for any} \quad \delta > 0,$

(iii) $\quad |r(\tau)| < 1 \quad \text{for } \tau > 0 \text{ and } \lim_{\tau \to \infty} r(\tau) = 0.$

These three assumptions are successively more stringent, and our estimates for the time discretization error will be correspondingly successively sharper. We shall treat the cases in the order (iii), (ii), (i) (Lemmas 4.1-4.3) and finish by considering the total error in the case (iii) (Theorem 4.2). Examples of (iii) are all subdiagonal Padé approximations of $e^{-\tau}$. An example of (ii) is given by (1.23), and (i) allows also diagonal Padé approximations. For the latter, we have $\lim_{\tau\to\infty} |r(\tau)| = 1$, and we shall then need two extra assumptions to assure that the factors $r(k\Lambda_j)^n$ are small for large n, namely the "inverse" assumption

(i') $$L(V, V) \leq Ch^{-2} \|V\|^2 \quad \text{for} \quad V \in S_h ,$$

and a boundedness condition for the mesh ratio,

(i") $$k/h^2 \leq K .$$

Throughout this paper, C and c will denote positive constants, not necessarily the same at different occurrences.

2. Error estimates based on the energy method.

For the purpose of estimating the error in the Galerkin procedure (1.7) we shall need some estimates for the projection error $\rho = u - P_1 u$ where u is the exact solution of (1.1) and P_1 is defined by (1.11).

Lemma 2.1.

There is a constant C such that $(\rho_t = \partial\rho/\partial t)$

(2.1) $$\left(\int_0^\infty \|\rho_t(t)\|_j^2 dt\right)^{\frac{1}{2}} \leq Ch^{s-j} \|v\|_{s+1}, \quad j = 0, -1, \quad 1 \leq s \leq \mu ,$$

(2.2) $$\left(\int_0^\infty t \|\rho_t(t)\|^2 dt\right)^{\frac{1}{2}} \leq Ch^s \|v\|_s , \quad 1 \leq s \leq \mu .$$

Proof.

Since P_1 commutes with time differentiation, we have by (1.12)

$$\|\rho_t\|_j = \|(I-P_1)u_t\|_j \leq Ch^{s-j} \|u_t\|_s \leq Ch^{s-j} \|u\|_{s+2} .$$

Using (1.2) and (1.4), we obtain

$$\Big(\int_0^\infty \|u\|_{s+2}^2 dt\Big)^{\frac{1}{2}} = \Big(\sum_j \beta_j^2 \lambda_j^{s+2} \int_0^\infty e^{-2\lambda_j t} dt\Big)^{\frac{1}{2}} = \frac{1}{\sqrt{2}} \|v\|_{s+1},$$

which proves (2.1). Similarly

$$\Big(\int_0^\infty t\|u\|_{s+2}^2 dt\Big)^{\frac{1}{2}} = \Big(\sum_j \beta_j^2 \lambda_j^{s+2} \int_0^\infty te^{-2\lambda_j t} dt\Big)^{\frac{1}{2}} = \tfrac{1}{2}\|v\|_s ,$$

which proves (2.2).

We shall now estimate the difference $\vartheta = U - P_1 u$ between the Galerkin solution and the elliptic projection of the exact solution. For the time being, we do not specify V in (1.7), and the estimate in the following lemma depends therefore on $\vartheta(0) = V - P_1 v$.

Lemma 2.2.

There is a constant C *such that*

$$\sup_{t \leq 0} \|\vartheta(t)\| + \Big(\int_0^\infty \|\vartheta(t)\|_1^2 dt\Big)^{\frac{1}{2}} \leq C[\|\vartheta(0)\| + h^s \|v\|_s] \tag{2.3}$$

for $2 \leq s \leq \mu+1$,

and

$$\sup_{t \geq 0} \|\vartheta(t)\|_1 \leq C[\|\vartheta(0)\|_1 + h^{s-1}\|v\|_s] \tag{2.4}$$

for $2 \leq s \leq \mu+1$.

Proof.

Recalling that $\rho = u - P_1 u$, the Galerkin equation can be written

$$(\vartheta_t, \chi) + L(\vartheta, \chi) = (\rho_t, \chi), \quad \text{for all } \chi \in S_h .$$

Hence, with $\chi = \vartheta$,

$$\tfrac{1}{2}\frac{d}{dt}\|\vartheta\|^2 + L(\vartheta,\vartheta) \leq \|\rho_t\|_{-1}\|\vartheta\|_1 \leq \tfrac{1}{2}L(\vartheta,\vartheta) + C\|\rho_t\|_{-1}^2,$$

and, after integration,

$$\|\vartheta(t)\|^2 + \int_0^t L(\vartheta, \vartheta)ds \leq \|\vartheta(0)\|^2 + C\int_0^t \|\rho_t\|_{-1}^2 ds .$$

The first inequality follows therefore by (2.1) with $j = -1$. Setting instead $\chi = \vartheta_t$, we obtain

$$\|\vartheta_t\|^2 + \tfrac{1}{2}\frac{d}{dt} L(\vartheta, \vartheta) = (\rho_t, \vartheta_t) \leq \tfrac{1}{2}\|\vartheta_t\|^2 + \tfrac{1}{2}\|\rho_t\|^2, \tag{2.5}$$

so that

$$L(\vartheta, \vartheta)(t) \leq L(\vartheta, \vartheta)(0) + \int_0^t \|\rho_t\|^2 ds .$$

The second inequality follows now by (2.1) with $j = 0$.

We shall now turn to the estimate for the error $U - u = \vartheta - \rho$.

Theorem 2.1.

Assume $V = Jv$ *where* J *satisfies* (1.9). *Then there is a constant* C *such that, for* $v \in \dot{H}^{\mu}$,

$$\|U(t) - u(t)\| \leq Ch^{\mu}\|v\|_{\mu} \quad \text{for} \quad t \geq 0 . \tag{2.6}$$

If, in addition J *satisfies*

$$\|v - Jv\|_1 \leq Ch^{\mu-1}\|v\|_{\mu}, \tag{2.7}$$

then we have

$$\|U(t) - u(t)\|_1 \leq Ch^{\mu-1}\|v\|_{\mu} .$$

Proof.

We have, by (2.3) with $s = \mu$,

$$\|\vartheta(t)\| \leq C[\|\vartheta(0)\| + h^{\mu}\|v\|_{\mu}] \leq Ch^{\mu}\|v\|_{\mu} .$$

Since, by (1.12) and (1.5),

$$\|\rho(t)\| \leq Ch^{\mu}\|u(t)\|_{\mu} \leq Ch^{\mu}\|v\|_{\mu} ,$$

the first result follows by the triangle inequality. Similarly, by (2.4) with $s = \mu$, (2.7), (1.12) and (1.5),

$$\|U(t)-u(t)\|_1 \leq \|\vartheta(t)\|_1 + \|\rho(t)\|_1 \leq C[\|\vartheta(0)\|_1 + h^{\mu-1}\|v\|_{\mu}]$$
$$\leq Ch^{\mu-1}\|v\|_{\mu} ,$$

which completes the proof of the theorem.

For less smooth v, correspondingly lower rates of convergence can be proved, provided some supplementary bounds for J are known. We mention only briefly the case of the L_2-projection P_0.

Theorem 2.2.

Let $V = P_0 v$. *Then there is a constant* C *such that for* $v \in \dot{H}^s$, $0 \leq s \leq \mu$,

$$\|U(t) - u(t)\| \leq Ch^s \|v\|_s \quad \textit{for} \quad t \geq 0 .$$

Proof.

We have, by (1.8) and (1.5),

$$\|U(t) - u(t)\| \leq \|P_0 v\| + \|v\| \leq 2\|v\|. \tag{2.8}$$

The result follows now immediately by interpolation between (2.6) and (2.8).

In the proof of Theorem 2.1, we did not use the full strength of Lemma 2.2 in that we only chose $s = \mu$ in (2.3) and (2.4). As we shall see in the next theorem, if the additional estimate (1.13) for the elliptic projection P_1 holds, then the fact that one can obtain a $O(h^{\mu})$ estimate for $\|\vartheta\|_1$ can be used to obtain a $O(h^{\mu})$ error estimate in the maximum norm. We shall then first assume that J satisfies the somewhat more restrictive condition (1.14).

Theorem 2.3.

Assume that (1.13) and (1.14) hold. Then there is a constant C such that for $v \in \mathring{H}^{\mu+1}$,

$$|U(t) - u(t)| \leq Ch^{\mu} \|v\|_{\mu+1} \quad \text{for } t \geq 0 .$$

Proof.

We have, by Sobolev's inequality, (2.4) with $s = \mu+1$, and (1.14),

$$|\vartheta(t)| \leq C\|\vartheta(t)\|_1 \leq Ch^{\mu} \|v\|_{\mu+1},$$

and also, by (1.13) and (1.5),

$$|\rho(t)| \leq Ch^{\mu} |u^{(\mu)}(t)| \leq Ch^{\mu} \|u(t)\|_{\mu+1} \leq Ch^{\mu} \|v\|_{\mu+1},$$

which again proves the result by the triangle inequality.

We shall see now that for t positive we may use the smoothing property (1.15) of the solution operator of the Galerkin problem to prove a $O(h^{\mu})$ maximum norm estimate also for the less restrictive condition (1.9) and under lesser regularity assumptions. The proof hinges on the following lemma.

Lemma 2.3.

There is a constant C such that, for $v \in \mathring{H}^{\mu}$,

$$\|\vartheta(t)\|_1 \leq Ct^{-\frac{1}{2}}[\|\vartheta(0)\| + h^{\mu} \|v\|_{\mu}].$$

Proof.

Multiplying (2.5) by $2t$, we obtain

$$2t\|\vartheta_t\|^2 + \frac{d}{dt}(tL(\vartheta, \vartheta)) \leq t\|\vartheta_t\|^2 + L(\vartheta, \vartheta) + t\|\rho_t\|^2 ,$$

and hence by integration

$$tL(\vartheta,\vartheta)(t) \leq \int_0^\infty L(\vartheta,\vartheta)dt + \int_0^\infty t\|\rho_t\|^2 dt .$$

The result now follows by Lemmas 2.1 and 2.2.

Theorem 2.4.

Assume that (1.13) and (1.9) hold. Then there is a constant C such that for $v \in \dot{H}^\mu$,

$$|U(t)-u(t)| \leq Ct^{-\frac{1}{2}}h^\mu\|v\|_\mu \quad \text{for} \quad t > 0 .$$

Proof.

We have, by Lemma 2.3,

$$|\vartheta(t)| \leq C\|\vartheta(t)\|_1 \leq Ct^{-\frac{1}{2}}[\|Jv-v\|+\|P_1v-v\|+h^\mu\|v\|_\mu]$$

$$\leq Ct^{-\frac{1}{2}}h^\mu\|v\|_\mu ,$$

and using (1.13) and (1.5),

$$|\rho(t)| \leq Ch^\mu\|u(t)\|_{\mu+1} \leq Ct^{-\frac{1}{2}}h^\mu\|v\|_\mu .$$

The result therefore follows again by the triangle inequality.

3. Error estimates based on eigenfunction expansions.

We begin this section by studying the eigenvalues $\{\Lambda_j\}_1^{N_h}$ and eigenfunctions $\{\Phi_j\}_1^{N_h}$ of the eigenvalue problem (1.16). As is well-known,

$$\Lambda_j = \min_{\mathcal{E}_j} \max_{\substack{v\in\mathcal{E}_j \\ \|v\|=1}} L(v,v) , \tag{3.1}$$

where $\mathcal{E}_j$ ranges over all j-dimensional subspaces of S_h.

Since $S_h \subset \dot{H}^1$, it follows from the corresponding characterization of the eigenvalues λ_j of the problem (1.3) that $\Lambda_j \geq \lambda_j$ for all j.

We shall need estimates for $\Lambda_j - \lambda_j$ and $\Phi_j - \varphi_j$ in terms of both h and j for the purpose of comparing the eigenfunction expansions (1.2) and (1.17) term by term. Such estimates are the contents of the next two lemmas.

Lemma 3.1.

There are positive δ_0 and c_0 such that, for $h^2\lambda_j \leq \delta_0$,

$$\Lambda_j - \lambda_j \leq c_0 \lambda_j (h^2 \lambda_j)^{\mu/2} .$$

Proof.

We shall prove the lemma with $c_0 = 4\tilde{c}$ and with δ_0 such that $\tilde{c}\,\delta_0^{\mu/2} = \frac{1}{4}$ where $\tilde{c}$ is the constant C in (1.12). Let Ψ_j denote the linear span of $\{\varphi_l\}_1^j$ and Σ_j the unit sphere in Ψ_j. We notice that, for $h^2\lambda_j \leq \delta_0$, $P_1\Psi_j$ is also j-dimensional. For if $P_1\Psi_j$ had smaller dimension, we would be able to find $w \in \Psi_j$ with $w \neq 0$ such that $P_1 w = 0$. But then

$$\|w\| = \|(I-P_1)w\| \leq \tilde{c} h^{\mu} \|w\|_{\mu} \leq \tilde{c} h^{\mu} \lambda_j^{\mu/2} \|w\| \leq \tfrac{1}{4}\|w\|,$$

which is a contradiction.

Let now $h^2\lambda_j \leq \delta_0$. Since $\mathcal{E}_j = P_1\Psi_j$ is j-dimensional, we have by (3.1) and (1.11) that

$$\Lambda_j \leq \max_{v \in \Sigma_j} \frac{L(P_1 v, P_1 v)}{\|P_1 v\|^2} \leq \max_{v \in \Sigma_j} \frac{L(v,v)}{\|P_1 v\|^2} \leq \lambda_j \max_{v \in \Sigma_j} \|P_1 v\|^{-2}.$$

Now, for $v \in \Sigma_j$,

$$\|(I-P_1)v\| \leq \tilde{c} h^{\mu} \|v\|_{\mu} \leq \tilde{c}(h^2\lambda_j)^{\mu/2} \leq \tfrac{1}{4},$$

and hence

$$\|P_1 v\|^2 \geq (1 - \|(I-P_1)v\|)^2 \geq 1 - 2\|(I-P_1)v\|$$
$$\geq (1 + 4\|(I-P_1)v\|)^{-1} \geq (1+4\,\tilde{c}\,(h^2\lambda_j)^{\mu/2})^{-1} ,$$

so that

$$\Lambda_j \leq \lambda_j(1 + 4\,\tilde{c}\,(h^2\lambda_j)^{\mu/2}) ,$$

which proves the lemma.

Lemma 3.2.

There are positive constants $\delta(\leq \delta_0)$ and c_1 such that, for $h\lambda_j \leq \delta$,

$$\|\Phi_j - \varphi_j\| \leq c_1(h\lambda_j)^{\mu} . \tag{3.2}$$

Proof.

Setting $\gamma_j = (P_1\varphi_j, \Phi_j)$, we have

$$\|\Phi_j - \varphi_j\| \leq \|(I-P_1)\varphi_j\| + \|P_1\varphi_j - \gamma_j\Phi_j\| + |1-\gamma_j| . \tag{3.3}$$

For the first term we have, for all j ,

$$\|(I-P_1)\varphi_j\| \leq \tilde{c}\,h^{\mu}\|\varphi_j\|_{\mu} = \tilde{c}\,(h^2\lambda_j)^{\mu/2} \leq C(h\lambda_j)^{\mu} ,$$

and for the last

$$|1-\gamma_j| = |\|\varphi_j\| - \|\gamma_j\Phi_j\|| \leq \|\varphi_j - \gamma_j\Phi_j\| \leq \|P_1\varphi_j - \gamma_j\Phi_j\| + C(h\lambda_j)^{\mu} .$$

It remains therefore to estimate the middle term in (3.3). Using the definitions, we find

$$(\Lambda_\ell - \lambda_j)(P_1\varphi_j, \Phi_\ell) = \lambda_j(\varphi_j - P_1\varphi_j, \Phi_\ell) . \tag{3.4}$$

Since it is well-known that $\lambda_{j+1} - \lambda_j$ is bounded below, we may find a positive $\gamma \le \frac{1}{2} \min_j |\lambda_{j+1} - \lambda_j|$. By Lemma 3.1, we have, for $h^2\lambda_j < \delta_0$ and $\ell < j$,

$$|\Lambda_\ell - \lambda_j| \ge 2\gamma - c_0\lambda_j(h^2\lambda_j)^{\mu/2} \ge 2\gamma - \tilde{c}_0(h\lambda_j)^\mu .$$

Hence, if we choose $\delta \le \delta_0$ so that $\tilde{c}_0\delta^\mu \le \gamma$, we have, for $h\lambda_j \le \delta$, $\ell < j$,

$$|\Lambda_\ell - \lambda_j| \ge \gamma . \tag{3.5}$$

Since $\Lambda_\ell \ge \lambda_j$, (3.5) also holds for $\ell > j$ and hence, using (3.4),

$$\|P_1\varphi_j - \gamma_j\Phi_j\|^2 = \sum_{\ell \ne j} (P_1\varphi_j, \Phi_\ell)^2 \le (\lambda_j/\gamma)^2 \sum_{\ell \ne j} (\varphi_j - P_1\varphi_j, \Phi_\ell)^2$$

$$\le (\lambda_j/\gamma)^2 \|(I - P_1)\varphi_j\|^2 \le C(\lambda_j h^\mu \lambda_j^{\mu/2})^2 \le C(h\lambda_j)^{2\mu} .$$

This completes the proof.

Remark.

Although sufficient for our purposes, the result of Lemma 3.1 is not best possible; for fixed j it can be proved that $\Lambda_j - \lambda_j = O(h^{2\mu-2})$ as $h \to 0$ (cf. e.g. [1]). We also do not claim that the dependence of our estimates on λ_j is optimal; all that is needed here is that the right hand sides of the estimates grow no faster than a power of λ_j. An estimate for $\|\Phi_j - \varphi_j\|$ with a lower power of λ_j than in (3.2) was stated in [11] but the proof there does not appear binding.

We now turn to the error estimates for the continuous time Galerkin problem (1.7) for positive t. The first result is an estimate in L_2.

Theorem 3.1.

Let $V = P_0 v$. *Then for* $t_0 > 0$ *there is a* $C(t_0)$ *such that, for* $v \in L_2$,

$$\|U(t) - u(t)\| \leq C(t_0)h^{\mu}\|v\| \quad \underline{\text{for}} \quad t \geq t_0 .$$

Proof.

By the representations (1. 2) and (1. 17), we find that we may write

$$(3.6) \qquad U(t) - u(t) = \sum_{j=1}^{5} \varepsilon_j(t) ,$$

where, with δ as in Lemma 3. 2 and j_h the first integer such that $h\lambda_j \geq \delta$,

$$(3.7) \quad \varepsilon_1(t) = -\sum_{j \geq j_h} \beta_j e^{-t\lambda_j} \varphi_j , \qquad \varepsilon_2(t) = \sum_{j \geq j_h} B_j e^{-t\Lambda_j} \Phi_j ,$$

$$\varepsilon_3(t) = \sum_{j<j_h} \beta_j (e^{-t\Lambda_j} - e^{-t\lambda_j})\varphi_j , \quad \varepsilon_4(t) = \sum_{j<j_h} B_j e^{-t\Lambda_j}(\Phi_j - \varphi_j),$$

$$\varepsilon_5(t) = \sum_{j<j_h} (B_j - \beta_j) e^{-t\Lambda_j} \varphi_j .$$

We shall estimate the norms of these five terms. For later use we estimate the first three in the maximum-norm and assume only $v \in L_1$. We shall then need

$$|\varphi_j| \leq C\|\varphi_j\|_1 = C\lambda_j^{\frac{1}{2}} ,$$

$$(3.8) \qquad |\Phi_j| \leq CL(\Phi_j, \Phi_j)^{\frac{1}{2}} = C\Lambda_j^{\frac{1}{2}} ,$$

and also

$$|\beta_j| = |(v, \varphi_j)| \leq \|v\|_{L_1} |\varphi_j| \leq C\lambda_j^{\frac{1}{2}} \|v\|_{L_1} ,$$

$$|B_j| = |(V, \Phi_j)| = |(v, \Phi_j)| \leq C\Lambda_j^{\frac{1}{2}} \|v\|_{L_1} ,$$

where we have used that $V = P_0 v$. We have therefore first, since $\delta \leq h\lambda_j$ for $j > j_h$,

$$|\varepsilon_1(t)| \leq C\|v\|_{L_1} \sum_{j \geq j_h} \lambda_j e^{-t\lambda_j} \leq Ch^{\mu}\|v\|_{L_1} \sum_j \lambda_j^{1+\mu} e^{-t_0\lambda_j}$$

$$= C(t_0)h^{\mu}\|v\|_{L_1} .$$

Similarly we obtain

$$|\varepsilon_2(t)| \leq C\|v\|_{L_1} \sum_{j \geq j_h} \Lambda_j e^{-t_0\Lambda_j} \leq C\|v\|_{L_1} \sum_{j \geq j_h} e^{-t_0\Lambda_j/2}$$

$$\leq Ch^{\mu}\|v\|_{L_1} \sum_j \lambda_j^{\mu} e^{-t_0\lambda_j/2} = C(t_0)h^{\mu}\|v\|_{L_1} .$$

Turning now to the terms with $h\lambda_j < \delta$ we obtain first by Lemma 3.1 that

$$|e^{-t\Lambda_j} - e^{-t\lambda_j}| \leq e^{-t\lambda_j} t|\Lambda_j - \lambda_j| \leq C(h^2\lambda_j)^{\mu/2}\lambda_j t e^{-t\lambda_j}$$

$$\leq C(h^2\lambda_j)^{\mu/2} e^{-t\lambda_j/2}$$

and hence

$$|\varepsilon_3(t)| \leq Ch^{\mu}\|v\|_{L_1} \sum_j \lambda_j^{1+\mu/2} e^{-t_0\lambda_j/2} = C(t_0)h^{\mu}\|v\|_{L_1} .$$

Using Lemma 3.2, we shall now estimate $\varepsilon_4(t)$ in L_2. We have

$$\|\varepsilon_4(t)\| \leq C\|v\|_{L_1} \sum_{j < j_h} \Lambda_j^{\frac{1}{2}} e^{-t\Lambda_j} \|\Phi_j - \varphi_j\|$$

$$\leq Ch^{\mu}\|v\|_{L_1} \sum_j \lambda_j^{\mu} e^{-t_0\lambda_j/2} = C(t_0)h^{\mu}\|v\|_{L_1} .$$

We shall finally estimate $\varepsilon_5(t)$ in the maximum-norm, assuming $v \in L_2$. Since $V = P_0 v$, we have

$$|B_j - \beta_j| = |(V, \Phi_j) - (v, \varphi_j)| = |(v, \Phi_j - \varphi_j)| \leq C\|v\|(h\lambda_j)^{\mu},$$

so that

$$|\varepsilon_5(t)| \leq Ch^{\mu}\|v\| \sum_j \lambda_j^{\mu+\frac{1}{2}} e^{-t_0\lambda_j} = C(t_0)h^{\mu}\|v\| .$$

Together these estimates prove the theorem since the maximum-norm dominates the L_2-norm and the L_2-norm the L_1-norm .

We shall now see that under the additional assumption (1.13) for P_1 we can in fact obtain a maximum-norm estimate when v is only in L_1. We shall need the following lemma.

Lemma 3.3.

Assume that (1.13) holds. Then there is a constant C such that, for $h\lambda_j \leq \delta$,

$$|\Phi_j - \varphi_j| \leq Ch^{\mu}\lambda_j^{\mu+1} .$$

Proof.

We shall first prove that, for $h\lambda_j \leq \delta$,

$$\|\Phi_j - P_1\varphi_j\|_1 \leq Ch^{\mu}\lambda_j^{\mu+1} . \tag{3.9}$$

For, with $\chi = \Phi_j - P_1\varphi_j$, we have, since $\chi \in S_h \subset \dot{H}^1$, that

$$\|\chi\|_1^2 \leq CL(\Phi_j - P_1\varphi_j, \chi) = C[L(\Phi_j, \chi) - L(\varphi_j, \chi)]$$

$$\leq C[\Lambda_j(\Phi_j, \chi) - \lambda_j(\varphi_j, \chi)] \leq C\|\Lambda_j\Phi_j - \lambda_j\varphi_j\|\|\chi\|_1.$$

Hence

$$\|\chi\|_1 \leq C\|\Lambda_j\Phi_j - \lambda_j\varphi_j\| \leq C[|\Lambda_j - \lambda_j| + \lambda_j\|\Phi_j - \varphi_j\|]$$

$$\leq C[\lambda_j(h^2\lambda_j)^{\mu/2} + \lambda_j(h\lambda_j)^{\mu}] \leq Ch^{\mu}\lambda_j^{\mu+1},$$

which proves (3.9). This now implies by (1.13) that

$$|\Phi_j - \varphi_j| \leq C\|\Phi_j - P_1\varphi_j\|_1 + |(I-P_1)\varphi_j| \leq Ch^{\mu}\lambda_j^{\mu+1},$$

which concludes the proof of the lemma.

Theorem 3.2.

Assume that (1.13) holds and let $V = P_0 v$. Then for $t_0 > 0$ there is a $C(t_0)$ such that for $v \in L_1$,

$$|U(t) - c(t)| \leq C(t_0)h^{\mu}\|v\|_{L_1} \quad \text{for} \quad t \geq t_0$$

Proof.

We have already proved the appropriate estimates for $\varepsilon_j(t)$, $j = 1, 2, 3$. This time we obtain, using Lemma 3.3,

$$|\varepsilon_4(t)| \leq C\|v\|_{L_1} \sum_{j<j_h} \Lambda_j^{\frac{1}{2}} e^{-t\Lambda_j} |\Phi_j - \varphi_j|$$

$$\leq Ch^{\mu}\|v\|_{L_1} \sum_j \lambda_j^{\mu+1} e^{-t_0\lambda_j/2},$$

and

$$|B_j - \beta_j| \leq \|v\|_{L_1} |\Phi_j - \varphi_j| \leq Ch^{\mu}\lambda_j^{\mu+1}\|v\|_{L_1},$$

so that

$$|\varepsilon_5(t)| \leq Ch^{\mu}\|v\|_{L_1} \sum_j \lambda_j^{\mu+3/2} e^{-t_0\lambda_j},$$

which completes the proof of the theorem.

It is clear from the proofs above that the technique could be used to treat also choices for V other than $P_0 v$. We shall consider only a case with the elliptic projection $P_1 v$ which, as is easily seen, can be represented for $v \in \dot{H}^1$ by

$$P_1 v = \sum_{j=1}^{N_h} \frac{L(v, \Phi_j)}{\Lambda_j} \Phi_j .$$

We denote by $\dot{W}_1^2$ the closure of C^∞ functions vanishing at $x = 0$ and $x = 1$ with respect to

$$\|v\|_{\dot{W}_1^2} = \int_0^1 (|v| + |v'| + |v''|)dx .$$

It is clear that $\dot{W}_1^2 \subset \dot{H}^1$ with the corresponding norm inequality.

Theorem 3. 3.

Assume that (1. 13) holds and let $V = P_1 v$. Then for $t_0 > 0$ there is a $C(t_0)$ such that, for $v \in \dot{W}_1^2$,

$$|U(t) - u(t)| \leq C(t_0) h^{\mu} \|v\|_{\dot{W}_1^2}, \quad \text{for} \quad t \geq t_0 .$$

Proof.

We can write (3. 6) and (3.7) as before with $\beta_j = (v, \varphi_j)$ but this time $B_j = L(v, \Phi_j)/\Lambda_j$. The estimates for $\varepsilon_1(t)$ and $\varepsilon_3(t)$ are unchanged. We have now

$$|B_j| \leq C\|v\|_1 L(\Phi_j, \Phi_j)^{\frac{1}{2}}/\Lambda_j = C\|v\|_1 (\Lambda_j)^{-\frac{1}{2}} \leq C\|v\|_{\dot{W}_1^2},$$

and $\varepsilon_2(t)$ and $\varepsilon_4(t)$ are therefore, by the same argument as before, estimated by

$$|\varepsilon_2(t)| + |\varepsilon_4(t)| \leq C(t_0) h^{\mu} \|v\|_{\dot{W}_1^2} .$$

It remains to consider $\varepsilon_5(t)$. We find easily

$$L(v,\Phi_j) = \Lambda_j(v,\varphi_j) + (\lambda_j - \Lambda_j)(v,\varphi_j) + (Lv,\Phi_j - \varphi_j),$$

so that, using Lemmas 3.1 and 3.3, for $j < j_h$,

$$|B_j - \beta_j| \leq C\lambda_j^{-1}\{|\Lambda_j - \lambda_j| \, \|v\| + |\Phi_j - \varphi_j| \, \|v\|_{\dot{W}_1^2}\}$$
$$\leq Ch^{\mu}\lambda_j^{\mu} \|v\|_{\dot{W}_1^2}.$$

The estimate for $\varepsilon_5(t)$ is now obtained in the same way as above,

$$|\varepsilon_5(t)| \leq C(t_0)h^{\mu}\|v\|_{\dot{W}_1^2},$$

which completes the proof of the theorem.

4. Completely discrete schemes.

We shall now turn to the error estimates for the completely discrete scheme (1.21). We notice that since $A_k(v,v)$ is positive definite, $A_k(v,w)$ defines an inner product in $\dot{H}^{\alpha}$, and we denote the corresponding norm

$$a_k(v) = A_k(v,v)^{\frac{1}{2}}. \tag{4.1}$$

Since $a(\tau)$ is bounded below, $a_k(\cdot)$ dominates the L_2 norm. We also notice that $B_k(v,w)$ is defined on $\dot{H}^{\alpha} \times \dot{H}^{\alpha}$ and that, by (1.20),

$$|B_k(v,w)| \leq a_k(v)a_k(w). \tag{4.2}$$

We shall assume in this section that $S_h \subset \dot{H}^{\alpha}$ and that

$$\inf_{\chi \in S_h}(\|v-\chi\| + h^{\alpha}\|v-\chi\|_{\alpha}) \leq Ch^s\|v\|_s, \quad \alpha \leq s \leq \mu. \tag{4.3}$$

We shall prove the following result.

Theorem 4.1.

Assume that $r(\tau)$ satisfies (1.19) and (1.20), with $a_\alpha \neq b_\alpha$. Assume further that $S_h \subset \dot{H}^\alpha$ and (4.3) holds. Let $\sigma > \max(\mu, 2\nu)$ and $\kappa = k/h^2 \geq c_0 > 0$. Then for $T>0$ there is a constant C such that, for $U_0 = v$, we have

$$(4.4) \qquad a_k(U_n - u(nk)) \leq C(\kappa^{\alpha/2} h^\mu + k^\nu) \|v\|_\sigma, \text{ for } t \leq T.$$

Proof.

We want to estimate first the difference between U_n and a certain projection into S_h of $u_n = u(nk)$. For this purpose, let for $v, w \in \dot{H}^\alpha$,

$$G_k(v,w) = A_k(v,w) - B_k(v,w) = \sum_{j=1}^{\infty} g(k\lambda_j) v_j w_j \,.$$

Here, by (1.20)

$$g(\tau) = a(\tau) - b(\tau) > 0 \qquad \text{for } \tau > 0$$

so that $G_k(v,w)$ defines an inner product in $\dot{H}^\alpha$. Let now Q be the projection in $\dot{H}^\alpha$ onto S_h with respect to this product. Then, with $g_k(v) = G_k(v,v)^{\frac{1}{2}}$,

$$\begin{aligned}(4.5) \qquad g_k((I-Q)v) &= \inf_{\chi \in S_h} g_k(v-\chi) \\ &\leq C \inf_{\chi \in S_h} (\|v-\chi\| + k^{\alpha/2} \|v-\chi\|_\alpha) \\ &\leq C\kappa^{\alpha/2} h^s \|v\|_s, \quad \alpha \leq s \leq \mu .\end{aligned}$$

We shall prove that this implies

$$(4.6) \qquad a_k((I-Q)v) \leq C\kappa^{\alpha/2} h^s \|v\|_s \quad \text{for } \alpha \leq s \leq \mu .$$

In fact, let $\tilde{v}$, ψ and $\tilde{\psi}$ be defined by

$$\tilde{v} = (I-Q)v, \quad \psi = \sum_{j=1}^{\infty} \frac{\tilde{v}_j}{g(k\lambda_j)} \varphi_j \quad \text{with} \quad \tilde{v}_j = (\tilde{v}, \varphi_j), \quad \tilde{\psi} = (I-Q)\psi .$$

We have

$$(4.7) \qquad \|\tilde{v}\|^2 = G_k(\tilde{v}, \psi) = G_k(\tilde{v}, \tilde{\psi}) \leq g_k(\tilde{v}) g_k(\tilde{\psi}) ,$$

and, by (4.5) with $s = \alpha =$ degree g,

$$g_k(\tilde{\psi}) \leq Ck^{\alpha/2} \|\psi\|_\alpha = C\Big(\sum_j \frac{(k\lambda_j)^\alpha}{g(k\lambda_j)^2} \tilde{v}_j^2\Big)^{\frac{1}{2}} \leq C\|\tilde{v}\| .$$

It follows from (4.7) and (4.5) that

$$\|\tilde{v}\| \leq Cg_k(\tilde{v}) \leq C\kappa^{\alpha/2} h^s \|v\|_s , \quad \alpha \leq s \leq \mu ,$$

and hence

$$a_k(\tilde{v}) \leq C(\|\tilde{v}\| + g_k(\tilde{v})) \leq C\kappa^{\alpha/2} h^s \|v\|_s$$

which proves (4.6).

We shall also need the estimate

$$(4.8) \qquad |A_k(u_{m+1}, w) - B_k(u_m, w)| \leq Ck^{s/2} \|u_m\|_s a_k(w)$$

for the truncation error in the time discretization. Here $u_m = u(mk)$ where u is the exact solution of (1.1) and $w \in \dot{H}^\alpha$, with $\alpha \leq s \leq 2\nu+2$. To prove (4.8), it is sufficient to consider $v, w \in \dot{H}^\infty$. We have then, with $u_{mj} = (u_m, \varphi_j)$,

$$A_k(u_{m+1}, w) - B_k(u_m, w) = \sum_j (a(k\lambda_j)e^{-k\lambda_j} - b(k\lambda_j))u_{mj} w_j .$$

The estimate now follows by Cauchy's inequality and the inequality

$$|a(\tau)^{\frac{1}{2}}(e^{-\tau} - r(\tau))| \leq C\tau^{s/2} \quad \text{for } \tau > 0 , \quad \alpha \leq s \leq 2\nu+2$$

(for large τ we have used $\beta-\alpha/2 \le \frac{\alpha}{2} \le \nu$).

We notice also that

$$(4.9) \quad \|u_{m+1}-u_m\|_\mu = \left(\sum_j \lambda_j^\mu (e^{-k\lambda_j}-1)^2 u_{mj}^2\right)^{\frac{1}{2}} \le k\|u_m\|_{\mu+2}$$

and that by (1.5), for $s \ge \sigma$,

$$\|u_m\|_s \le C(mk)^{-\frac{1}{2}(s-\sigma)}\|v\|_\sigma .$$

Set now $\vartheta_m = U_m - Qu_m$, $\rho_m = (I-Q)u_m$ so that $e_m = U_m - u_m = \vartheta_m - \rho_m$. We have, from (4.6) and (1.5),

$$(4.10) \quad \begin{aligned} a_k(\rho_m) = a_k((I-Q)u_m) &\le C\kappa^{\alpha/2}h^\mu\|u_m\|_\mu \\ &\le C\kappa^{\alpha/2}h^\mu\|v\|_\mu , \end{aligned}$$

which is bounded from above by the right side of (4.4). It remains to prove a similar estimate for ϑ_m. By our definitions and the Galerkin equation (1.21) we have, for $\chi \in S_h$,

$$\begin{aligned} A_k(\vartheta_{m+1},\chi) = B_k(\vartheta_m,\chi) &+ B_k(\rho_{m+1}-\rho_m,\chi) \\ &- [A_k(u_{m+1},\chi) - B_k(u_m,\chi)]. \end{aligned}$$

Setting here $\chi = \vartheta_{m+1}$, we obtain, using (4.1), (4.2) and (4.8),

$$\begin{aligned} a_k(\vartheta_{m+1})^2 \le a_k(\vartheta_m)a_k(\vartheta_{m+1}) &+ a_k(\rho_{m+1}-\rho_m)a_k(\vartheta_{m+1}) \\ &+ Ck^{\nu+1}\|u_m\|_{2\nu+2}a_k(\vartheta_{m+1}) . \end{aligned}$$

Noticing that, by (4.6) and (4.9),

$$a_k(\rho_{m+1}-\rho_m) = a_k((I-Q)(u_{m+1}-u_m))$$

$$\leq C\kappa^{\alpha/2} h^{\mu} \|u_{m+1}-u_m\|_{\mu} \leq C\kappa^{\alpha/2} k h^{\mu} \|u_m\|_{\mu+2},$$

this yields by summation, for $\max(\mu, 2\nu) < \sigma \leq \max(\mu, 2\nu)+2$,

$$a_k(\vartheta_n) \leq a_k(\vartheta_1) + C(\kappa^{\alpha/2} h^{\mu} + k^{\nu}) k \sum_{m=1}^{n-1} \|u_m\|_{\max(\mu, 2\nu)+2}$$

$$\leq a_k(\vartheta_1) + C(\kappa^{\alpha/2} h^{\mu} + k^{\nu})(k \sum_{m=1}^{n-1} (mk)^{-\frac{1}{2}(\max(\mu, 2\nu) + 2 - \sigma)}) \|v\|_{\sigma}$$

$$\leq a_k(\vartheta_1) + C(\kappa^{\alpha/2} h^{\mu} + k^{\nu}) \|v\|_{\sigma} . \tag{4.11}$$

In order to estimate $a_k(\vartheta_1)$ we notice that since $e_0 = 0$ we have by (4.8), recalling that $2\nu \geq \alpha$,

$$A_k(e_1, \vartheta_1) = A_k(e_1, \vartheta_1) - B_k(e_0, \vartheta_1) \leq Ck^{\nu} \|v\|_{2\nu} a_k(\vartheta_1),$$

and hence

$$a_k(\vartheta_1)^2 = A_k(e_1, \vartheta_1) + A_k(\rho_1, \vartheta_1) \leq (a_k(\rho_1) + Ck^{\nu} \|v\|_{2\nu}) a_k(\vartheta_1),$$

so that using (4.10) we may conclude

$$a_k(\vartheta_1) \leq C(\kappa^{\alpha/2} h^{\mu} + k^{\nu}) \|v\|_{\sigma} .$$

Together with (4.11) this completes the proof.

We now turn to the error estimates for positive t and for non-smooth initial data, using the eigenfunction expansion (1.25). In the next three lemmas we consider, under the assumptions (i), (ii), (iii) of the introduction, the error between the solutions of the semi-discrete and completely discrete problems. These results can then be combined with the various estimates of Section 3 to yield complete error estimates. An example of this is given in Theorem 4.2 below.

Lemma 4.1.

Assume that (iii) holds. Then for each $t_0 > 0$ there is a constant $C(t_0)$ such that, for $nk \geq t_0$,

$$|U_n - U(nk)| \leq C(t_0) k^{\nu} \|V\|_{\ell_1}, \quad \text{where} \quad \|V\|_{\ell_1} = \max_j \frac{|(V, \Phi_j)|}{|\Phi_j|}.$$

Proof.

By (1.25) and (1.17) we may write

$$U_n - U(nk) = \sum_{m=1}^{3} \tilde{\varepsilon}_{m,n}$$

where, with j_k the smallest integer such that $k\Lambda_{j_k} \geq 1$,

$$\tilde{\varepsilon}_{1,n} = \sum_{j<j_k} B_j (r(k\Lambda_j)^n - e^{-nk\Lambda_j}) \Phi_j,$$

$$\tilde{\varepsilon}_{2,n} = -\sum_{j \geq j_k} B_j e^{-nk\Lambda_j} \Phi_j,$$

$$\tilde{\varepsilon}_{3,n} = \sum_{j \geq j_k} B_j r(k\Lambda_j)^n \Phi_j.$$

We first estimate $\tilde{\varepsilon}_{1n}$. By our assumptions (1.19) and (iii) we have, for $0 \leq \tau \leq 1$,

$$|r(\tau) - e^{-\tau}| \leq C\tau^{\nu+1} \quad \text{and} \quad |r(\tau)| \leq e^{-c\tau},$$

and hence, for these τ,

$$|r(\tau)^n - e^{-n\tau}| = |r(\tau) - e^{-\tau}| \left| \sum_{j=0}^{n-1} r(\tau)^{n-1-j} e^{-j\tau} \right|$$

$$\leq Cn\tau^{\nu+1} e^{-c(n-1)\tau} \leq C\tau^{\nu} e^{-cn\tau}.$$

Recalling (3.8) it follows for $nk \geq t_0$ that

$$(4.12) \qquad |\tilde{\varepsilon}_{1n}| \leq C\|v\|_{\ell_1} \sum_{j<j_k} |\Phi_j|^2 (k\Lambda_j)^{\nu} e^{-cnk\Lambda_j}$$

$$\leq Ck^{\nu}\|v\|_{\ell_1} \sum_{j=1}^{\infty} \Lambda_j^{\nu+1} e^{-ct_0\Lambda_j}$$

$$\leq C(t_0)k^{\nu}\|v\|_{\ell_1} .$$

Similarly, for $nk \geq t_0$,

$$(4.13) \qquad |\tilde{\varepsilon}_{2n}| \leq C\|v\|_{\ell_1} \sum_{j \geq j_k} |\Phi_j|^2 e^{-nk\Lambda_j}$$

$$\leq C\|v\|_{\ell_1} \sum_{j} \Lambda_j (k\Lambda_j)^{\nu} e^{-t_0\Lambda_j}$$

$$\leq C(t_0)k^{\nu}\|v\|_{\ell_1} .$$

Finally, we have for $nk \geq t_0$,

$$|\tilde{\varepsilon}_{3n}| \leq C\|v\|_{\ell_1} \sum_{j \geq j_k} \Lambda_j |r(k\Lambda_j)^n|$$

$$\leq C(t_0)k^{\nu}\|v\|_{\ell_1} n^{\nu+1} \sum_{j \geq j_k} k\Lambda_j |r(k\Lambda_j)^n| .$$

By (iii) the degree of the numerator is less than that of the denominator in $r(\tau)$ and it follows that there is a $c > 0$ such that

$$|r(\tau)| \leq \frac{1}{1+c\tau} \qquad \text{for } \tau \geq 1.$$

Hence, using the fact that $\Lambda_j \geq \lambda_j \geq cj^2 \geq cj$, we obtain that

$$\sum_{j \geq j_k} k\Lambda_j |r(k\Lambda_j)^n| \leq C \sum_{j \geq j_k} (1 + ck\Lambda_j)^{-(n-1)}$$

$$\leq C \sum_{j=1}^{\infty} (1+c\max(kj,1))^{-(n-1)}$$

$$\leq C \int_0^{\infty} (1+c\max(kx,1))^{-(n-1)} dx \leq C(t_0) \frac{n}{(1+c)^n}.$$

It follows that

$$|\tilde{\varepsilon}_{3n}| \leq C(t_0) k^\nu \|V\|_{\ell_1} \frac{n^{\nu+2}}{(1+c)^n} \leq C(t_0) k^\nu \|V\|_{\ell_1}$$

which together with (4.12) and (4.13) completes the proof.

Notice that, for $V = P_0 v$,

$$\|V\|_{\ell_1} = \max_j \frac{|(v, \Phi_j)|}{|\Phi_j|} \leq \|v\|_{L_1}. \tag{4.14}$$

Lemma 4.2.

Assume that (ii) holds. Then for each $t_0 > 0$ there is a constant $C(t_0)$ such that, for $nk \geq t_0$,

$$\|U_n - U(nk)\| \leq C(t_0) k^\nu \|V\|.$$

Proof.

In exactly the same way as before we have (4.12) and (4.13) from which clearly follows

$$\|\tilde{\varepsilon}_{jn}\| \leq C(t_0) k^\nu \|V\|, \quad j = 1,2. \tag{4.15}$$

It remains to consider $\tilde{\varepsilon}_{3n}$. This time, by (ii) there is a $\gamma > 0$ such that

$$\sup_{\tau \geq 1} |r(\tau)| = e^{-\gamma}.$$

We obtain by Parseval's relation that, for $nk \geq t_0$,

$$\|\tilde{\varepsilon}_{3n}\|^2 = \sum_{j \geq j_k} B_j^2 r(k\Lambda_j)^{2n} \leq e^{-2n\gamma}\|V\|^2 \tag{4.16}$$

$$\leq C(t_0)k^{2\nu}\|V\|^2$$

which concludes the proof of the lemma.

Lemma 4.3.

Assume that (i), (i'), (i") hold. Then for each $t_0 > 0$ there is a constant $C(t_0)$ such that, for $nk \geq t_0$,

$$\|U_n - U(nk)\| \leq C(t_0)k^{\nu}\|V\| .$$

Proof.

In exactly the same way as above we have (4.15). In order to prove that under the present assumptions also the conclusion in (4.16) holds we only have to notice that by (i') we have, for all j,

$$\Lambda_j = L(\Phi_j, \Phi_j) \leq Ch^{-2}$$

so that, by (i"), $k\Lambda_j \leq CK$, and that by (i) here is a $\gamma > 0$ such that

$$\sup_{1 \leq \tau \leq CK} |r(\tau)| = e^{-\gamma} .$$

This completes the proof of the lemma.

We finish by stating an example of a complete error estimate.

Theorem 4.2.

Assume that (1.13) and (iii) hold and that $V = P_0 v$. Then for each $t_0 > 0$ there is a constant $C(t_0)$ such that, for $nk \geq t_0$,

$$|U_n - u(nk)| \leq C(t_0)(h^{\mu} + k^{\nu})\|v\|_{L_1} .$$

Proof.

Immediate consequence of Theorem 3.2, Lemma 4.1 and (4.14).

References

1. J. H. Bramble and J. E. Osborn, Rate of convergence estimates for non-selfadjoint eigenvalue approximations, Math. Comp. 27 (1973), 525-549 .

2. J. H. Bramble and V. Thomée, Discrete time Galerkin methods for a parabolic boundary value problem, to appear in Ann. Mat. Pura Appl.

3. J. Douglas, Jr. and T. Dupont, Galerkin methods for parabolic equations, SIAM J. Numer. Anal. 7 (1970), 575-626.

4. J. Douglas, Jr., T. Dupont, and L. Wahlbin, Optimal L_∞ error estimates for Galerkin approximations of two point boundary value problems, to appear.

5. T. Dupont, Some L^2 error estimates for parabolic Galerkin methods, The Mathematical Foundations of the Finite Element Method with Applications to Partial Differential Equations. Ed. K. Aziz and I. Babuska. Academic Press, New York and London 1972, 491-504.

6. G. Fix and N. Nassif, On finite element approximations to time dependent problems, Numer. Math. 19 (1972), 127-135.

7. H. -P. Helfrich, Lokale Fehlerabschätzungen für das Galerkinverfahren zur Lösung von Evolutionsgleichungen, to appear.

8. H. O. Kreiss, V. Thomée, and O. Widlund, Smoothing of initial data and rates of convergence for parabolic difference equations, Comm. Pure Appl. Math. 23 (1970), 241-259.

9. O. A. Ladyženskaja, V. A. Solonnikov, N. N. Ural'ceva, Linear and quasilinear equations of parabolic type, Translations of mathematical monographs, Vol. 23, American Mathematical Society, Providence, R. I., 1968.

10. H. S. Price and R. S. Varga, Error bounds for semidiscrete Galerkin approximations of parabolic problems with applications to petroleum reservoir mechanics, Numerical Solution of Field Problems in Continuum Physics, American Mathematical Society, Providence, R. I., 1970, 74-94.

11. G. Strang and G. J. Fix, An analysis of the finite element method, Prentice-Hall, Englewood Cliffs, N. J., 1973.

12. V. Thomée and L. Wahlbin, Convergence rates of parabolic difference schemes for non-smooth data, to appear in Math. Comp.

13. L. Wahlbin, On maximum norm error estimates for Galerkin approximations to one dimensional second order parabolic boundary value problems, to appear.

14. M. Wheeler, L_∞ estimates of optimal order for Galerkin methods for one dimensional second order parabolic and hyperbolic equations, SIAM J. Numer. Anal. 10 (1973), 908-913.

15. M. F. Wheeler, An optimal L_∞ error estimate for Galerkin approximations to solutions of two point boundary value problems, SIAM J. Numer. Anal. 10 (1973), 914-917.

16. M. Zlámal, Finite element methods for parabolic equations I, to appear.

Department of Mathematics
Chalmers University of Technology
and
The University of Göteborg
Göteborg, Sweden

On a Finite Element Method for Solving the Neutron Transport Equation

P. LASAINT AND P. A. RAVIART

Introduction.

Let Ω be a convex open set in the (x, y)-plane with boundary Γ. Denote by $\underline{n} = (n_x, n_y)$ the outward unit vector normal to Γ.

Let Q be the unit disk in the (μ, ν)-plane. We consider the following problem: Find a function $u = u(x, y, \mu, \nu)$ such that

$$\text{(1.1)} \qquad \mu \frac{\partial u}{\partial x} + \nu \frac{\partial u}{\partial y} + \sigma u = f \quad \text{in } \Omega \times Q,$$

$$\text{(1.2)} \qquad u(x, y, \mu, \nu) = 0 \quad \text{if } (x, y) \in \Gamma, \quad (\mu n_x + \nu n_y)(x, y) < 0 .$$

Equation (1.1) is the neutron transport equation: The function $u(x, y, \mu, \nu)$ represents the flux of neutrons at the point (x, y) in the angular direction (μ, ν), σ is the nuclear cross section, and f stands for the scattering, the fission and the inhomogeneous source terms. The boundary condition (1.2) simply means that no neutrons are entering the system from outside.

In this paper, we shall be only concerned with the spatial discretization of problem (1.1), (1.2). Thus, we shall assume that the angular direction (μ, ν) is fixed and we shall consider the reduced problem: Given a function f defined over Ω, find a function u defined over Ω such that

$$
(1.3) \qquad \begin{aligned} &\underline{\underline{m}} \cdot \operatorname{grad} u + \sigma u = f \text{ in } \Omega , \\ &u = 0 \quad \text{on} \quad \Gamma_- , \end{aligned}
$$

where $\underline{\underline{m}} = (\mu, \nu)$ and

$$
(1.4) \qquad \Gamma_- = \{(x,y) \in \Gamma \mid \underline{\underline{m}} \cdot \underline{\underline{n}}(x,y) < 0\} .
$$

This paper will be devoted to the numerical approximation of problem (1.3) by a finite element method using triangular or quadrilateral elements which has been recently introduced by Reed and Hill [17] and which appears to be very effective in practice. Other finite element methods for solving the neutron transport equation have been introduced by several authors (cf. for instance [10], [14], [15], [16]). We refer to [12] for a mathematical discussion of some of them.

An outline of the paper is as follows. In §2, we study a discontinuous Galerkin method for ordinary differential equations using polynomials of degree k. This Galerkin method is shown to be strongly A-stable and of order $2k+1$. In §3, we introduce the finite element method as a generalization of the discontinuous Galerkin method of §2. We prove the existence and uniqueness of the approximate solution and we give an algorithm for computing this approximate solution. In §4, we derive general error bounds in the L_2-norm. Finally, we give in §5 a superconvergence result.

Note that problem (1.3) is a simple but important example of a first-order hyperbolic problem. In fact, the finite element method studied in this paper provides an effective way for numerically solving such problems. For other finite element methods for solving first order systems of partial differential equations, we refer to [11], [13].

For the sake of simplicity, we have confined ourselves to polygonal domains Ω. It is probably an easy matter to handle general curved domains by using curved isoparametric elements and the analysis given in [5], [6].

2. A Discontinuous Galerkin Method for Ordinary Differential Equations.

We begin by studying the numerical solution of the ordinary differential equation

$$\text{(2.1)}\qquad \begin{aligned} u'(x) &= f(x, u(x)), \quad x \geq x_0 , \\ u(x_0) &= u_0 , \end{aligned}$$

on a finite interval $[x_0, x_0+a]$ by a discontinuous Galerkin method. For continuous Galerkin methods and related collocation methods, we refer for instance to Axelsson [1], de Boor and Swartz [2], Hulme [9].

Let $x_n = x_0 + nh$, $0 \leq n \leq N$ $(Nh = a)$ be a uniform mesh for the sake of simplicity. Then we may approximate u by a function u_h which, on each subinterval $[x_n, x_{n+1}]$, reduces to a polynomial of degree $\leq k$. We require that u_h satisfies on each subinterval $[x_n, x_{n+1}]$, $0 \leq n \leq N-1$:

$$\text{(2.2)}\qquad \begin{aligned} &(u_h(x_{n+}) - u_h(x_{n-}))v(x_n) \\ &+ \int_{x_n}^{x_{n+1}} \{u_h'(x) - f(x, u_h(x))\}v(x)dx = 0 \\ &\qquad\qquad\qquad \text{for all } v \in P_k \end{aligned}$$

with the initial condition

$$\text{(2.3)}\qquad u_h(x_{0-}) = u_0 ,$$

where P_k denotes the space of all polynomials of degree $\leq k$. Notice that the function u_h is in general discontinuous at the mesh points x_n.

To obtain a computational form of (2.2)-(2.3), we replace the integral in (2.2) by an interpolatory quadrature formula

$$\text{(2.4)}\qquad \int_{x_n}^{x_{n+1}} \varphi(x)dx = h \sum_{i=1}^{k+1} b_i \varphi(x_{n,i}) + O(h^{p+1}) ,$$

(2.5) $$x_{n,i} = x_n + \xi_i h \,, \quad 1 \le i \le k+1 \,, \quad \xi_1 = 0 \,,$$

where b_i and ξ_i are the weights and abscissae for $[0,1]$. Notice that $k+1 \le p \le 2k+1$. Then (2.2) becomes

(2.6) $$\begin{aligned} &(u_h(x_{n+}) - u_h(x_{n-}))v(x_n) \\ &+ h \sum_{i=1}^{k+1} b_i \{u_h'(x_{n,i}) - f(x_{n,i}, u_h(x_{n,i}))\}\, v(x_{n,i}) = 0 \\ &\qquad\qquad \text{for all } v \in P_k \,. \end{aligned}$$

Let us now show that the discrete Galerkin method (2.3), (2.6) is equivalent to some implicit Runge-Kutta method. We define

(2.7) $$\begin{aligned} u_n &= u_h(x_{n-}), \\ u_{n,1} &= u_h(x_{n+}) = u_h(x_{n,1}) \,, \\ u_{n,i} &= u_h(x_{n,i}) \,, \quad 2 \le i \le k+1 \,. \end{aligned}$$

We introduce the Lagrange interpolation coefficients

(2.8) $$\ell_i(x) = \prod_{\substack{j=2 \\ j \ne i}}^{k+1} \frac{x - \xi_j}{\xi_i - \xi_j} \,, \quad 2 \le i \le k+1 \,.$$

<u>Lemma 1.</u>

<u>The discrete Galerkin method</u> (2.3), (2.6) <u>is equivalent to the following implicit Runge-Kutta method</u>

(2.9) $$\begin{aligned} u_{n,i} &= u_n + h \sum_{j=1}^{k+1} a_{ij} f(x_{n,j}, u_{n,j}), \quad 1 \le i \le k+1 \,, \\ u_{n+1} &= u_n + h \sum_{j=1}^{k+1} b_j f(x_{n,j}, u_{n,j}) \,, \end{aligned}$$

<u>where</u>

(2.10) $$\begin{aligned} a_{i1} &= b_1 \,, \quad 1 \le i \le k+1 \,, \\ a_{ij} &= \int_0^{\xi_i} \ell_j(x)dx - b_1 \ell_j(\xi_1), \quad 1 \le i \le k+1, \ 2 \le j \le k+1. \end{aligned}$$

Proof.

Let us introduce the basis $\{v_i\}_{1 \leq i \leq k+1}$ for the space P_k defined by

$$v_i(x_{n,j}) = \delta_{ij} \ , \quad 1 \leq i,j \leq k+1 \ .$$

By replacing successively in (2.6) v by v_i, we find that an equivalent form of (2.6) is given by

$$\text{(2.11)} \quad \begin{aligned} & u_h(x_{n+}) - u_h(x_{n-}) + hb_1[u_h'(x_{n,1}) - f(x_{n,1}, u_h(x_{n,1}))] = 0 \\ & u_h'(x_{n,i}) - f(x_{n,i}, u_h(x_{n,i})) = 0 \ , \quad 2 \leq i \leq k+1 \ . \end{aligned}$$

In the subinterval $[x_n, x_{n+1}]$, we have $u_h' \in P_{k-1}$ so that

$$u_h'(x) = \sum_{j=2}^{k+1} \ell_j\left(\frac{x-x_n}{h}\right) u_h'(x_{n,j})$$

and by (2.11)

$$\text{(2.12)} \quad u_h'(x) = \sum_{j=2}^{k+1} \ell_j\left(\frac{x-x_n}{h}\right) f(x_{n,j}, u_h(x_{n,j})) \ .$$

Taking $x = x_n = x_{n,1}$ in (2.12), substituting this expression into the 1st equation (2.11) and using (2.7), we obtain

$$\text{(2.13)} \quad u_{n,1} = u_n + h\, b_1\{f(x_{n,1}, u_{n,1}) - \sum_{j=2}^{k+1} \ell_j(\xi_1) f(x_{n,j}, u_{n,j})\}.$$

On the other hand, we may write for $2 \leq i \leq k+1$

$$u_h(x_{n,i}) = u_h(x_{n,1}) + \int_{x_{n,1}}^{x_{n,i}} u_h'(x)dx$$

and by (2.7), (2.12), (2.13)

$$\text{(2.14)} \quad \begin{aligned} u_{n,i} = u_n & + h\{b_1 f(x_{n,1}, u_{n,1}) \\ & + \sum_{j=2}^{k+1} [\int_0^{\xi_i} \ell_j(x)dx - b_1 \ell_j(\xi_1)] f(x_{n,j}, u_{n,j})\} \ . \end{aligned}$$

Similarly, we have

$$u_h(x_{n+1}-) = u_h(x_{n,1}) + \int_{x_n}^{x_{n+1}} u_h'(x)dx$$

and then

$$u_{n+1} = u_n + h\{b_1 f(x_{n,1}, u_{n,1}) + \sum_{j=2}^{k+1} [\int_0^1 \ell_j(x)dx - b_1\ell_j(\xi_1)] f(x_{n,j}, u_{n,j})\} .$$

By noticing that

$$\int_0^1 \ell_j(x)dx = \sum_{i=1}^{k+1} b_i \ell_j(\xi_i) = b_1\ell_j(\xi_1) + b_j ,$$

we get

$$u_{n+1} = u_n + h \sum_{j=1}^{k+1} b_j f(x_{n,j}, u_{n,j}) . \tag{2.15}$$

The equations (2.13)-(2.15) are identical to the equations (2.9), (2.10). We then have proved that the discrete Galerkin method leads to the one-step method (2.9), (2.10). Conversely, the Runge-Kutta method (2.9), (2.10) can be clearly viewed as a discrete Galerkin method. ■

Theorem 1.

The discrete Galerkin method (2.3), (2.6) *is a one-step method of order* p .

Proof.

Following Butcher [3], Crouzeix [7], we know that the conditions

$$\sum_{j=1}^{k+1} b_j \xi_j^\ell = \frac{1}{\ell+1} , \quad 0 \le \ell \le p-1 , \tag{2.16}$$

$$\sum_{j=1}^{k+1} a_{ij} \xi_j^\ell = \frac{\xi_i^{\ell+1}}{\ell+1} , \quad 0 \le \ell \le k-1 , \quad 1 \le i \le k+1 \tag{2.17}$$

$$(2.18)\quad \sum_{i=1}^{k+1} b_i a_{ij} \xi_i^{\ell} = \frac{1}{\ell+1} b_j(1-\xi_j^{\ell+1}),\ k+\ell \leq p-1,\ 1 \leq j \leq k+1,$$

are sufficient for the Runge-Kutta method (2.9) to be of order p . Let us show that these conditions hold in the present case.

First, conditions (2.16) simply mean that the interpolatory quadrature formula (2.4) is exact for all polynomials of degree $\leq p-1$.

Next, consider conditions (2.17). Using (2.8), we may write

$$x^{\ell} = \sum_{j=2}^{k+1} \ell_j(x)\xi_j^{\ell}, \qquad 0 \leq \ell \leq k-1 ,$$

so that

$$\xi_1^{\ell} = \sum_{j=2}^{k+1} \ell_j(\xi_1)\xi_j^{\ell}, \qquad 0 \leq \ell \leq k-1 ,$$

$$\frac{\xi_i^{\ell+1}}{\ell+1} = \sum_{j=2}^{k+1} \left(\int_0^{\xi_i} \ell_j(x)dx\right)\xi_j^{\ell}, \quad 0 \leq \ell \leq k-1,\ 1 \leq i \leq k+1 .$$

Using (2.10), we have

$$\sum_{j=1}^{k+1} a_{ij}\xi_j^{\ell} = b_1\left(\xi_1^{\ell} - \sum_{j=2}^{k+1} \ell_j(\xi_1)\xi_j^{\ell}\right) + \sum_{j=2}^{k+1}\left(\int_0^{\xi_i} \ell_j(x)dx\right)\xi_j^{\ell}$$

and by the previous relations

$$\sum_{j=1}^{k+1} a_{ij}\xi_j^{\ell} = \frac{\xi_i^{\ell+1}}{\ell+1}, \quad 0 \leq \ell \leq k-1 , \quad 1 \leq i \leq k+1 .$$

Finally, let us show that conditions (2.18) hold. We begin by noticing that

$$(2.19)\quad \sum_{i=1}^{k+1} b_i a_{i1}\xi_i^{\ell} = b_1 \sum_{i=1}^{k+1} b_i \xi_i^{\ell} = \frac{b_1}{\ell+1}, \quad 0 \leq \ell \leq p-1 .$$

On the other hand, following Crouzeix [7], we may write for any continuous function φ

$$\int_0^1 x^{\ell}\left(\int_0^x \varphi(y)dy\right)dx = \frac{1}{\ell+1}\int_0^1 (1-x^{\ell+1})\varphi(x)dx \,. \tag{2.20}$$

Taking $\varphi \in P_{k-1}$, we obtain for $k+\ell \leq p-1$

$$\begin{aligned}\int_0^1 x^{\ell}\left(\int_0^x \varphi(y)dy\right)dx &= \sum_{i=1}^{k+1} b_i \xi_i^{\ell} \int_0^{\xi_i} \varphi(y)dy \\ &= \sum_{i=1}^{k+1} b_i \xi_i^{\ell} \sum_{j=2}^{k+1} \left(\int_0^{\xi_i} \ell_j(y)dy\right)\varphi(\xi_j)\end{aligned}$$

and by (2.10)

$$\int_0^1 x^{\ell}\left(\int_0^x \varphi(y)dy\right)dx = \sum_{i,j=1}^{k+1} b_i a_{ij} \xi_i^{\ell} \varphi(\xi_j), \quad \varphi \in P_{k-1}, \quad k+\ell \leq p-1 \,. \tag{2.21}$$

Similarly, we get

$$\int_0^1 (1-x^{\ell+1})\varphi(x)dx = \sum_{j=1}^{k+1} b_j (1-\xi_j^{\ell+1})\varphi(\xi_j), \quad \varphi \in P_{k-1}, \quad k+\ell \leq p-1 \,. \tag{2.22}$$

Hence, combining (2.19)-(2.22), we have for all $\varphi \in P_{k-1}$ and for $k + \ell \leq p-1$

$$\sum_{j=1}^{k+1} \left[\sum_{i=1}^{k+1} b_i a_{ij} \xi_i^{\ell} - \frac{1}{\ell+1} b_j (1 - \xi_j^{\ell+1})\right]\varphi(\xi_j) = 0 \,.$$

This implies

$$\sum_{i=1}^{k+1} b_i a_{ij} \xi_i^{\ell} = \frac{1}{\ell+1} b_j (1-\xi_j^{\ell+1}), \quad k+\ell \leq p-1, \quad 2 \leq j \leq k+1.$$

■

In order to investigate the stability properties of the one-step method (2.9), we consider the differential equation

$$u' = \lambda u \tag{2.23}$$

where λ is a complex constant with $Re(\lambda) < 0$.

Lemma 2.

Applied to the differential equation (2.23), *the one-step method* (2.9), (2.10) *gives*

(2.24) $$u_{n+1} = R(\lambda h)\, u_n$$

where $R(z) = \frac{P(z)}{Q(z)}$ *is the quotient of two polynomials* $P(z)$ *and* $Q(z)$ *of degree* $\leq k$ *and* $\leq k+1$, *respectively.*

Proof.

Applied to (2.23), the one-step method (2.9) becomes

(2.25) $$u_{n,i} = u_n + \lambda h \sum_{j=1}^{k+1} a_{ij}\, u_{n,j}\ , \qquad 1 \leq i \leq k+1$$

(2.26) $$u_{n+1} = u_n + \lambda h \sum_{j=1}^{k+1} b_j\, u_{n,j}\ .$$

Using obvious notations, we may write equations (2.25) in the form

$$(I - \lambda h\, [a_{ij}])[u_{n,i}] = u_n[1]$$

where the identity matrix I and $[a_{ij}]$ are $(k+1)\times(k+1)$-matrices. Since $a_{i1} = b_1$, $1 \leq i \leq k+1$, we get from Cramer's rule

$$u_{n,i} = \frac{P_i(\lambda h)}{Q(\lambda h)}\, u_n\ , \qquad 1 \leq i \leq k+1$$

where $P_1(z)$ is a polynomial of degree k whose leading coefficient is $b_1^{-1} \det [a_{ij}]$, $P_i(z)$, $2 \leq i \leq k+1$, are polynomials of degree $\leq k-1$ and where $Q(z)$ is a polynomial of degree $k+1$ whose leading coefficient is $\det[a_{ij}]$.

Using (2. 26), we obtain

$$u_{n+1} = \frac{P(\lambda h)}{Q(\lambda h)} u_n$$

where

$$P(z) = Q(z) - z \sum_{j=1}^{k+1} P_j(z) .$$

Clearly, in $P(z)$, the coefficient of z^{k+1} vanishes. The lemma is then proved. ∎

Let us now recall the following definition: A one-step method is <u>strongly A-stable</u> if

$$(2.27) \quad \begin{aligned} &|R(z)| < 1 \quad &&\text{for } \operatorname{Re}(z) < 0 , \\ &|R(z)| \to 0 \quad &&\text{as } \operatorname{Re}(z) \to -\infty. \end{aligned}$$

Theorem 2.

The Galerkin method (2. 2), (2. 3) is a strongly A-stable one-step method of order $2k+1$.

Proof.

Consider first the discrete Galerkin method (2. 3), (2. 6) associated with the Gauss-Radau abscissae ξ_i, $1 \leq i \leq k+1$ ($\xi_1 = 0$). Then, we have $p = 2k+1$ in (2. 4). By Theorem 1, this discrete Galerkin method is a one-step method of order $2k+1$ so that

$$R(z) = \exp(z) + O(z^{2k+2}) .$$

Moreover, by Lemma 2, $R(z)$ is the quotient of two polynomials $P(z)$ and $Q(z)$ of degree $\leq k$ and $\leq k+1$, respectively. Then, necessarily, $R(z)$ is the subdiagonal $(k+1, k)$ Padé rational approximation of $\exp(z)$. Using a result of Axelsson [1], we know that such a Padé approximation satisfies conditions (2. 27). Hence, the discrete Galerkin method (2. 3), (2. 6) associated with the Gauss-Radau abscissae is a strongly A-stable one-step method of order $2k+1$.

Now, it is a simple but lengthy matter to prove that the Galerkin method (2.2), (2.3) and the Gauss-Radau discrete Galerkin method (2.3), (2.6) are one-step methods of the same order $2k+1$. Moreover, these two methods coincide when applied to the differential equation (2.23). This completes the proof of the theorem. ∎

3. A Finite Element Method for the Neutron Transport Equation.

Consider now our neutron transport problem (1.3). First, we need some notations. Let us denote by $L_2(\Omega)$ the space of real-valued functions v which are square integrable over Ω. We provide $L_2(\Omega)$ with the usual norm

$$\|v\|_{0,\Omega} = \left(\int_\Omega |v(x)|^2 dx\right)^{\frac{1}{2}} . \tag{3.1}$$

Given any integer $m \geq 0$, let

$$H^m(\Omega) = \{v \in L_2(\Omega) \mid \partial^\alpha v \in L_2(\Omega),\ |\alpha| \leq m\} \tag{3.2}$$

be the usual Sobolev space provided with the norm

$$\|v\|_{m,\Omega} = \Big(\sum_{|\alpha| \leq m} \|\partial^\alpha v\|^2_{0,\Omega}\Big)^{\frac{1}{2}} \tag{3.3}$$

In (3.2), (3.3), $\alpha = (\alpha_1, \alpha_2) \in \mathbb{N}^2$ is a multi-index, $|\alpha| = \alpha_1 + \alpha_2$, and

$$\partial^\alpha = \left(\frac{\partial}{\partial x_1}\right)^{\alpha_1}\left(\frac{\partial}{\partial x_2}\right)^{\alpha_2} .$$

We shall also use the following semi-norm

$$|v|_{m,\Omega} = \Big(\sum_{|\alpha| = m} \|\partial^\alpha v\|^2_{0,\Omega}\Big)^{\frac{1}{2}} \tag{3.4}$$

Let us introduce the operator

$$A = \underline{\underline{m}} \cdot \text{grad} + \sigma = \mu \frac{\partial}{\partial x} + \nu \frac{\partial}{\partial y} + \sigma \tag{3.5}$$

and the space

$$D(A) = \{ v \in L_2(\Omega) \mid \underline{\underline{m}} \cdot \text{grad}\, v \in L_2(\Omega) \}. \tag{3.6}$$

Then, as a consequence of [8], we have the following result.

Theorem 3.

Assume that $\sigma \in L_\infty(\Omega)$ and $f \in L_2(\Omega)$. Then, problem (1.3) has a unique strong solution $u \in D(A)$.

With the substitution

$$u = \exp(\lambda(\frac{x}{\mu} + \frac{y}{\nu}))w,$$

equation (1.3) becomes

$$\underline{\underline{m}} \cdot \text{grad}\, w + (\sigma + \lambda)w = \exp(-\lambda(\frac{x}{\mu} + \frac{y}{\nu}))f .$$

Thus, by eventually changing $\sigma(x, y)$ into $\sigma(x, y) + \lambda$, we can restrict ourselves to the case where σ is positive. More precisely, we shall assume in the sequel that

$$M \geq \sigma(x, y) \geq \alpha > 0 \quad \text{a.e. in } \Omega . \tag{3.7}$$

Let us now generalize the one-dimensional discontinuous Galerkin method of §2 to our two-dimensional neutron transport problem. For the sake of simplicity, we shall assume in the following that $\bar{\Omega}$ is a polygon. In order to approximate problem (1.3), we first construct a triangulation $\mathcal{T}_h$ of $\bar{\Omega}$ with triangles and convex quadrilaterals K with diameters $\leq h$. With any $K \in \mathcal{T}_h$, we associate the finite-dimensional space P_K of real-valued functions defined on K such that

$$P_K \subset H^1(K) . \tag{3.8}$$

We then consider the finite-dimensional space

$$(3.9) \qquad V_h = \{v \mid v \in L_2(\Omega), \quad v|_K \in P_K \quad \text{for all } K \in \mathfrak{T}_h\}.$$

It is worthwhile to notice that in general <u>a function</u> $v \in V_h$ <u>does not satisfy any continuity requirement at the interelement boundaries.</u>

Let $K \in \mathfrak{T}_h$ and let ∂K be the boundary of K. We set

$$(3.10) \qquad \begin{aligned} \partial_- K &= \{(x, y) \in \partial K \mid \underline{\underline{m}} \cdot \underline{\underline{n}}(x, y) < 0\}, \\ \partial_+ K &= \{(x, y) \in \partial K \mid \underline{\underline{m}} \cdot \underline{\underline{n}}(x, y) > 0\}, \end{aligned}$$

where $\underline{\underline{n}} = (n_x, n_y)$ is the outward unit vector normal to the boundary ∂K.

Then, the finite element approximation of problem (1.3) that we shall consider here can be stated as follows. <u>Find a function</u> $u_h \in V_h$ <u>such that for all</u> $K \in \mathfrak{T}_h$

$$(3.11) \qquad -\int_{\partial_- K} \underline{\underline{m}} \cdot \underline{\underline{n}}\,(u_h - \xi_h) v \, ds + \int_K (Au_h - f) v \, dx dy = 0 \qquad \text{for all } v \in P_K$$

<u>where</u>

$$(3.12) \qquad \xi_h = \begin{cases} 0 & \underline{\text{on}} \quad \partial_- K \cap \Gamma_- , \\ \underline{\text{outward trace of}}\ u_h & \underline{\text{on}}\ \partial_- K \setminus (\partial_- K \cap \Gamma_-). \end{cases}$$

This method is clearly a direct generalization of the discontinuous Galerkin method (2.2), (2.3).

Before proving existence and uniqueness of the solution $u_h \in V_h$, we shall show that there exists an ordering of the elements of $\mathfrak{T}_h$ well suited for numerically solving equations (3.11), (3.12).

<u>Lemma 3.</u>

<u>There exists an ordering</u> $K_1, K_2, \ldots, K_I$ <u>of the elements of</u> $\mathfrak{T}_h$ <u>such that, for all</u> $i = 1, \ldots, I$, <u>each side of</u>

$\partial_- K_i$ is either a subset of Γ_- or a subset of $\partial_+ K_j$ for some $j < i$.

Proof.

Let us introduce first some notations. We shall say that K is a boundary element if at least one side of ∂K is a subset of Γ, and that K is a semi-boundary element if one and only one vertex of K belongs to Γ. Let us consider Γ_- and let us number clockwise the corresponding

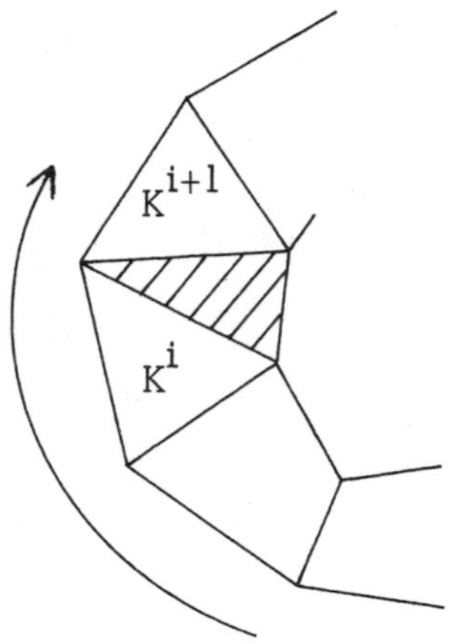

Figure 1

boundary elements $K^1, K^2, \ldots, K^s$. Two consecutive boundary elements K^i and K^{i+1} can have a common side or not. In the latter case (cf. Figure 1), there exists at least one semi-boundary element located between K^i and K^{i+1}. Then, we shall say that a side of K^i (resp. K^{i+1}) is semi-common with K^{i+1} (resp. K^i) if it is a subset of the union of the semi-boundary elements located between K^i and K^{i+1}.

Next, we show that there exists at least one boundary element K such that $\partial_- K \subset \Gamma_-$. To prove this, let us assume on the contrary that $\partial_- K^i \not\subset \Gamma_-$ for all $i = 1, \ldots, s$ and let us show that this assumption leads to a contradiction. Consider the first boundary element K^1 and use the notation of Figure 2.

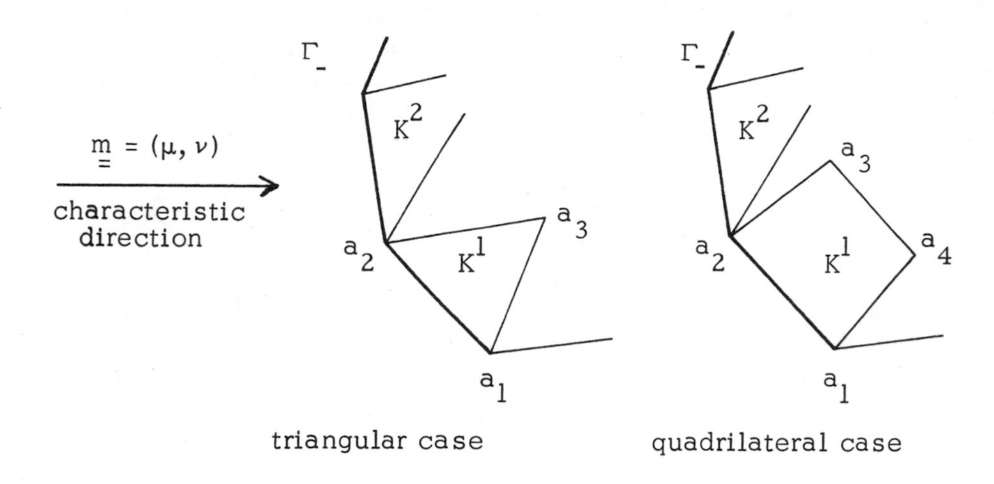

Figure 2

In the triangular case (resp. in the quadrilateral case), the side $[a_1, a_3]$ (resp. $[a_1, a_4]$) of K^1 is a subset of ∂_+K^1. Otherwise, K^1 would not be the first boundary element of Γ_-. Then, the side $[a_2, a_3]$ of K^1 which is common or semi-common with K^2 belongs to ∂_-K^1. Otherwise, we should get $\partial_-K^1 = [a_1, a_2] \subset \Gamma_-$ which is excluded. Therefore, the side of K^2 which is common or semi-common with K^1 belongs to ∂_+K^2. More generally, we get for every $i = 1, \ldots, s-1$ the following property: the side of K^i which is common or semi-common with K^{i+1} is a subset of ∂_-K^i and therefore the side of K^{i+1} which is common or semi-common with K^i is a subset of ∂_+K^{i+1}. Now consider the last boundary element K^s and use the notations of

Figure 3. In the triangular case (resp. in the quadrilateral case), the side $[a_1, a_3]$ (resp. $[a_1, a_4]$) of K^s is a subset of $\partial_+ K^s$. Moreover, the side $[a_2, a_3]$ is a subset of $\partial_+ K^s$. Otherwise, K^s would not be the last boundary element of Γ_-. Thus, we get $\partial_- K^s = [a_1, a_2] \subset \Gamma_-$ which has been excluded. The existence of a boundary element K such that $\partial_- K \subset \Gamma_-$ is then proved.

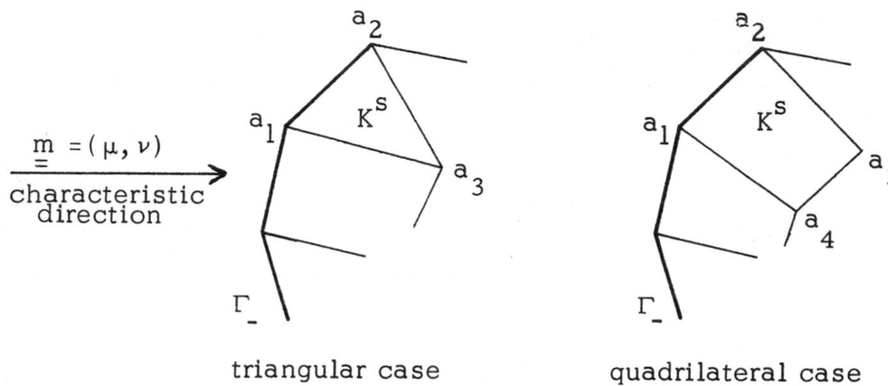

Figure 3

Now, choose for K_1 a boundary element of Γ_- such that $\partial_- K_1 \subset \Gamma_-$ and define $\Omega_1 = \Omega \setminus \Omega \cap K_1$, $\Gamma_{1-} = \partial_- \Omega_1$. Note that each side of Γ_{1-} is either a subset of Γ_- or a subset of $\partial_+ K_1$. By the previous argument, there exists a boundary element K_2 of Γ_{1-} such that $\partial_- K_2 \subset \Gamma_{1-}$, etc. Repeating this process, we take into account all the elements of $\mathfrak{T}_h$, and we obtain an ordering $K_1, K_2, \ldots, K_I$ of the elements of $\mathfrak{T}_h$ such that the desired property holds. ∎

This proof suggests an ordering algorithm for the element of $\mathfrak{T}_h$ which is effectively used in practice.

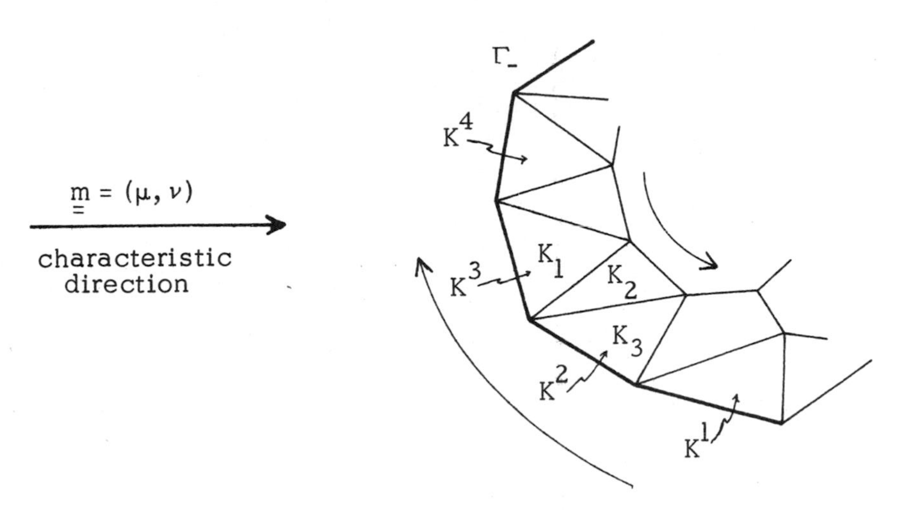

Figure 4

Consider the sequence $K^1, K^2, \ldots, K^s$ of boundary elements of Γ_-. For K_1 we choose the first element K of this sequence which satisfies $\partial_- K \subset \Gamma_-$. Let K^p be this element ($p = 3$ in Figure 4). From $K_1 = K^p$, we then number counterclockwise $K_1, K_2, \ldots, K_r$ the boundary and semi-boundary elements located between K^p and K^1 which satisfy the following condition: for all $i = 1, \ldots, r$, each side of $\partial_- K_i$ is either a subset of Γ_- or a subset of $\partial_+ K_j$ for some $j < i$ ($r = 3$ in Figure 4). Next, we replace the set Ω by $\Omega \backslash \Omega \cap (\bigcup_{i=1}^{r} K_i)$ and we repeat the process, etc.

We are now able to prove

Theorem 4.

Assume that $f \in L_2(\Omega)$ *and that condition* (3.7) *holds. Then, there exists a unique function* $u_h \in V_h$ *which satisfies equations* (3.11) *and* (3.12) *for all* $K \in \mathcal{T}_h$.

Proof.

Clearly, the finite element method (3.11), (3.12) is equivalent to an $N \times N$ linear system of equations with $N = \dim V_h$. Then, it is sufficient to prove the uniqueness of the solution u_h. Thus, let us assume that $f = 0$ and let us show that necessarily $u_h = 0$. Let $K_1, K_2, \ldots, K_I$ be an ordering of the elements $K \in \mathfrak{T}_h$ such that the condition of Lemma 3 holds. If $u_h = 0$ in $K_1 \cup K_2 \cup \ldots \cup K_{i-1}$, then $\xi_h = 0$ on $\partial_- K_i$ and equation (3.11) becomes, in $K = K_i$,

$$-\int_{\partial_- K} \underline{\underline{m}} \cdot \underline{\underline{n}} \; u_h v \, ds + \int_{K_i} (A u_h) v \, dx dy = 0$$

for all $v \in P_{K_i}$

Taking $v = u_h$ and using Green's formula

$$\int_{K_i} (\underline{\underline{m}} \cdot \operatorname{grad} u_h) u_h \, dx dy = \tfrac{1}{2} \int_{\partial K_i} \underline{\underline{m}} \cdot \underline{\underline{n}} \; u_h^2 \, ds ,$$

we get

$$\int_{\partial_+ K_i} \underline{\underline{m}} \cdot \underline{\underline{n}} \; u_h^2 \, ds - \int_{\partial_- K_i} \underline{\underline{m}} \cdot \underline{\underline{n}} \; u_h^2 \, ds + \int_{K_i} \sigma \, u_h^2 \, dx dy = 0 .$$

Using (3.7) and (3.10), we obtain $u_h = 0$ in K_i. Therefore, using an inductive argument, we get $u_h = 0$ in Ω. ∎

In practice, the computation of the approximate solution $u_h \in V_h$ goes along the following lines:

(i) Find an ordering $K_1, K_2, \ldots, K_I$ of the elements $K \in \mathfrak{T}_h$ which satisfies the condition of Lemma 3, for instance by using the previous algorithm;

(ii) Compute successively u_h in $K_1, K_2, \ldots, K_I$. The computation of u_h in each K_i has a local character and involves the numerical solution of a $d_i \times d_i$ linear system where $d_i = \dim P_{K_i}$.

In other words, by using an ordering of $\mathfrak{T}_h$ such that the condition of Lemma 3 holds, the $N \times N$ matrix of the approximate problem becomes block triangular and the ith diagonal block is a $d_i \times d_i$ matrix associated with the ith element K_i.

Note that, in many practical problems, the geometry of Ω and the triangulation $\mathfrak{T}_h$ are so simple that step (i) becomes obvious.

4. General Error Bounds.

Let us now derive some estimates for the error $u_h - u$ when the solution u of problem (1.3) is smooth enough. We begin with

Lemma 4.

For any $K \in \mathfrak{T}_h$, *any* $v \in P_K$ *and any function* $\eta \in L_2(\partial_- K)$ *we have the estimate*

$$\frac{1}{2}\int_{\partial_+ K} \underline{\underline{m}} \cdot \underline{\underline{n}} \; (u_h - v)^2 ds + \frac{1}{2}\int_{\partial_- K} \underline{\underline{m}} \cdot \underline{\underline{n}} \; (\xi_h - \eta)^2 ds$$

$$- \frac{1}{2}\int_{\partial_- K} \underline{\underline{m}} \cdot \underline{\underline{n}} \; ((u_h - v) - (\xi_h - \eta))^2 ds + \int_K \sigma (u_h - v)^2 dxdy$$

(4.1)

$$= \int_{\partial_+ K} \underline{\underline{m}} \cdot \underline{\underline{n}} \; (u - v)(u_h - v) ds + \int_{\partial_- K} \underline{\underline{m}} \cdot \underline{\underline{n}} \; (u - \eta)(u_h - v) ds$$

$$+ \int_K (u - v) A^* (u_h - v) dxdy$$

where A^* *is the formal adjoint of the operator* A, *i.e.*,

$$A^* = -\underline{\underline{m}} \cdot \operatorname{grad} + \sigma . \tag{4.2}$$

<u>Proof.</u>

Given $v \in P_K$ and $\eta \in L_2(\partial_- K)$, we set:

$$w = u_h - v \in P_K, \quad \zeta = \xi_h - \eta. \tag{4.3}$$

Consider the expression

$$X_h = -\int_{\partial_- K} \underline{\underline{m}} \cdot \underline{\underline{n}} \; (w-\zeta) w \, ds + \int_K (Aw) w \, dx dy. \tag{4.4}$$

First, using Green's formula, we obtain

$$X_h = \tfrac{1}{2} \int_{\partial K} \underline{\underline{m}} \cdot \underline{\underline{n}} \; w^2 ds - \int_{\partial_- K} \underline{\underline{m}} \cdot \underline{\underline{n}} \; (w-\zeta) w \, ds + \int_K \sigma w^2 dx \, dy.$$

Since

$$(w-\zeta) w = \tfrac{1}{2} (w^2 - \zeta^2 + (w-\zeta)^2),$$

we get

$$\begin{aligned} X_h &= \tfrac{1}{2} \int_{\partial_+ K} \underline{\underline{m}} \cdot \underline{\underline{n}} \; w^2 ds + \tfrac{1}{2} \int_{\partial_- K} \underline{\underline{m}} \cdot \underline{\underline{n}} \; \zeta^2 ds \\ &\quad - \tfrac{1}{2} \int_{\partial_- K} \underline{\underline{m}} \cdot \underline{\underline{n}} \; (w-\zeta)^2 ds + \int_K \sigma w^2 dx \, dy \end{aligned} \tag{4.5}$$

On the other hand, using (3.11), we obtain

$$X_h = \int_{\partial_- K} \underline{\underline{m}} \cdot \underline{\underline{n}} \; (v-\eta) w \, ds + \int_K (f - Av) w \, dx dy$$

and therefore

$$X_h = \int_{\partial_- K} \underline{\underline{m}} \cdot \underline{\underline{n}} \; (v-\eta) w \, ds + \int_K A(u-v) w \, dx dy.$$

Since $u \in D(A)$, we may write

$$\int_K A(u-v)w\,dx\,dy = \int_K (u-v)A^* w\,dxdy + \int_{\partial K} \underline{\underline{m}} \cdot \underline{\underline{n}}\,(u-v)w\,ds$$

so that

$$X_h = \int_{\partial_+ K} \underline{\underline{m}} \cdot \underline{\underline{n}}\ (u-v)w\,ds + \int_{\partial_- K} \underline{\underline{m}} \cdot \underline{\underline{n}}\ (u-\eta)w\,ds + \int_K (u-v)A^* w\,dxdy\,. \tag{4.6}$$

By combining (4.3), (4.5) and (4.6), we get the desired estimate. ∎

In order to get explicit error bounds, we need to define more precisely the finite-dimensional spaces P_K. Let K be an element of $\mathcal{T}_h$. If K is a triangle, there exists an affine invertible mapping F_K which maps a reference triangle $\hat{K}$ onto K ($\hat{K}$ is usually chosen as a unit, isosceles, right triangle). If K is a nondegenerate convex quadrilateral, there exists a biaffine invertible mapping F_K which maps the reference element $\hat{K} = [-1,+1]^2$ onto K. Note that this mapping F_K becomes affine when K is a parallelogram.

In both cases, let $\hat{P} \subset H^1(\hat{K})$ be a finite-dimensional space of real-valued functions defined on the reference element $\hat{K}$. We shall always assume in the following that

$$P_K = \{p \mid p = \hat{p} \circ F_K^{-1}\,,\ \hat{p} \in \hat{P}\}\,. \tag{4.7}$$

We shall make constant use of the one-to-one correspondence

$$\hat{v} \mapsto v = \hat{v} \circ F_K^{-1}\,, \qquad v \mapsto \hat{v} = v \circ F_K$$

between the functions $\hat{v}$ defined on $\hat{K}$ and the functions v defined on K.

For any integer $m \geq 0$, let P_m denote the space of all polynomials of degree $\leq m$ in the two variables x, y and let Q_m denote the space of all polynomials of the form

$$p(x, y) = \sum_{i,j=0}^{m} c_{ij} x^i y^j , \quad c_{ij} \in \mathbb{R} .$$

We shall need

Hypothesis H.1.

There exists an integer $k \geq 0$ such that:

(4.8) $P_k \subset \hat{P}$ if $\hat{K}$ is the reference triangle,

(4.9) $Q_k \subset \hat{P}$ if $\hat{K}$ is the reference quadrilateral $[-1, +1]^2$.

Let us now introduce the following geometric parameters:

$$\begin{aligned} & h(K) = \text{diameter of } K , \\ (4.10) \quad & \rho(K) = \sup\{\text{diameter of the circles contained in } K\}, \\ & \theta_i(K) \ (1 \leq i \leq 4) = \text{angles of } K \text{ if } K \text{ is a quadrilateral}. \end{aligned}$$

Hypothesis H.2.

There exists a constant $\sigma > 1$ independent of h such that

(4.11) $$h(K) \leq \sigma \rho(K) \quad \text{for all} \quad K \in \mathfrak{T}_h .$$

Moreover, there exists a constant γ independent of h with $0 < \gamma < 1$ such that

(4.12) $$\max_{1 \leq i \leq 4} |\cos \theta_i(K)| \leq \gamma \quad \text{for all quadrilateral} \quad K \in \mathfrak{T}_h .$$

Given a reference element $\hat{K}$, we define $\hat{\pi}$ to be the orthogonal projection operator on $L_2(\hat{K})$ to $\hat{P}$. For any $K \in \mathcal{T}_h$, we define $\pi_K \in \mathcal{L}(L_2(K); P_K)$ by

$$(4.13) \qquad \widehat{\pi_K v} = \hat{\pi}\,\hat{v} \qquad \text{for all} \quad v \in L_2(K)\,.$$

Then, for any $v \in L_2(\Omega)$, we define $\pi_h v$ to be the function in V_h such that

$$(4.14) \qquad \pi_h v\big|_K = \pi_K v \qquad \text{for all} \quad K \in \mathcal{T}_h\,.$$

Let us now state some standard results which can be easily proved by using the techniques of Ciarlet and Raviart [4], [5].

Lemma 5.

Assume that Hypothesis H.2 holds. Then, there exists a constant $C > 0$ independent of $K \in \mathcal{T}_h$ such that for all $p \in P_K$

$$(4.15) \qquad |p|_{1,K} \leq C(h(K))^{-1}\|p\|_{0,K}\,,$$

$$(4.16) \qquad \|p\|_{0,K'} \leq C(h(K))^{-\frac{1}{2}}\|p\|_{0,K}\,,$$

where K' is any side of K and $\|p\|_{0,K'} = (\int_{K'} |p|^2 ds)^{\frac{1}{2}}$.

Lemma 6.

Assume that Hypotheses H.1, H.2 and (4.13) hold. Then, there exists a constant $C > 0$ independent of $K \in \mathcal{T}_h$ such that for all $v \in H^{k+1}(K)$

$$(4.17) \qquad |v - \pi_K v|_{m,K} \leq C(h(K))^{k+1-m}\|v\|_{k+1,K}\,, \qquad m = 0, 1,$$

$$(4.18) \qquad \|v - \pi_K v\|_{0,K'} \leq C(h(K))^{k+\frac{1}{2}}\|v\|_{k+1,K}$$

where K' is any side of K.

Let $K_1, K_2, \ldots, K_I$ be a fixed ordering of the elements of $\mathcal{T}_h$ which satisfies the condition of Lemma 3. For all $i = 1, \ldots, I$, we set

$$\Omega_i = \widehat{\overset{\circ}{\bigcup_{j=1}^{i} K_j}} \tag{4.19}$$

and we define $\partial_+\Omega_i$ and $\partial_-\Omega_i$ in the usual way. Note that $\partial_-\Omega_i \subset \Gamma_-$.

Theorem 5.

Assume that Hypotheses H.1 and H.2 hold. Assume in addition that the solution u of problem (1.3) belongs to $H^{k+1}(\Omega)$. Then, there exists a constant $C > 0$ independent of h such that for all $i = 1, \ldots, I$

$$\|u_h - u\|_{0,\Omega_i} \le C h^k \|u\|_{k+1,\Omega_i}, \tag{4.20}$$

$$\left(\int_{\partial_+\Omega_i} \underline{\underline{m}} \cdot \underline{\underline{n}} \, (u_h - u)^2 ds\right)^{\frac{1}{2}} \le C h^k \|u\|_{k+1,\Omega_i}, \tag{4.21}$$

$$\left(-\sum_{j=1}^{i} \int_{\partial_- K_j} \underline{\underline{m}} \cdot \underline{\underline{n}} \, (u_h - \xi_h)^2 ds\right)^{\frac{1}{2}} \le C h^k \|u\|_{k+1,\Omega_i}. \tag{4.22}$$

Proof.

For any $K \in \mathcal{T}_h$, we define

$$\eta_h = \begin{cases} 0 & \text{on } \partial_- K \cap \Gamma_-, \\ \text{outward trace of } \pi_h u & \text{on } \partial_- K \setminus (\partial_- K \cap \Gamma_-). \end{cases} \tag{4.23}$$

We start from equation (4.1) with $v = \pi_h u$, $\eta = \eta_h$ and we estimate the corresponding right hand side member. First, we have[†]

$$\left|\int_K (u - \pi_h u) A^* (u_h - \pi_h u) dxdy\right| \le c_1 \|u - \pi_h u\|_{0,K} \|u_h - \pi_h u\|_{1,K}$$

[†] In the sequel, we shall denote by c_i various constants independent of h.

and by (4.15), (4.17)

$$(4.24)\qquad \left| \int_K (u-\pi_h u) A^* (u_h - \pi_h u)\,dxdy \right| \leq c_2 h^k \|u\|_{k+1,K} \|u_h - \pi_h u\|_{0,K}\,.$$

Next, using (4.16) and (4.18), we obtain

$$(4.25)\qquad \left| \int_{\partial_+ K} \underline{\underline{m}} \cdot \underline{\underline{n}}\, (u - \pi_h u)(u_h - \pi_h u)\,ds \right| \leq c_3 h^k \|u\|_{k+1,K} \|u_h - \pi_h u\|_{0,K}\,.$$

Similarly, we get

$$(4.26)\qquad \left| \int_{\partial_- K} \underline{\underline{m}} \cdot \underline{\underline{n}}\, (u - \eta_h)(u_h - \pi_h u)\,ds \right| \leq c_4 h^k \|u\|_{k+1,\mathcal{D}_K} \|u_h - \pi_h u\|_{0,K}$$

where $\mathcal{D}_K$ is the union of the elements of $\mathcal{T}_h$ which have a side contained in $\partial_- K$.

Thus, combining (4.1) with $v = \pi_h u$, $\eta = \eta_h$, (4.24), (4.25), (4.26) and using (3.7), we obtain

$$\tfrac{1}{2}\int_{\partial_+ K} \underline{\underline{m}} \cdot \underline{\underline{n}}\, (u_h - \pi_h u)^2 ds - \tfrac{1}{2}\int_{\partial_- K} \underline{\underline{m}} \cdot \underline{\underline{n}}\, ((u_h - \pi_h u) - (\xi_h - \eta_h))^2 ds + \alpha \|u_h - \pi_h u\|^2_{0,K}$$

$$\leq -\tfrac{1}{2}\int_{\partial_- K} \underline{m} \cdot \underline{\underline{n}}\, (\xi_h - \eta_h)^2 ds + c_5 h^k \|u\|_{k+1,K \cup \mathcal{D}_K} \|u_h - \pi_h u\|_{0,K}\,.$$

Summing over all the elements K_j, $1 \leq j \leq i$, and using (3.12), (4.23), we get

$$\tfrac{1}{2}\int_{\partial_+ \Omega_i} \underline{\underline{m}} \cdot \underline{\underline{n}}\, (u_h - \pi_h u)^2\, ds$$

$$(4.27)\qquad - \tfrac{1}{2}\sum_{j=1}^{i} \int_{\partial_- K_j} \underline{\underline{m}} \cdot \underline{\underline{n}}\, ((u_h - \pi_h u) - (\xi_h - \eta_h))^2 ds + \alpha \|u_h - \pi_h u\|^2_{0,\Omega_i}$$

$$\leq c_6 h^k \|u\|_{k+1,\Omega_i} \|u_h - \pi_h u\|_{0,\Omega_i}\,.$$

From (4.17) and (4.27), we deduce

$$\|u_h - u\|_{0,\Omega_i} \leq \|u_h - \pi_h u\|_{0,\Omega_i} + \|\pi_h u - u\|_{0,\Omega_i}$$

$$\leq \frac{c_6}{\alpha} h^k \|u\|_{k+1,\Omega_i} + c_7 h^{k+1} \|u\|_{k+1,\Omega_i}$$

so that (4.20) holds.

Next, we have by (4.27)

$$\left(\int_{\partial_+\Omega_i} \underset{=}{m}\cdot\underset{=}{n} \; (u_h - \pi_h u)^2 ds\right)^{\frac{1}{2}} \leq c_8 h^k \|u\|_{k+1,\Omega_i}$$

and by (4.18)

$$\left(\int_{\partial_+\Omega_i} \underset{=}{m}\cdot\underset{=}{n} \; (\pi_h u - u)^2 ds\right)^{\frac{1}{2}} \leq c_9 h^{k+\frac{1}{2}} \|u\|_{k+1,\Omega_i} .$$

This proves inequality (4.21).

Similarly, we have by (4.27)

$$\left(-\sum_{j=1}^{i} \int_{\partial_-\Omega_i} \underset{=}{m}\cdot\underset{=}{n} \; ((u_h - \pi_h u) - (\xi_h - \eta_h))^2 ds\right)^{\frac{1}{2}}$$

$$\leq c_{10} h^k \|u\|_{k+1,\Omega_i}$$

and by (4.18)

$$\left(-\sum_{j=1}^{i} \int_{\partial_- K_j} \underset{=}{m}\cdot\underset{=}{n} \; (\pi_h u - u)^2 ds\right)^{\frac{1}{2}} \leq c_{11} h^{k+\frac{1}{2}} \|u\|_{k+1,\Omega_i} ,$$

$$\left(-\sum_{j=1}^{i} \int_{\partial_- K_j} \underset{=}{m}\cdot\underset{=}{n} \; (\eta_h - u)^2 ds\right)^{\frac{1}{2}} \leq c_{12} h^{k+\frac{1}{2}} \|u\|_{k+1,\Omega_i} .$$

This implies inequality (4.22). ■

5. A Superconvergence Result.

Let us notice that the error estimates of Theorem 5 are not optimal in the exponent of the parameter h. In fact, numerical calculations have shown that these error bounds could not be improved *in general*. However, the one-dimensional results of §2 clearly indicate that better estimates must hold in some special cases. Indeed, we shall prove in this section that the rate of convergence of our finite element method is $O(h^{k+1})$ <u>when all the elements $K \in \mathfrak{T}_h$ are rectangles and when</u> $\hat{P} = Q_k$. In the sequel, we shall confine ourselves to this particular case.

On the interval $[-1,+1]$, let $-1 < \theta_1 < \theta_2 < \ldots < \theta_{k+1} = 1$ denote the $(k+1)$ Gauss-Radau quadrature abscissae. In the reference square $\hat{K} = [-1,+1]^2$, we consider the points $\hat{a}_{ij}$ with coordinates (θ_i, θ_j), $1 \leq i, j \leq k+1$.

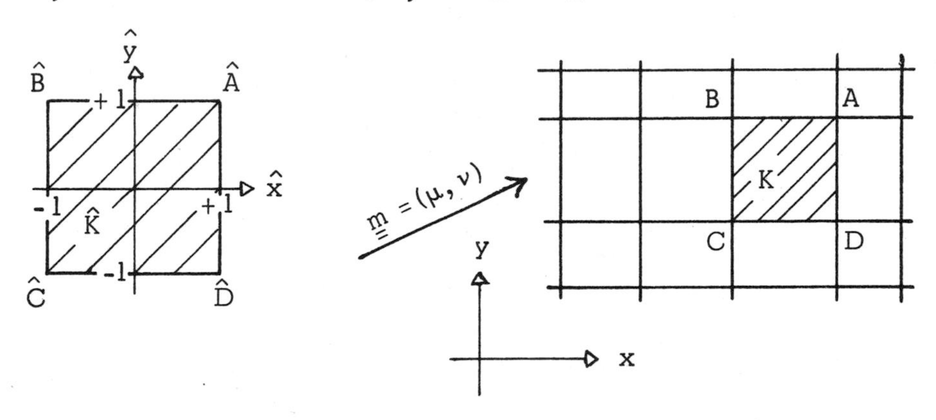

Figure 5

Just for convenience, we shall assume that the sides of the rectangles $K \in \mathfrak{T}_h$ are parallel to the (x, y) axes and that the coefficients μ, ν are > 0. Given a rectangle K with vertices A, B, C, D as in Figure 5, we denote by F_K the affine invertible mapping such that $A = F_K(\hat{A}), \ldots, D = F_K(\hat{D})$. Given a function $v \in C^0(K)$, we define $r_K v$ as the unique polynomial in Q_k which interpolates v at the points $a_{ij} = F_K(\hat{a}_{ij})$, $1 \leq i, j \leq k+1$. Then, for any $v \in C^0(\Omega)$, we define $r_h v$ to be the function in V_h such that

$$(5.1) \qquad r_h v|_K = r_K v \quad \text{for all } K \in \mathfrak{T}_h .$$

We provide $L_\infty(\Omega)$ with the following norm

$$\|v\|_{0,\infty,\Omega} = \sup\{ |v(x)| ; x \in \Omega \}.$$

Given any integer $m \geq 0$, let

$$W_\infty^m(\Omega) = \{v \in L_\infty(\Omega) \mid \partial^\alpha v \in L_\infty(\Omega),\ |\alpha| \leq m\}$$

be the Sobolev space provided with the norm

$$\|v\|_{m,\infty,\Omega} = \max\{\|\partial^\alpha v\|_{0,\infty,\Omega};\ |\alpha| \leq m\}.$$

Using [4], for instance, one can easily prove

Lemma 7.

Assume that Hypothesis H.2 holds. Then there exists a constant $c > 0$ independent of $K \in \mathfrak{T}_h$ such that

(5.2) $\|v - r_K v\|_{0,K} \leq c(h(K))^{k+1}\|v\|_{k+1,K}$ for all $v \in H^{k+1}(K)$,

(5.3) $\|v - r_K v\|_{0,K'} \leq c(h(K))^{k+3/2}\|v\|_{k+1,\infty,K}$ for all $v \in W_\infty^{k+1}(K)$ where K' is any side of $\partial_+ K$.

We are now able to prove

Theorem 6.

Assume that all the elements $K \in \mathfrak{T}_h$ are rectangles, that $\hat{P} = Q_k$, and that Hypothesis H.2 holds. Assume, in addition, that the solution u of problem (1.3) belongs to $H^{k+2}(\Omega) \cap W_\infty^{k+1}(\Omega)$. Then, there exists a constant $C > 0$ independent of h such that, for all $i = 1, \ldots, I$,

(5.4) $$\|u_h - u\|_{0,\Omega_i} \leq Ch^{k+1}\|u\|_{k+2,\Omega_i},$$

(5.5) $$\left(\int_{\partial_+\Omega_i} \underline{\underline{m}}\cdot\underline{\underline{n}}\,(u_h - u)^2 ds\right)^{\frac{1}{2}} \leq Ch^{k+1}(\|u\|_{k+2,\Omega_i} + \|u\|_{k+1,\infty,\Omega_i}).$$

Proof.

For any $K \in \mathcal{T}_h$, we now define

$$(5.6) \qquad \eta_h = \begin{cases} 0 & \text{on } \partial_- K \cap \Gamma_- , \\ \text{outward trace of } r_h u & \text{on } \partial_- K \setminus (\partial_- K \cap \Gamma_-). \end{cases}$$

We start from equation (4.1) with $v = r_h u$, $\eta = \eta_h$. The corresponding right hand side may be written in the form

$$(5.7) \qquad X_K(u, u_h - r_h u) = Z_K(u, u_h - r_h u) + \int_K \sigma(u - r_h u)(u_h - r_h u)dxdy$$

where

$$(5.8) \qquad \begin{aligned} Z_K(u, w) = & \int_{\partial_+ K} \underline{\underline{m}} \cdot \underline{\underline{n}} \, (u - r_h u) w \, ds \\ & + \int_{\partial_- K} \underline{\underline{m}} \cdot \underline{\underline{n}} \, (u - \eta_h) w \, ds \\ & - \int_K (u - r_h u) \, \underline{\underline{m}} \cdot \text{grad } w \, dx \, dy . \end{aligned}$$

We now use the following essential lemma which will be proved later.

Lemma 8.

With the same assumptions as in Theorem 6, there exists a constant $C > 0$ independent of $K \in \mathcal{T}_h$ such that for all $w \in Q_k$

$$(5.9) \qquad |Z_k(u, w)| \le C(h(K))^{k+1} \|u\|_{k+2,K} \|w\|_{0,K} .$$

Using (5.2), (5.7) and (5.9), we obtain for all $K \in \mathcal{T}_h$

$$(5.10) \qquad |X_K(u, u_h - r_h u)| \le c_1 h^{k+1} \|u\|_{k+2,K} \|u_h - r_h u\|_{0,K} .$$

Thus, combining (4.1) with $v = r_h u$, $\eta = \eta_h$, (5.10) and using (3.7), we get

$$\tfrac{1}{2}\int_{\partial_+ K} \underline{\underline{m}}\cdot\underline{\underline{n}}\,(u_h - r_h u)^2 ds + \alpha \|u_h - r_h u\|^2_{0,K}$$
$$\leq -\tfrac{1}{2}\int_{\partial_- K} \underline{\underline{m}}\cdot\underline{\underline{n}}\,(\xi_h - \eta_h)^2 ds$$
$$+ c_1 h^{k+1} \|u\|_{k+2,K} \|u_h - r_h u\|_{0,K}.$$

Summing over all the elements K_j, $1 \leq j \leq i$, we obtain

$$(5.11)\quad \tfrac{1}{2}\int_{\partial_+ \Omega_i} \underline{\underline{m}}\cdot\underline{\underline{n}}\,(u_h - r_h u)^2 ds + \alpha \|u_h - r_h u\|^2_{0,\Omega_i} \leq c_1 h^{k+1} \|u\|_{k+2,\Omega_i} \|u_h - r_h u\|_{0,\Omega_i}.$$

Thus, the estimates (5.4) and (5.5) are simple consequences of inequality (5.11) and Lemma 7. ∎

Proof of Lemma 8.

Consider a rectangle $K \in \mathcal{T}_h$ with vertices A, B, C, D (cf. Figure 5). Let us denote by Δx (resp. Δy) the length of the side AB (resp. BC). We may write

$$(5.12)\quad Z_K(u,w) = \mu Z_{K,x}(u,w) + \nu\, Z_{K,y}(u,w)$$

with

$$Z_{K,x}(u,w) = \int_D^A (u - r_h u) w\,dy - \int_C^B (u - \eta_h) w\,dy - \int_K (u - r_h u) \frac{\partial w}{\partial x}\,dxdy,$$

$$Z_{K,y}(u,w) = \int_B^A (u - r_h u) w\,dx - \int_C^D (u - \eta_h) w\,dx - \int_K (u - r_h u) \frac{\partial w}{\partial y}\,dxdy.$$

By using the one-to-one correspondence $v \mapsto \hat{v} = v \circ F_K$, we get

$$(5.13)\quad Z_{K,x}(u,w) = \frac{\Delta y}{2}\, \hat{Z}_{\hat{x}}(\hat{u},\hat{w})$$

with

$$\hat{Z}_{\hat{x}}(\hat{u},\hat{w}) = \int_{-1}^{+1} (\hat{u}(1,\hat{y}) - \hat{r}\hat{u}(1,\hat{y}))\hat{w}(1,\hat{y})d\hat{y}$$

$$- \int_{-1}^{+1} (\hat{u}(-1,\hat{y}) - \hat{\eta}(\hat{y}))\,\hat{w}(-1,\hat{y})d\hat{y}$$

$$- \int_{-1}^{+1}\int_{-1}^{+1} (\hat{u} - \hat{r}\hat{u}) \frac{\partial \hat{w}}{\partial \hat{y}}\, d\hat{x}d\hat{y},$$

where $\hat{r}\hat{u}$ is the polynomial in Q_k which interpolates $\hat{u}$ at the points $\hat{a}_{ij}$, $1 \leq i,j \leq k+1$, and where $\hat{\eta}$ is the polynomial of degree $\leq k$ which interpolates the function $\hat{y} \mapsto \hat{u}(-1,\hat{y})$ at the points θ_i, $1 \leq i \leq k+1$.

Clearly

$$\hat{Z}_{\hat{x}}(\hat{u},\hat{w}) = 0 \qquad \text{for all} \quad \hat{u},\hat{w} \in Q_k .$$

Now, when $\hat{u} = \hat{x}^{k+1}$, we have

$$\hat{u}(1,\hat{y}) = \hat{r}\hat{u}(1,\hat{y}) = 1, \quad \hat{u}(-1,\hat{y}) = \hat{\eta}(\hat{y}) = (-1)^{k+1} .$$

Moreover, $\hat{r}\hat{u}$ does not depend on $\hat{y}$ and then, for all $\hat{w} \in Q_k$, the function $\hat{x} \mapsto (\hat{u} - \hat{r}\hat{u})(\hat{x}) \frac{\partial \hat{w}}{\partial \hat{x}}(\hat{x},\hat{y})$ is a polynomial of degree $\leq 2k$ which vanishes at the $(k+1)$ Gauss-Radau points θ_i. Therefore,

$$\int_{-1}^{+1} (\hat{u} - \hat{r}\hat{u})\hat{w}\, d\hat{x} = 0 \qquad \text{for all} \quad \hat{w} \in Q_k .$$

Thus, when $\hat{u} = \hat{x}^{k+1}$, we get

$$\hat{Z}_{\hat{x}}(\hat{u},\hat{w}) = 0 \qquad \text{for all} \quad \hat{w} \in Q_k .$$

On the other hand, when $\hat{u} = \hat{y}^{k+1}$, $\hat{r}\hat{u}$ is independent of $\hat{x}$ so that we obtain by integration by parts

$$\int_{-1}^{+1}\int_{-1}^{+1} (\hat{u} - \hat{r}\hat{u}) \frac{\partial \hat{w}}{\partial \hat{x}}\, d\hat{x}d\hat{y} = \int_{-1}^{+1} (\hat{u}(1,\hat{y}) - \hat{r}\hat{u}(1,\hat{y}))\hat{w}(1,\hat{y})d\hat{y}$$

$$- \int_{-1}^{+1} (\hat{u}(-1,\hat{y}) - \hat{\eta}(\hat{y}))\hat{w}(-1,\hat{y})d\hat{y}.$$

This gives again

$$\hat{Z}_{\hat{x}}(\hat{u}, \hat{w}) = 0 \quad \text{for all } \hat{w} \in Q_k .$$

Therefore, we have proved that

$$\hat{Z}_{\hat{x}}(\hat{u}, \hat{w}) = 0 \quad \text{for all } \hat{u} \in P_{k+1} \text{ and all } \hat{w} \in Q_k .$$

Then, for fixed $\hat{w} \in Q_k$, the linear functional $\hat{u} \mapsto \hat{Z}_{\hat{x}}(\hat{u}, \hat{w})$ is continuous on $H^{k+2}(\hat{K})$, has norm $\leq c_1 \|\hat{w}\|_{0,\hat{K}}$ and vanishes over P_{k+1}. By the Bramble-Hilbert lemma in the form given in [4, Lemma 6], we get for all $\hat{u} \in H^{k+2}(\hat{K})$ and all $\hat{w} \in Q_k$

$$|\hat{Z}_{\hat{x}}(\hat{u}, \hat{w})| \leq c_2 |\hat{u}|_{k+2,\hat{K}} \|\hat{w}\|_{0,\hat{K}} .$$

Going back to the element K by using the correspondence $\hat{v} \mapsto v = \hat{v} \circ F_K^{-1}$ and (5.13), we obtain for all $u \in H^{k+2}(K)$ and all $w \in Q_k$

$$(5.14) \qquad |Z_{K,x}(u,w)| \leq c_3 (h(K))^{k+1} \|u\|_{k+2,K} \|w\|_{0,K} .$$

Likewise, we get

$$(5.15) \qquad |Z_{K,y}(u,w)| \leq c_4 (h(K))^{k+1} \|u\|_{k+2,K} \|w\|_{0,K} .$$

Then, combining (5.12), (5.14) and (5.15), we obtain the desired inequality (5.9). ∎

Note that the error estimates of Theorem 6 are now optimal in the exponent of the parameter h. However, as the one-dimensional results of §1 suggest, we conjecture that, for any rectangle $K \in \mathfrak{T}_h$, there exist some points of $\partial_+ K$ where even more precise error bounds hold. Unfortunately, we have not been able to prove the existence of such points.

References.

1. Axelsson, O., A class of A-stable methods, B.I.T. 9 (1969), 185-199.

2. de Boor, C., and B. Swartz, Collocation at Gaussian points, SIAM J. Numer. Anal. 10 (1973), 582-606.

3. Butcher, J. C., Implicit Runge-Kutta processes, Math. Comp. 18 (1964), 50-64.

4. Ciarlet, P. G., and P.-A. Raviart, General Lagrange and Hermite interpolation in $\mathbb{R}^n$ with applications to finite element methods, Arch. Rat. Mech. Anal. 46 (1972), 177-199.

5. Ciarlet, P. G., and P.-A. Raviart, Interpolation theory over curved elements, with applications to finite element methods, Comp. Meth. Appl. Mech. Eng. 1 (1972), 217-249.

6. Ciarlet, P. G., and P.-A. Raviart, The combined effect of curved boundaries and numerical integration in isoparametric finite element methods, The Mathematical Foundations of the Finite Element Method with Applications to Partial Differential Equations (A.K. Aziz, Ed.), 409-474, Academic Press. New York, 1972.

7. Crouzeix, M., Thesis, to appear.

8. Friedrichs, K. O., Symmetric positive differential equations, Comm. Pure Appl. Math. 11 (1958), 333-418.

9. Hulme, B. L., Discrete Galerkin and related one-step methods for ordinary differential equations, Math. Comp. 26 (1972), 881-891.

10. Kaper, H. G., G. K. Leaf, and A. J. Lindeman, Application of finite element techniques for the numerical solution of the neutron transport and diffusion equations, Proc. Conf. on Transport Theory, 2nd Conf-710107, Los Alamos (1971).

11. Lesaint, P., Finite element methods for symmetric hyperbolic equations, Numer. Math. 21 (1973), 244-255.

12. Lesaint, P., Finite element methods for the transport equation, to appear in RAIRO, Série Mathématiques.

13. Lesaint, P., Thesis, to appear.

14. Lesaint, P., and J. Gérin-Roze, Isoparametric finite element methods for the neutron transport equation, to appear.

15. Miller, W. F. Jr., E. E. Lewis, and E. C. Rossow, The application of phase-space finite elements to the two-dimensional transport equation in x-y geometry, to appear in Nucl. Sci. Eng.

16. Ohnishi, T., Application of finite element solution technique to neutron diffusion and transport equations, Proc. Conf. on New Developments in Reactor Mathematics and Applications, Conf.-710302, Idaho Falls (1971).

17. Reed, W. H., and T. R. Hill, Triangular mesh methods for the neutron transport equation, to appear in Proc. Amer. Nucl. Soc.

FINITE ELEMENTS AND THE NEUTRON TRANSPORT EQUATION

P. Lesaint
Service de Mathématiques Appliquées,
Centre d'Etudes de Limeil, B. P. 27
94190 VILLENEUVE-St. Georges,
France

P. -A. Raviart
Analyse Numerique, Tour 55-65,
Université de Paris VI
4, Place Jussieu,
75230, Paris CEDEX 05,
France

A Mixed Finite Element Method for the Biharmonic Equation

P. G. CIARLET AND P. A. RAVIART

Introduction.

Consider the problem

$$(1.1) \qquad \Delta^2 u = f \quad \text{in } \Omega ,$$

$$(1.2) \qquad u = \frac{\partial u}{\partial \nu} = 0 \quad \text{on } \Gamma ,$$

where Ω is a bounded and connected subset of $\mathbb{R}^n$, with boundary Γ, f is a given function which throughout this paper is assumed to belong to the space $L^2(\Omega)$, and $\frac{\partial}{\partial \nu}$ is the exterior normal derivative along Γ.

The most straightforward discretization of problem (1.1)-(1.2) consists in using a conforming method, in which the approximate solution lies in a subspace of the space $H_0^2(\Omega)$. However, constructing such subspaces requires the use of fairly sophisticated finite elements. One way to avoid the related computational difficulty is to use a non-conforming method, in which the approximate solution lies in a finite-dimensional space which is "only" a subspace of $H^1(\Omega)$, or even a subspace of $L_2(\Omega)$. For a survey about the above methods as well as references, see [4].

There are other approaches based on a different variational principle (the "standard" variational formulation of problem (1.1)-(1.2) is mentioned at the beginning of §2), which fall themselves into two categories: Either the variational principle is explicitly dependent upon a given triangulation of the set Ω, in which case the method is called hybrid, or otherwise, it is called a mixed method. It is the object of this paper to study a method of the latter type. For general theorems related to mixed methods, the papers of Oden [12, 13] are particularly relevant. For the numerical analysis of hybrid methods, see the paper of Brezzi [3].

In practice, problem (1.1)-(1.2) is found in plate problems and in hydrodynamics problems. In the first case, it is of particular interest to have good approximations of all the second partial derivatives of the solution u whose physical interpretation are moments. To achieve this goal, there is a particularly well-suited variational principle, which is known as the complementary energy principle, and Johnson [8, 9] has made the numerical analysis of the associated finite element methods, with triangular and rectangular finite elements.

In this paper, we study an approximate method related to yet another variational principle, in which the Laplacian Δu of the solution plays a special role. As such, this approach is more appropriate for hydrodynamics problems, where $-\Delta u$ represents the vorticity.

Our starting point is the work of Glowinski [7], who studied a related method. In particular, Glowinski has observed the interesting fact that for specific choices of subspaces, the method is identical to the usual 13-point finite difference approximation to the operator Δ^2. Glowinski has proved convergence, however without orders of convergence; likewise, Mercier [10] has also studied a similar method, again proving convergence without orders of convergence.

A main feature of such methods is that the numerical work is reduced to solving a sequence of Dirichlet problems for the operator $-\Delta$. This is achieved by using a Uzawa method for solving the saddle-point equations of the associated Lagrangian. As a consequence, such a method may be applied as long as a good finite-element program for solving second-order problems is available, which is indeed quite realistic an objective for solving fourth-order problems (for related ideas, see Bossavit [1], Smith [14]), and actually, numerical results obtained by Glowinski at the I. R. I. A. show that the method is quite efficient, from a numerical standpoint.

Let us now briefly describe the contents of the following sections. In §2, we describe a variational principle which may be associated with problem (1.1)-(1.2); in §3, we describe the associated discrete problem, and in §4, the error analysis is made: We show that, with C^0-finite elements associated with piecewise polynomials of degree k, one obtains the asymptotic error estimate:

$$\|u-u_h\|_{H^1(\Omega)} + \|\Delta u + \varphi_h\|_{L_2(\Omega)} \leq \mathcal{K} \|u\|_{H^{k+2}(\Omega)} h^{k-1},$$

where (u_h, φ_h) is the solution of the discrete problem, so that, as a special case, we get an $O(h)$ convergence with polynomials of degree only 2. Thus, an interesting feature is that a global L_2-estimate of the Laplacian Δu is obtained.

As regards the above error estimates, particular mention should be made of recent results of Brezzi [2] concerning the abstract approximation of saddle-points by variational methods. His results are similar, but they are more adapted to hybrid methods; see Brezzi [3].

Throughout this paper, we will use the following notations:

$$|\cdot|_{m,\Omega} = \left(\int_\Omega \sum_{|\alpha|=m} |\partial^\alpha \cdot|^2 dx\right)^{\frac{1}{2}}$$

$$\|\cdot\|_{m,\Omega} = \left(\int_\Omega \sum_{|\alpha|\leq m} |\partial^\alpha \cdot|^2 dx\right)^{\frac{1}{2}}.$$

Also, we shall say that the boundary Γ of an open subset Ω of $\mathbb{R}^n$ is <u>sufficiently smooth</u> if it is Lipschitz-continuous in the sense of Nečas [11, page 14].

2. The continuous problem.

We first recall that the "standard" variational formulation of problem (1.1)-(1.2) consists in finding the unique function u in the space $H_0^2(\Omega)$ which satisfies

$$\forall v \in H_0^2(\Omega) , \quad \int_\Omega \Delta u \, \Delta v \, dx = \int fv dx . \tag{2.1}$$

In order to derive another variational formulation of this problem (cf. Theorem 1), we need a few definitions. First we let

$$\underline{\underline{W}} = H_0^1(\Omega) \times L^2(\Omega) . \tag{2.2}$$

Then the bilinear form $a(\cdot,\cdot)$ defined on $\underline{\underline{W}} \times \underline{\underline{W}}$ by

$$a(\underline{\underline{u}}, \underline{\underline{v}}) = \int_\Omega \varphi\psi dx \tag{2.3}$$

for all $(\underline{\underline{u}}, \underline{\underline{v}}) = ((u,\varphi), (v,\psi)) \in \underline{\underline{W}} \times \underline{\underline{W}}$ is continuous, the space $\underline{\underline{W}}$ being equipped with the product norm

$$\underline{\underline{v}} = (v,\psi) \in \underline{\underline{W}} \rightarrow |v|_{1,\Omega} + |\psi|_{0,\Omega} . \tag{2.4}$$

Likewise, the linear form $f(\cdot)$, defined on $\underline{\underline{W}}$ by

$$f(\underline{\underline{v}}) = \int_\Omega fv dx \tag{2.5}$$

for all $\underline{\underline{v}} = (v,\psi) \in \underline{\underline{W}}$, is continuous.

Next, we let

$$M = H^1(\Omega) , \tag{2.6}$$

so that the bilinear form $b(\cdot,\cdot)$ defined on $\underline{\underline{W}} \times M$ by

$$b(\underline{\underline{v}},\mu) = \int_\Omega \operatorname{grad} v \operatorname{grad} \mu \, dx - \int_\Omega \psi\mu \, dx \tag{2.7}$$

for all $(\underline{\underline{v}} = (v,\psi),\mu) \in \underline{\underline{W}} \times M$ is also continuous.

Finally, we let

$$\underline{\underline{V}} = \{\underline{\underline{v}} = (v,\psi) \in \underline{\underline{W}};\ \forall\, \mu \in M,\ b(\underline{\underline{v}},\mu) = 0\}\ , \tag{2.8}$$

so that $\underline{\underline{V}}$ is a Hilbert space since it is a closed subspace of the space $\underline{\underline{W}}$.

Lemma 1.

The semi-norm

$$\underline{\underline{v}} = (v,\psi) \in \underline{\underline{V}} \to |\psi|_{0,\Omega} \tag{2.9}$$

is a norm over the space $\underline{\underline{V}}$, which is equivalent to the product norm defined in (2.4).

Proof.

Let (v,ψ) be any element in the space $\underline{\underline{V}}$. Since $H_0^1(\Omega) \subset H^1(\Omega)$, the particular choice $\mu = v$ in (2.8) yields

$$|v|^2_{1,\Omega} = \int_\Omega \text{grad } v \cdot \text{grad } v\, dx = \int_\Omega \psi v\, dx \leq |\psi|_{0,\Omega} |v|_{0,\Omega}.$$

By Friedrich's inequality, there exists some constant C solely dependent upon the set Ω such that $|v|_{0,\Omega} \leq C|v|_{1,\Omega}$ for all $v \in H_0^1(\Omega)$. Therefore, we have

$$|v|_{1,\Omega} \leq C|\psi|_{0,\Omega}\ , \tag{2.10}$$

for all $(v,\psi) \in \underline{\underline{V}}$, which finishes the proof. ∎

We next define problem (P): Find a function $\underline{\underline{u}} \in \underline{\underline{V}}$ such that

$$\forall\ \underline{\underline{v}} \in \underline{\underline{V}}\ ,\quad a(\underline{\underline{u}},\underline{\underline{v}}) = f(\underline{\underline{v}}) \tag{2.11}$$

Theorem 1.

The problem (P) has a unique solution $\underline{\underline{u}} = (u,\varphi) \in \underline{\underline{V}}$. In addition, if the boundary Γ is sufficiently smooth, then $u \in H_0^2(\varphi)$, $\varphi = -\Delta u$, and u is also the unique solution of problem (2.1).

Proof.

Since $a(\underline{\underline{v}}, \underline{\underline{v}}) = |\psi|^2_{0,\Omega}$, the bilinear form is $\underline{\underline{V}}$-elliptic in view of Lemma 1, and therefore the existence and uniqueness of the solution $\underline{\underline{u}} = (u, \varphi)$ of problem (P) follows from the Lax-Milgram lemma.

Given an arbitrary function ψ in the space $L_2(\Omega)$, it follows from Lemma 1 that there is at most one function $v \in H^1_0(\Omega)$ such that $\underline{\underline{v}} = (v, \psi)$ belongs to the space $\underline{\underline{V}}$. If this is the case, then we have in particular for all $\mu \in H^1_0(\Omega)$:

$$\int_\Omega \text{grad } v \text{ grad } \mu \, dx = \int_\Omega \psi\mu \, dx .$$

Then if the boundary Γ is sufficiently smooth, the function $v \in H^1_0(\Omega)$, solution of the above problem, is also in the space $H^2(\Omega)$ [11], and

$$-\Delta v = \psi ,$$

the equality holding in the space $L_2(\Omega)$. Therefore, for any function $\mu \in H^1(\Omega)$, we may apply Green's formula

$$\int_\Omega \text{grad } v \text{ grad } \mu \, dx + \int_\Omega \Delta v \, \mu \, dx = \int_\Gamma \frac{\partial v}{\partial \nu} \mu \, d\gamma$$

which in conjunction with the relation $b(\underline{\underline{v}}, \mu) = 0$ and the equality $-\Delta v = \psi$ shows that $\frac{\partial v}{\partial \nu} = 0$ on Γ, i.e., $v \in H^2_0(\Omega)$.

In other words, if the boundary Γ is sufficiently smooth, then the space $\underline{\underline{V}}$ can be alternately described as those pairs $(v, \psi) \in H^2_0(\Omega) \times L_2(\Omega)$ which satisfy the equation $-\Delta v = \psi$.

Going back to problem (P), we see that $\underline{\underline{u}} = (u, \varphi) \in \underline{\underline{V}}$ is such that $u \in H^2_0(\varphi)$ and $-\Delta u = \varphi$ in $L_2(\Omega)$ so that, using the above characterization of the space $\underline{\underline{V}}$, the equations (2.11) become identical to equations (2.1). ∎

Since the space $\underline{\underline{V}}$ is a subspace of the space $\underline{\underline{W}}$, defined by the constraint:

$$\underline{\underline{v}} \in \underline{\underline{V}} \Leftrightarrow \forall \mu \in M , \ b(\underline{\underline{v}}, \mu) = 0 ,$$

we may use a standard procedure in optimization theory which consists in replacing the variational problem (2.11) by an unconstrained problem, at the expense however of having to find a Lagrange multiplier μ^*, in addition to the function u.

More precisely, with the bilinear forms $a(\cdot,\cdot)$ and $b(\cdot,\cdot)$ and the linear form $f(\cdot)$ being defined as in (2.3), (2.7) and (2.5), respectively, we define problem (P'): Find $(\underline{\underline{u}},\mu^*) \in \underline{\underline{W}} \times M$ such that

$$\forall\, \underline{\underline{v}} \in \underline{\underline{W}}, \quad a(\underline{\underline{u}},\underline{\underline{v}}) - f(\underline{\underline{v}}) + b(\underline{\underline{v}},\mu^*) = 0, \tag{2.12}$$

$$\forall\, \mu \in M, \quad b(\underline{\underline{u}},\mu) = 0 . \tag{2.13}$$

The relationship between problems (P) and (P') is described in the following theorem.

Theorem 2.

If $(\underline{\underline{u}},\mu^*)$ is a solution of problem (P'), then it is the only solution, and $\underline{\underline{u}}$ is the solution of problem (P).

Conversely, if the boundary Γ is sufficiently smooth, and if the solution $\underline{\underline{u}}$ of problem (P) is such that $u \in H^3(\Omega)$, then $(\underline{\underline{u}}, -\Delta u)$ is the solution of problem (P').

Proof.

If $(\underline{\underline{u}},\mu^*)$ is a solution of problem (P'), then $\underline{\underline{u}} \in \underline{\underline{V}}$, in view of (2.13). Next, since $\mu^* \in M$, it follows that $b(\underline{\underline{v}},\mu^*) = 0$ for all $\underline{\underline{v}} \in \underline{\underline{V}}$ and therefore $a(\underline{\underline{u}},\underline{\underline{v}}) - f(\underline{\underline{v}}) = 0$ for all $\underline{\underline{v}} \in \underline{\underline{V}}$, i.e., $\underline{\underline{u}}$ is the solution of problem (P).

As a consequence, the function $\underline{\underline{u}}$ is uniquely determined (cf. Theorem 1). To show that this is also the case for μ^*, let $\mu^* \in H^1(\Omega)$ be such that

$$\forall\, \underline{\underline{v}} \in \underline{\underline{W}}, \quad b(\underline{\underline{v}},\mu^*) = 0 .$$

Then $\mu^* = 0$ since in particular $\int_\Omega \psi\,\mu^* dx = 0$ for all $\psi \in L_2(\Omega)$.

Conversely, let $\underline{\underline{u}} = (u, \varphi)$ be the solution of problem (P), with $\underline{\underline{u}} \in H^3(\Omega)$. Since $\underline{\underline{u}} \in \underline{\underline{V}}$, the equations (2.13) are satisfied. To show that the equations (2.12) are also satisfied, let $\mu^* = -\Delta u$. Then for any element $\underline{\underline{v}} = (v, \psi) \in \underline{\underline{W}}$, we may write, using the relation $\varphi = -\Delta u$ (cf. Theorem 1):

$$b(\underline{\underline{v}}, \mu^*) = -\int_\Omega \operatorname{grad} v \operatorname{grad} \Delta u \, dx + \int_\Omega \psi \Delta u \, dx$$

$$= -\int_\Omega \operatorname{grad} v \operatorname{grad} \Delta u \, dx - \int_\Omega \psi \varphi \, dx .$$

Since for all $v \in \mathcal{D}(\Omega)$, we have

$$-\int_\Omega \operatorname{grad} v \operatorname{grad} \Delta u \, dx = \int_\Omega \Delta v \, \Delta u \, dx = \int_\Omega f v \, dx ,$$

we deduce that the same relation holds for all functions $v \in H_0^1(\Omega)$, and therefore

$$b(\underline{\underline{v}}, \mu^*) = \int_\Omega f v \, dx - \int_\Omega \psi \varphi \, dx = f(\underline{\underline{v}}) - a(\underline{\underline{u}}, \underline{\underline{v}}) ,$$

which finishes the proof. ∎

Remark 1.

It follows from Theorems 1 and 2 that the function $-\Delta u$ is both the second argument φ of the solution $\underline{\underline{u}} = (u, \varphi)$ of problem (P) and the Lagrange multiplier of problem (P'). ∎

3. The discrete problem.

Given three finite-dimensional spaces X_h, Y_h and M_h, which satisfy the inclusions

$$(3.1) \quad X_h \subset H_0^1(\Omega) , \quad Y_h \subset L_2(\Omega), \quad M_h \subset M = H^1(\Omega) ,$$

we define the spaces

$$(3.2) \quad \underline{\underline{W}}_h = X_h \times Y_h ,$$

and

$$\underset{=}{V}_h = \{\underset{=}{v}_h = (v_h, \psi_h) \in \underset{=}{W}_h ;\ \forall \mu_h \in M_h \ ,\ b(\underset{=}{v}_h, \mu_h) = 0\}. \tag{3.3}$$

Lemma 2.

If the inclusion

$$X_h \subset M_h \tag{3.4}$$

holds, the semi-norm $|\cdot|_{0,\Omega}$ *defined in* (2.9) *is also a norm over the space* $\underset{=}{V}_h$, *which is equivalent to the product norm defined in* (2.4).

Proof.

In view of the inclusion (3.4), we may parallel the proof of Lemma 1. In particular, we have

$$|v_h|_{1,\Omega} \leq C|\psi_h|_{0,\Omega} \tag{3.5}$$

for all $(v_h, \psi_h) \in \underset{=}{V}_h$, the constant C being the same as that of inequality (2.10). ∎

Then we define *problem* (P_h): *Find a function* $\underset{=}{u}_h = (u_h, \varphi_h) \in \underset{=}{V}_h$ *such that*

$$\forall \underset{=}{v}_h \in \underset{=}{V}_h \ , \quad a(\underset{=}{u}_h, \underset{=}{v}_h) = f(\underset{=}{v}_h) \ , \tag{3.6}$$

where the bilinear form $a(\cdot\,,\cdot)$ *and the linear form* $f(\cdot)$ *are defined as in* (2.3) *and* (2.5), *respectively.*

Remark 2.

Since the space $\underset{=}{V}_h$ is *not* contained in the space $\underset{=}{V}$ in general, the above discrete method may be viewed as a *nonconforming* method. ∎

With a proof similar to the first part of the proof of Theorem 1, we prove:

Theorem 3.

If the inclusion (3.4) *holds, problem* (P_h) *has a unique solution* $\underline{\underline{u}}_h = (u_h, \varphi_h) \in \underline{\underline{V}}_h$. ■

We next define *problem* (P_h'): *Find* $(u_h, \mu_h^*) \in \underline{\underline{W}}_h \times M_h$ *such that*

$$(3.7) \qquad \forall\, \underline{\underline{v}}_h \in \underline{\underline{W}}_h\,, \quad a(\underline{\underline{u}}_h, \underline{\underline{v}}_h) - f(\underline{\underline{v}}_h) + b(\underline{\underline{v}}, \mu_h^*) = 0\,,$$

$$(3.8) \qquad \forall\, \mu_h \in M_h, \quad b(\underline{\underline{u}}_h, \mu_h) = 0\,.$$

Remark 3.

In contrast to problem (P_h), problem (P_h') corresponds to a *conforming* method, since the unknown lies in the finite-dimensional subspace $\underline{\underline{W}}_h \times M_h$ of the space $\underline{\underline{W}} \times M$. ■

As regards the existence and uniqueness of the solution for problem (P_h'), we have:

Theorem 4.

If the inclusions

$$(3.9) \qquad X_h \subset M_h \subset Y_h$$

hold, problem (P_h') *has a unique solution* (u_h, μ_h^*). *The function* $u_h \in \underline{\underline{V}}_h$ *is also the unique solution of problem* (P_h).

Proof.

Since the relations (3.7)-(3.8) represent a linear system in a finite-dimensional space, it suffices to prove uniqueness. If $(\underline{\underline{u}}_h, \mu_h^*)$ is a solution of problem (P_h'), then $\underline{\underline{u}}_h \in V_h$ is also a solution of problem (P_h), and therefore, $\underline{\underline{u}}_h$ is uniquely determined, since the inclusion $X_h \subset M_h$ holds. (cf. Theorem 3). Likewise, the inclusion $M_h \subset Y_h$ shows that the function μ_h^* is uniquely determined (the proof is the same as that of Theorem 2). ■

Remark 4.

The inclusions (3.9) are indeed natural, since they are the discrete analogs of the inclusions

$$H_0^1(\Omega) \subset H^1(\Omega) \subset L_2(\Omega) . \qquad \blacksquare$$

One reason for introducing problem (P_h') is that it yields one way of actually computing the solution of problem (P_h), by solving the linear system represented by relations (3.7)-(3.8).

4. Error bounds.

Throughout this section, the boundary Γ is assumed to be sufficiently smooth. Our first result shows that the problem of estimating the error is reduced to a problem in approximation theory.

Theorem 5.

Assume that $u \in H^3(\Omega)$ *and that the inclusion*

$$X_h \subset M_h \tag{4.1}$$

holds. Then there exists a constant $\mathbb{C}$ *independent of the subspaces* X_h, Y_h *and* M_h *such that*

$$\begin{aligned} &|u-u_h|_{1,\Omega} + |\Delta u+\varphi_h|_{0,\Omega} \\ &\le \mathbb{C}\Big(\inf_{(v_h,\psi_h)\in \underline{\underline{V}}_h} \{ |u-v_h|_{1,\Omega} + |\Delta u+\psi_h|_{0,\Omega}\} \\ &\qquad + \inf_{\mu_h \in M_h} \|\Delta u+\mu_h\|_{1,\Omega} \Big), \end{aligned} \tag{4.2}$$

where $\underline{\underline{u}} = (u, -\Delta u)$ *is the solution of problem* (P), *and* $\underline{\underline{u}}_h = (u_h, \varphi_h)$ *is the solution of problem* (P_h).

Proof.

For ease of notation, the subscripts Ω will be omitted in the proof. We let $\underline{\underline{v}}_h = (v_h, \psi_h)$ be an arbitrary element in the space $\underline{\underline{V}}_h$, and we let μ_h be an arbitrary element in the space M_h. Since $u \in H^3(\Omega)$, then $\mu^* = -\Delta u$ (cf. Theorem 2), so that from relations (2.13) we obtain in particular

$$a(\underline{\underline{u}}, \underline{\underline{u}}_h - \underline{\underline{v}}_h) - f(\underline{\underline{u}}_h - \underline{\underline{v}}_h) - b(\underline{\underline{u}}_h - \underline{\underline{v}}_h, \Delta u + \mu_h) = 0 .$$

On the other hand, relations (3.6) yield

$$a(\underline{\underline{u}}_h, \underline{\underline{u}}_h - \underline{\underline{v}}_h) = f(\underline{\underline{u}}_h - \underline{\underline{v}}_h) ,$$

and hence,

$$a(\underline{\underline{u}} - \underline{\underline{u}}_h, \underline{\underline{u}}_h - \underline{\underline{v}}_h) = b(\underline{\underline{u}}_h - \underline{\underline{v}}_h, \Delta u + \mu_h) .$$

As a consequence, we obtain the inequality

$$|a(\underline{\underline{u}} - \underline{\underline{u}}_h, \underline{\underline{u}}_h - \underline{\underline{v}}_h)| \leq \beta |\varphi_h - \psi_h|_0 \|\Delta u + \mu_h\|_1 ,$$

valid for some constant β solely dependent upon the constant C occuring in inequality (3.5), which we may apply in view of the inclusion (4.1). Therefore we have

$$\begin{aligned} |\varphi_h - \psi_h|_0^2 &= a(\underline{\underline{u}} - \underline{\underline{v}}_h, \underline{\underline{u}}_h - \underline{\underline{v}}_h) + a(\underline{\underline{u}} - \underline{\underline{u}}_h, \underline{\underline{u}}_h - \underline{\underline{v}}_h) \\ &\leq |\Delta u + \psi_h|_0 |\varphi_h - \psi_h|_0 \\ &\qquad + \beta |\varphi_h - \psi_h|_0 \|\Delta u + \mu_h\|_1 . \end{aligned}$$

Combining the above inequality with the inequality

$$\begin{aligned} &|u - u_h|_1 + |\Delta u + \varphi_h|_0 \\ &\quad \leq |u - v_h|_1 + |\Delta u + \psi_h|_0 + (1 + C)|\varphi_h - \psi_h|_0 , \end{aligned}$$

we eventually obtain

$$|u-u_h|_1 + |\Delta u + \varphi_h|_0$$
$$\leq |u - v_h|_1 + (2+C)|\Delta u + \psi_h|_0 + \beta(1+C)\|\Delta u + \mu_h\|_1,$$

from which the inequality of (4.2) follows. ∎

Remark 5.

The norm which appears in the left-hand side of inequality (4.2) is a natural analog, in the present case, of the $\|\cdot\|_{2,\Omega}$ norm in which the error would be estimated for a conforming method. ∎

To apply Theorem 5, we must therefore evaluate the expression

$$\inf_{\mu_h \in M_h} \|\Delta u + \mu_h\|_{1,\Omega} ,$$

which is quite a standard problem for specific finite element subspaces. On the other hand, the evaluation of the quantity

$$\inf_{(v_h,\psi_h)\in \underline{\underline{V}}_h} \{|u - v_h|_{1,\Omega} + |\Delta u + \psi_h|_{0,\Omega}\}$$

is not straightforward since the functions v_h and ψ_h do not vary independently from each other. In the next result, we give a simple criterion which reduces this problem to a similar, but unconstrained, problem.

Theorem 6.

Assume that the inclusion

$$(4.3) \qquad Y_h \subset M_h$$

holds, and let $\alpha(h)$ be such that

$$(4.4) \qquad \forall\, \mu_h \in M_h , \qquad |\mu_h|_{1,\Omega} \leq \alpha(h)|\mu_h|_{0,\Omega} .$$

Then we have

$$\inf_{(v_h, \psi_h) \in \underline{\underline{V}}_h} \{ |u - v_h|_{1,\Omega} + |\Delta u + \psi_h|_{0,\Omega} \}$$

(4. 5)

$$\leq (1 + \alpha(h)) \inf_{v_h \in X_h} |u - v_h|_{1,\Omega} + 2 \inf_{\mu_h \in M_h} |\Delta u - \mu_h|_{0,\Omega}$$

Proof.

Let $\underline{\underline{v}}_h = (v_h, \psi_h)$ be an arbitrary element in $\underline{\underline{V}}_h$ and let μ_h be an arbitrary element in M_h. Since the inclusion (4. 3) holds, it follows that $\nu_h = \mu_h + \psi_h \in M_h$. Then on the one hand $b(\underline{\underline{v}}_h, \nu_h) = 0$ since $\underline{\underline{v}}_h \in \underline{\underline{V}}_h$, and on the other hand, $b(\underline{\underline{u}}, \nu_h) = 0$, since $\frac{\partial u}{\partial \nu} = 0$ on Γ. Therefore $b(\underline{\underline{u}} - \underline{\underline{v}}_h, \nu_h) = 0$, i. e.,

$$\int_\Omega (\Delta u + \psi_h) \nu_h \, dx = -\int \operatorname{grad}(u - v_h) \operatorname{grad} \nu_h \, dx ,$$

and thus,

$$\left| \int_\Omega (\Delta u + \psi_h) \nu_h dx \right| \leq |u - v_h|_1 |\nu_h|_1 \leq \alpha(h) |u - v_h|_1 |\nu_h|_0 .$$

As a consequence,

$$|\nu_h|_0^2 = \int_\Omega (\mu_h - \Delta u) \nu_h \, dx + \int_\Omega (\Delta u + \psi_h) \nu_h \, dx$$
$$\leq |\Delta u - \mu_h|_0 |\nu_h|_0 + \alpha(h) |u - v_h|_1 |\nu_h|_0 ,$$

so that we obtain the inequality

$$|\Delta u + \psi_h|_0 \leq |\Delta u - \mu_h|_0 + |\nu_h|_0$$
$$\leq 2 |\Delta u - \mu_h|_0 + \alpha(h) |u - v_h|_1 ,$$

from which we deduce the inequality (4. 5). ∎

As an immediate consequence of Theorems 5 and 6, we have

Theorem 7.

Assume that the inclusions

$$X_h \subset M_h \quad \text{and} \quad Y_h \subset M_h \tag{4.6}$$

hold, and that $u \in H^3(\Omega)$. Then there exists a constant $\mathscr{D}$ independent of the subspaces X_h, Y_h and M_h such that

$$\begin{aligned} &|u - u_h|_{1,\Omega} + |\Delta u + \varphi_h|_{0,\Omega} \\ &\leq \mathscr{D}((1+\alpha(h)) \inf_{v_h \in X_h} |u - v_h|_{1,\Omega} \\ &\qquad + \inf_{\mu_h \in M_h} \|\Delta u - \mu_h\|_{1,\Omega}). \end{aligned} \tag{4.7}$$

where $\alpha(h)$ is defined as in (4.4). ∎

To apply the above results, we assume that the open set Ω is polyhedral, i.e., $\bar{\Omega}$ is a polyhedron in $\mathbb{R}^n$, so that, in particular, its boundary Γ is sufficiently smooth in the sense understood here. We establish a triangulation $\mathfrak{T}_h$ over the set $\bar{\Omega}$, i.e., we write

$$\bar{\Omega} = \bigcup_{k \in \mathfrak{T}_h} K,$$

where the finite elements K satisfy the usual geometrical conditions that their interior shall be pairwise disjoint, and that the intersection of any two distinct finite elements K and K' is either empty or is a face of both K and K'. In addition, for all the triangulations considered, all the finite elements K are the image $F_K(\hat{K})$ of a reference finite element $\hat{K}$ through an affine mapping F_K.

We consider finite-dimensional spaces X_h, Y_h and M_h defined by

$$Y_h = M_h = \{v_h \in C^0(\bar{\Omega}); \forall K \in \mathfrak{T}_h, v_h|_K \in P_K\} \tag{4.8}$$

$$X_h = \{v_h \in M_h \,;\, v_h = 0 \text{ on } \Gamma\}, \tag{4.9}$$

where

$$P_K = \{v : K \to \mathbb{R} \,;\, v = \hat{v} \circ F_K^{-1}, \quad \forall \hat{v} \in \hat{P}\},$$

and $\hat{P}$ is a given finite-dimensional space of functions $\hat{v} : \hat{K} \to \mathbb{R}$, which satisfies the inclusion

$$P_k \subset \hat{P}. \tag{4.10}$$

Here, P_k denotes the set of all polynomials of degree $\leq k$ in the n variables $x_1, x_2, \ldots, x_n$.

We then consider a regular family $(\mathcal{T}_h)$ of triangulations in the sense of [5], in that there exists a constant σ independent of h such that

$$\max_{K \in \mathcal{T}_h} \left(\frac{h(K)}{\rho(K)}\right) \leq \sigma, \tag{4.11}$$

where $h(K)$ = diameter of K, and $\rho(K)$ = sup{diameter of inscribed spheres in K}. Also, we assume that for some constant τ,

$$h = \max_{K \in \mathcal{T}_h} h(K) \leq \frac{1}{\tau} \min_{K \in \mathcal{T}_h} h(K). \tag{4.12}$$

We first evaluate the constant $\alpha(h)$ which appears in (4.4).

Lemma 3.

Let there be given spaces M_h defined as in (4.8), which are associated with a regular family of triangulations. Then there exists a constant A independent of h such that

$$\forall \mu_h \in M_h, \quad |\mu_h|_{1,\Omega} \leq \frac{A}{h} |\mu_h|_{0,\Omega}. \tag{4.13}$$

Proof.

Let $\hat{K}$ be a reference finite element, and let $K \in \mathcal{T}_h$ be given. Then for any h there exists an affine transformation:

$$F_K: x \in \mathbb{R}^n \to F_K(x) = B_K x + b_K ,$$

where B_K is an invertible element of $\mathcal{L}(\mathbb{R}^n)$ and b_K a vector in $\mathbb{R}^n$, such that $K = F_K(\hat{K})$, and any function $v \in P_K$ is such that [5, equation (4.16)]

$$|v|_{1,K} \leq \|B_K^{-1}\| \, |\det(B_K)|^{\frac{1}{2}} |\hat{v}|_{1,\hat{K}} ,$$

where $\hat{v} = v \cdot F_K \in \hat{P}$. The space $\hat{P}$ being finite-dimensional, there exists a constant $\hat{c}$ such that

$$\forall \, \hat{v} \in \hat{P}, \quad |\hat{v}|_{1,\hat{K}} \leq \hat{c} \, |\hat{v}|_{0,\hat{K}} .$$

We also have, for all functions $v \in P_K$ [5, equation (4.15)]

$$|\hat{v}|_{0,\hat{K}} \leq (\det(B_K))^{-\frac{1}{2}} \, |v|_{0,K} .$$

Using next Lemma 2 of [5] and inequalities (4.11) and (4.12), we obtain

$$\|B_K^{-1}\| \leq \frac{h(\hat{K})}{\rho(K)} \leq \frac{\sigma}{\tau} \, \frac{h(\hat{K})}{h} ,$$

so that, combining all the above inequalities, we obtain

$$\forall \, v \in P_K , \quad |v|_{1,K} \leq \left(\frac{\sigma \hat{c}}{\tau} \, h(\hat{K})\right) \frac{|v|_{0,K}}{h} ,$$

from which the inequality (4.13) follows. ∎

We now have our main result.

Theorem 8.

We assume that $u \in H_0^2(\Omega) \cap H^{k+2}(\Omega)$ *for some integer* $k \geq 2$. *Let there be given spaces* Y_h, M_h *and* X_h *defined as in* (4.8)-(4.9), *which are associated with a regular family of triangulations. It is assumed that the inclusions of* (4.6) *and* (4.10) *hold.*

Then there exists a constant $\mathcal{K}$ *independent of* h *such that*

$$(4.14)\quad |u-u_h|_{1,\Omega} + |\Delta u + \varphi_h|_{0,\Omega} \leq \mathcal{K}\,(|u|_{k+1,\Omega} + |u|_{k+2,\Omega})\,h^{k-1}.$$

Proof.

In view of the inclusions $P_k \subset P_K$ for all $K \in \mathcal{T}_h$ and all h, and since the families are regular, it is well known (see [5, Theorem 5], for example) that for some constants C_1, C_2, C_3 independent of h,

$$(4.15)\quad \inf_{v_h \in X_h} |u - v_h|_{1,\Omega} \leq C_1 |u|_{k+1,\Omega}\, h^k \;,$$

$$(4.16)\quad \inf_{\mu_h \in M_h} \|\Delta u - \mu_h\|_{1,\Omega} \leq C_2 |\Delta u|_{k,\Omega}\, h^{k-1} \leq C_3 |u|_{k+2,\Omega}\, h^{k-1} \;.$$

Then inequality (4.14) follows from inequalities (4.7), (4.13), (4.15) and (4.16). ∎

Remark 6.

Using an extension of the duality technique of Aubin-Nitsche, one can also prove that with the same assumptions as in Theorem 8

$$(4.17)\quad |u - u_h|_{0,\Omega} \leq \mathcal{K}'(|u|_{k+1,\Omega} + |u|_{k+2,\Omega})h^k \;,$$

for another constant $\mathcal{K}'$ independent of h. The proof will be found in [6]. ∎

5. A concluding remark.

Because of the numerical complexity involved with standard conforming finite elements for solving fourth-order problems, even in the case of polygonal boundaries, it seems unrealistic to handle curved boundaries with the associated curved isoparametric finite elements.

With some numerical evidence to support this statement, it is suggested here that one proper way to handle fourth-order problems on curved domains is to use the method analogous to that described here, with the corresponding isoparametric finite elements.

References.

1. Bossavit, A., Une méthode de décomposition de l'opérateur biharmonique, Electricité de France, Note HI 585/2, 1971.

2. Brezzi, F., Théorèmes d'existence et unicité et d'approximation numérique pour des problèmes de point de selle. To appear.

3. Brezzi, F., Sur la méthode des éléments finis hybrides pour le problème biharmonique. To appear.

4. Ciarlet, P. G., Quelques méthodes d'éléments finis pour le problème d'une plaque encastrée, to appear in Proceedings of the Colloque International sur les Méthodes de Calcul Scientifique et Technique, I. R. I. A., Le Chesnay, 1973.

5. Ciarlet, P. G., and Raviart, P.-A., General Lagrange and Hermite interpolation in R^n with applications to finite element methods, Arch. Rat. Mech. Anal. 46 (1972), 177-199.

6. Ciarlet, P. G., and Raviart, P.-A., La Méthode des Eléments Finis pour les Problèmes aux Limites Elliptiques. To appear.

7. Glowinski, R., Approximations externes, par éléments finis de Lagrange d'ordre un et deux, du problème de Dirichlet pour l'opérateur biharmonique. Methodes itératives de résolution des problèmes approchés, Topics in Numerical Analysis (J. J. H. Miller, Editor), pp. 123-171. Academic Press, London, 1973.

8. Johnson, C., Convergence of another mixed finite-element method for plate bending problems, Report No. 27, Chalmers Institute of Technology and the University of Göteborg, 1972.

9. Johnson, C., On the convergence of a mixed finite-element method for plate bending problems, Numer. Math. 21 (1973), 43-62.

10. Mercier, B., Numerical solution of the biharmonic problem by mixed finite elements of class C^0, Report, Laboratorio di Analisi Numerica del C. N. R., Pavia, 1973.

11. Nečas, J., Les Méthodes Directes en Théorie des Equations Elliptiques, Masson, Paris, 1967.

12. Oden, J. T., Generalized conjugate functions for mixed finite element approximations of boundary-value problems, The Mathematical Foundations of the Finite Element Method (A. K. Aziz, Editor), pp. 629-670, Academic Press, New York, 1973.

13. Oden, J. T., Some contributions to the mathematical theory of mixed finite element approximations, Tokyo Seminar on Finite Elements, Tokyo, 1973.

14. Smith, J., On the approximate solution of the first boundary value problem for $\nabla^4 u = f$, SIAM J. Numer. Anal. 10 (1973), 967-982.

Added in proof:

It was brought to the authors' attention that the following references are also particularly relevant:

Oden, J. T.; and J. N. Reddy, On dual-complementary variational principles in mathematical physics, Int. J. Engng. Sci., 12 (1974), 1-29.

Reddy, J. N., A Mathematical Theory of Complementary-Dual Variational Principles and Mixed Finite Element Approximations of Linear Boundary Value Problems in Continuum Mechanics, Ph. D. Dissertation, The University of Alabama in Huntsville, Huntsville, 1973.

Analyse Numérique, Tour 55-65
Université de Paris VI
4, Place Jussieu
75230 Paris CEDEX 05
France

A Dissipative Galerkin Method for the Numerical Solution of First Order Hyperbolic Equations

LARS B. WAHLBIN

1. Introduction

The purpose of this paper is to show a connection between dissipative finite difference operators and a certain Galerkin-type method, see (1.6) below, for the numerical solution of first order one-dimensional hyperbolic problems, and to discuss the extension of this Galerkin method to certain equations which are of third order in the space derivative.

Consider the 1-periodic initial value problem for the real-valued function $u = u(x,t)$, given by

$$(1.1)\qquad \begin{cases} u_t = a(x)u_x , & 0 \le t \le T , \quad x \in \mathbb{R} , \\ u(x,0) = v_0(x) , & x \in \mathbb{R} . \end{cases}$$

Here $a(x)$ and $v_0(x)$ are real-valued and periodic with period 1. It is assumed that $a(x)$, $v_0(x)$ and $u(x,t)$ are smooth functions.

For the numerical treatment of the problem (1.1) we shall use certain piecewise polynomial spaces. Let μ and k be integers, $\mu - 1 > k \ge 0$, and let $\{i \cdot h\}$, $i = 0,\dots,h^{-1} \in \mathbb{Z}$ be a uniform partition of $I = [0,1]$ depending on the parameter h. Let (supressing k and h in the notation)

$S^{\mu} = \{\chi(x),\ x \in I$: the periodic extension of χ lies in $C^k(\mathbb{R})$ and $\chi\big|_{(ih,(i+1)h)}$ is a polynomial of degree $\leq \mu-1$, $i = 0,\ldots,h^{-1}-1.\}$.

As one motivation for the method (1.6) below, consider the question of L_2-optimality for (continuous-in-time) Galerkin methods. Take first the ordinary Galerkin method; i.e., define the Galerkin approximation as a differentiable map $U:[0,T] \to S^{\mu}$ such that

$$(1.2)\quad \begin{cases} (U_t - a(x)U_x, \chi) = 0\ , & \chi \in S^{\mu}\ ,\quad 0 \leq t \leq T\ , \\ U(0) \in S^{\mu}\ . & \end{cases}$$

Here $(f,g) = \int_I f(x)g(x)dx$. As is well known, see e.g. Fix and Nassif [5] and Thomée and Wendroff [11], when S^{μ} is a space of smoothest splines of order μ $(k = \mu-2)$, then the method (1.2) gives optimal error in L_2, i.e.,

$$(1.3)\quad \| U(t)(\cdot) - u(\cdot,t) \|_{L_2(I)} \leq C_1 h^{\mu}\ ,\quad 0 \leq t \leq T\ ,$$

provided the initial data satisfy the condition

$$(1.4)\quad \| U(0) - v_0 \|_{L_2(I)} \leq C_0 h^{\mu}\ .$$

However, Dupont [3] has shown that, for S^{μ} the space of Hermite cubics $(\mu = 4,\ k = 1)$ and $a(x) \equiv 1$, (1.3) never holds no matter how $U(0)$ is chosen (for v_0 nonconstant, analytic and 1-periodic). The best estimate that can be given for general spaces S^{μ} is

$$\| U(t)(\cdot) - u(\cdot,t) \|_{L_2(I)} \leq C_1 h^{\mu-1}\ ,$$

cf. Swartz and Wendroff [8].

Dendy [2] has proposed a method which defined $U : [0,T] \to S^{\mu}$ by

$$(1.5.a)\quad (U_t - a(x)U_x,\ \chi - \chi_x) = 0,\ \chi \in S^{\mu},\quad 0 \leq t \leq T\ ,$$

(1.5.b) $\quad U(0) \in S^{\mu}$.

He has shown that this method gives an error which is optimal in L_2 for any space S^{μ}, provided the conditions (i) and (ii) below hold:

(i) There exists a positive constant a_0 such that
$a(x) \geq a_0 > 0\ , \quad x \in \mathbb{R}$.

(ii) $\|U(0)-W\|_{H^1(I)} \leq C_0 h^{\mu}$, where $W \in S^{\mu}$ is the projection of v_0 into S^{μ} defined by $(W-v_0, \chi) + ((W - v_0)_x, \chi_x) = 0\ , \quad \chi \in S^{\mu}$.

In Dupont and Wahlbin [4], the condition (i) was dispensed with, and L_2-optimality was shown for all spaces S^{μ} under a condition similar to (ii) for initial data for the method (1.5) in the form

$$(1.5.a)' \qquad (U_t - a(x)\, U_x, K\chi - a(x)\, \chi_x) = 0\ ,$$

K a constant $\geq \|a_x\|_{L_\infty(I)} + 1$.

In order to obtain L_2-optimality for all spaces S^{μ} with the sole constraint (1.4) on the initial data, the following method was considered in Wahlbin [12]:

$$(1.6) \qquad \begin{cases} (U_t - a(x)\, U_x\ , \chi - h\chi_x) = 0\ , & \chi \in S^{\mu},\ 0 \leq t \leq T, \\ U(0) \in S^{\mu}\ . & \end{cases}$$

Under the condition (i) above on the coefficient $a(x)$, L_2-optimality holds if the data fulfills (1.4). This was shown in [12] in the generality of the quasilinear equation $u_t = a(x,t,u)u_x + f(x,t,u)$.

Consider for a moment the equation $u_t = u_x$. Trivial calculations lead to the following relations, where

$\|\cdot\| = \|\cdot\|_{L_2(I)}$,

$$\text{for (1.2)}: \ \tfrac{1}{2}\frac{d}{dt}\|U\|^2 = 0 .$$

$$\text{for (1.5)}: \ \tfrac{1}{2}\frac{d}{dt}\{\|U\|^2 + \|U_x\|^2\} + [\|U_t\|^2 + \|U_x\|^2 - 2(U_t, U_x)] = 0 .$$

$$\text{for (1.6)}: \ \tfrac{1}{2}\frac{d}{dt}\{\|U\|^2 + h^2\|U_x\|^2\} + h\cdot[\|U_t\|^2 + \|U_x\|^2 - 2(U_t, U_x)] = 0 .$$

Since in the spaces S^{μ} , $U_t \not\equiv U_x$ in general, this shows that a "dissipative mechanism" is present in the methods (1.5) and (1.6), but not in the ordinary Galerkin method.

In Section 2, the method (1.6) is considered for the simple equation $u_t = u_x$ extended to $\mathbb{R}$, and using a space of smoothest splines on $\mathbb{R}$. Applying the analysis in Thomée [9], [10], the method is interpreted as a semidiscrete finite difference method and it is shown that the corresponding finite difference operator is accurate of order $2\mu - 1$ and dissipative of order 2μ. Thus the finite difference operator satisfies the Kreiss stability condition - this condition coincides in this case with a criterion for stability in L_∞ given in Brenner and Thomée [1].

In Section 3, the extension of the method (1.6) to a model equation which is of third order in the space derivative is discussed.

2. The Dissipativity of the Galerkin Operator (1.6).

In this section, consider the pure initial value problem

$$(2.1) \qquad \begin{cases} u_t = u_x , & t \geq 0 , \quad x \in \mathbb{R} , \\ u(x,0) = v_0(x) , & x \in \mathbb{R} , \end{cases}$$

with no demands on periodicity.

We first describe briefly the analogy between the method (1.6) applied with smoothest splines on $\mathbb{R}$ and a certain finite difference operator. For details, see Thomée [9], or [10] for a special case. We define the set $\mathcal{S}^{\mu}$ of smoothest splines of order μ on $\mathbb{R}$ as follows. Let χ_0 be the characteristic function of the interval $[-\frac{1}{2}, \frac{1}{2}]$ and let $\varphi = \chi_0 * \chi_0 * \cdots * \chi_0$, $\mu-1$ times. Let for $j \in \mathbb{Z}$ and h given,

$$\varphi_j(x) = \varphi(x h^{-1} - j) .$$

Then $\mathcal{S}^{\mu}$ is the set $\{\Sigma_j c_j \varphi_j\}$ where $|c_j|$ grows at most like a power of $|j|$ as $|j|$ tends to infinity. By meshpoints we shall mean the set $\{j \cdot h\}$, $j \in \mathbb{Z}$. Note that for μ odd, the discontinuities in the $(\mu-1)^{st}$ derivative of functions in $\mathcal{S}^{\mu}$ occur at points $(j+\frac{1}{2})h$, $j \in \mathbb{Z}$.

Let

$$U(t)(x) = \sum_j c_j(t)\, \varphi_j(x) .$$

With $(f, g) = \int_{\mathbb{R}} f(x)g(x)dx$, the method (1.6) now reads

$$\begin{aligned} &\sum_j c_j'(t)\, [(\varphi_j, \varphi_\ell) + h(\varphi_j', \varphi_\ell)] \\ &= \sum_j c_j(t)\, [(\varphi_j', \varphi_\ell) - h(\varphi_j', \varphi_\ell')] , \qquad \ell \in \mathbb{Z} . \end{aligned} \tag{2.2}$$

The quantities $(\varphi_j, \varphi_\ell)$, $(\varphi_j', \varphi_\ell)$ and $(\varphi_j', \varphi_\ell')$ depend only on $\ell - j$ and hence the system (2.2) is a convolution equation. Apply the discrete Fourier transform. Let

$$\tilde{c}(\theta, t) = \sum_j c_j(t) \exp(-ij\theta) ,$$

$$g_{\mu,0}(\theta) = h^{-1} \sum_\ell (\varphi_0, \varphi_\ell) \exp(-i\ell\theta) ,$$

$$g_{\mu,1}(\theta) = -i \sum_\ell (\varphi'_0, \varphi_\ell) \exp(-i\ell\theta) ,$$

$$g_{\mu,2}(\theta) = h \sum_\ell (\varphi_0', \varphi_\ell') \exp(-i\ell\theta) ,$$

these quantities are independent of h. Then

$$(2.3) \qquad \frac{d}{dt}\tilde{c}(\theta,t)[h g_{\mu,0}(\theta) + i h g_{\mu,1}(\theta)] = \tilde{c}(\theta,t)[i g_{\mu,1}(\theta) - g_{\mu,2}(\theta)] .$$

Let

$$(2.4) \qquad p(\theta) = \frac{i g_{\mu,1}(\theta) - g_{\mu,2}(\theta)}{g_{\mu,0}(\theta) + i g_{\mu,1}(\theta)} .$$

Then (2.3) implies that

$$(2.5) \qquad \tilde{c}(\theta,t) = \exp(t h^{-1} p(\theta))\,\tilde{c}(\theta,0) .$$

If $U(0)$ is chosen as the interpolant of v_0 (coinciding with v_0 at mesh points), then the equation (2.5) means that the Galerkin solution coincides at meshpoints with the semi-discrete finite difference solution $u_h(x,t)$ to the problem (2.1) given as

$$(2.6) \qquad u_h(jh,t) = \sum_\ell f_\ell(th^{-1})\, v_0(jh - \ell h), \quad t \geq 0, \quad j \in \mathbb{Z}$$

with the coefficients f_ℓ defined by the relation

$$(2.7) \qquad \exp(t h^{-1} p(\theta)) = \sum_\ell f_\ell(th^{-1}) \exp(-i \ell \theta).$$

Recall that a semidiscrete finite difference operator defined by (2.6) and (2.7) with general periodic $p(\theta)$, independent of h, is accurate of order r for the problem (2.1) if

$$(2.8) \qquad p(\theta) = i\theta + O(\theta^{r+1}), \quad \theta \to 0 ,$$

and dissipative of order $2s$ if there exists a constant $c > 0$ such that

$$(2.9) \qquad \operatorname{Re} p(\theta) \leq -c\,\theta^{2s} , \quad |\theta| \leq \pi .$$

We shall prove the following result.

Theorem 2.1.

The operator given by (2.4), (2.6) and (2.7) is accurate of order 2μ-1 and dissipative of order 2μ .

The theoretical interest of this result lies with the fact that it shows (1.6) to satisfy the Kreiss condition for stability in L_2, see e.g. Richtmyer and Morton [7, Section 5.4], as well as the Brenner and Thomée [1] condition for the stability in the maximum norm for the model problem (2.1):

> Consider a finite difference operator of the form (2.6), (2.7). Then the following two conditions are equivalent:
>
> (i). Given $T > 0$ there exists a constant C independent of h such that for all $v_0 \in L_\infty$ and $0 \le t \le T$,
>
> $$\sup_{j \in \mathbb{Z}} |u_h(jh, t)| \le C \sup_{j \in \mathbb{Z}} |v_0(jh)| .$$
>
> (ii). $r + 1 = 2s$ (with possibly $\infty = \infty$ in the obvious sense).

Note also that the fact that the finite difference operator is accurate of order 2μ-1 means that the error in the Galerkin solution is $O(h^{2\mu-1})$ at meshpoints whereas the over-all error is $O(h^{\mu})$ in L_2 (and in L_∞).

The rest of this section except Remark 2.1 is devoted to the proof of Theorem 2.1.

Proof of Theorem 2.1.

We must verify (2.8) and (2.9) for $R + 1 = 2s = 2\mu$ and for

$$(2.10) \qquad p = i\, g_{\mu,1}(g_{\mu,0} + g_{\mu,2})/G + (g^2_{\mu,1} - g_{\mu,0} g_{\mu,2})/G$$

where

$$(2.11) \qquad G(\theta) = g^2_{\mu,0}(\theta) + g^2_{\mu,1}(\theta) .$$

An application of the Poisson summation formula shows that, for $\sigma = 0, 1, 2,$

$$(2.12) \qquad \begin{aligned} g_{\mu,\sigma}(\theta) &= \sum_\ell (\theta + 2\pi\ell)^\sigma \hat{\varphi}^2(\theta + 2\pi\ell) \\ &= \theta^\sigma \hat{\varphi}^2(\theta) + R_{\mu,\sigma}(\theta) \end{aligned}$$

where

$$(2.13) \qquad \hat{\varphi}(\theta) = \left(\frac{2 \sin(\frac{1}{2}\theta)}{\theta} \right)^\mu ,$$

and, as $\theta \to 0$,

$$(2.14) \qquad R_{\mu,\sigma}(\theta) = \begin{cases} O(\theta^{2\mu}), & \sigma = 0, 2, \\ O(\theta^{2\mu+1}), & \sigma = 1 . \end{cases}$$

Consider first the real part of $p(\theta)$. We have by (2.10), (2.12), and (2.13),

$$(2.15) \qquad \operatorname{Re} p(\theta) = -(2 \sin(\tfrac{1}{2}\theta))^{4\mu} q(\theta)/G(\theta)$$

where

$$(2.16) \qquad q(\theta) = q_2(\theta)\, q_0(\theta) - q_1^2(\theta) ,$$

$$(2.17) \qquad q_i(\theta) = \sum_\ell (\theta + 2\pi\ell)^{i-2\mu} , \qquad i = 0, 1, 2.$$

We shall prove that there exist two positive constants c_1 and c_2 such that

$$(2.18) \qquad c_1 \theta^{-2\mu} \geq q(\theta) \geq c_2 \theta^{-2\mu} , \quad 0 < |\theta| \leq \pi .$$

Since $q(\theta)$ is even it suffices to consider $0 < \theta \leq \pi$. Note that then $2\pi - \theta \geq \theta > 0$ and hence

$$q_1(\theta) = \sum_{\ell=0}^{\infty} \{(\theta + 2\pi\ell)^{1-2\mu} + (\theta - 2\pi - 2\pi\ell)^{1-2\mu}\}$$

$$\geq 0 .$$

Thus

$$q_1(\theta) = \theta^{1-2\mu} + \sum_{\ell=1}^{\infty} \frac{(\theta + 2\pi\ell)^{2\mu-1} + (\theta - 2\pi\ell)^{2\mu-1}}{(\theta^2 - (2\pi\ell)^2)^{2\mu-1}} \geq 0$$

so that with $r(\theta)$ a nonnegative bounded function we have

$$\theta^{1-2\mu} \geq q_1(\theta) = \theta^{1-2\mu} - r(\theta) \geq 0 . \tag{2.19}$$

Next note that there exist positive constants c_j^i, $j = 1,2$, $i = 0,2$, such that

$$\theta^{i-2\mu} + c_1^i \geq q_i(\theta) \geq \theta^{i-2\mu} + c_2^i , \quad i = 0,2. \tag{2.20}$$

By (2.16), (2.19), and (2.20) we have

$$(\theta^{-2\mu} + c_1^0)(\theta^{2-2\mu} + c_1^2) - (\theta^{1-2\mu} - r(\theta))^2$$

$$\geq q(\theta)$$

$$\geq (\theta^{-2\mu} + c_2^0)(\theta^{2-2\mu} + c_2^2) - (\theta^{1-2\mu})^2$$

or with positive $\tilde{c}_1, \tilde{c}_2$,

$$\tilde{c}_1 \theta^{-2\mu} \geq q(\theta) \geq \tilde{c}_2 \theta^{-2\mu} , \qquad 0 < \theta \leq \pi .$$

This is the desired result (2.18).

Since the denominator G in (2.15) is strictly positive we have with positive c_1, c_2

(2.21) $$-c_2\theta^{2\mu} \leq \operatorname{Re} p(\theta) \leq -c_1\theta^{2\mu}, \qquad |\theta| \leq \pi .$$

Consider next the imaginary part of $p(\theta)$. From (2.10), (2.12), and (2.14) we get

$$\operatorname{Im} p(\theta) = \frac{\theta(\hat{\varphi}^2 + O(\theta^{2\mu}))(\hat{\varphi}^2(1+\theta^2) + O(\theta^{2\mu}))}{(\hat{\varphi}^2 + O(\theta^{2\mu}))^2 + (\theta\hat{\varphi}^2 + O(\theta^{2\mu+1}))^2}$$

(2.22) $$= \frac{\theta(1 + O(\theta^{2\mu}))((1+\theta^2) + O(\theta^{2\mu}))}{1 + \theta^2 + O(\theta^{2\mu})}$$

$$= \theta(1 + O(\theta^{2\mu})) \qquad \text{as} \quad \theta \to 0 .$$

The theorem follows from (2.21) and (2.22).

Remark 2.1.

The same proof can be carried through for the ordinary Galerkin method (this is done in [9]) and the method (1.5). For the ordinary Galerkin method one finds an order of accuracy 2μ and no dissipation. For the method (1.5) one finds that $p(\theta)$ depends on h also:

$$p = i\,\frac{g_{\mu,1}(h^2 g_{\mu,0} + g_{\mu,2})}{h^2 g_{\mu,0}^2 + g_{\mu,1}^2} + \frac{h(g_{\mu,1}^2 - g_{\mu,2}g_{\mu,0})}{h^2 g_{\mu,0}^2 + g_{\mu,1}^2}$$

so that

$$\operatorname{Im} p(\theta) = \theta\Big(1 + O\Big(\frac{\theta^{2\mu}}{h^2+\theta^2}\Big)\Big) .$$

$$\operatorname{Re} p(\theta) \sim -\frac{h\,\theta^{2\mu}}{h^2+\theta^2} .$$

Modifying the definition (2.8) to cover the case when $p(\theta)$ depends on h we see that the order of accuracy for the finite difference operator corresponding to (1.5) is $2\mu - 2$.

3. Discussion of Higher Order Equations.

We first consider two typical examples of equations which are of second order in the space derivative. Our interest is in the question of the accuracy in L_2 of the ordinary Galerkin method.

Parabolic problems present no difficulties. The L_2-optimality for any space S^μ is well-known, see Wheeler [13]. Consider next the 1-periodic problem for the simple Schrödinger equation,

$$(3.1)\qquad \begin{cases} u_t = i\, u_{xx}\,, & 0 \le t \le T\,, \quad x \in \mathbb{R}\,, \\ u(x,0) = v_0(x)\,, & x \in \mathbb{R}\,. \end{cases}$$

For this problem we consider complex-valued solutions, and hence $(f,g) = \int_I f(x)\, \overline{g(x)}\, dx$ below. Again the ordinary Galerkin method is optimal in L_2 for all spaces S^μ. We give a simple proof based on Nitsche [6] and [13].

Theorem 3.1.

Let $U : [0,T] \to S^\mu$ *be defined by*

$$(U_t, \chi) + i(U_x, \chi_x) = 0\,, \qquad \chi \in S^\mu\,,$$

$$U(0) \in S^\mu\,, \quad \|U(0) - v_0\|_{L_2(I)} \le C_0\, h^\mu\,.$$

Then there exists a constant C_1 *such that*

$$\|U(t)(\cdot) - u(\cdot, t)\|_{L_2(I)} \le C_1\, h^\mu\,, \qquad 0 \le t \le T\,.$$

Proof.

Let $W(t) \in S^\mu$ be defined by $\rho = u - W$,

$$(\rho_x, \chi_x) + (\rho, \chi) = 0\,, \qquad \chi \in S^\mu\,.$$

The following estimate is proved in [6] for u sufficiently smooth:

$$\|\rho_t\|_{L_2(I)} + \|\rho\|_{L_2(I)} \leq C h^{\mu} . \tag{3.2}$$

Set $\vartheta = U - W$. Then

$$(\vartheta_t, \chi) + i(\vartheta_x, \chi_x) = (\rho_t, \chi) - i(\rho, \chi) .$$

Let $\chi = \vartheta$ here. Taking real parts gives

$$\tfrac{1}{2} \frac{d}{dt} \|\vartheta\|^2_{L_2(I)} \leq \|\vartheta\|^2_{L_2(I)} + \tfrac{1}{2}(\|\rho_t\|^2_{L_2(I)} + \|\rho\|^2_{L_2(I)}) .$$

The theorem follows from Gronwall's lemma, the triangle inequality, and (3.2).

Consider now equations which are of third order in the space derivative. We take as a model 1-periodic problem the following based on the Korteweg-deVries equation,

$$\begin{aligned} &u_t + u_{xxx} + uu_x = 0 , \quad 0 \leq t \leq T , \quad x \in \mathbb{R} , \\ &u(x, 0) = v_0(x) , \quad x \in \mathbb{R} . \end{aligned} \tag{3.3}$$

In (3.3), only real valued functions are involved, and it is assumed that v_0 is periodic and smooth and that a solution exists which is sufficiently smooth for all our purposes.

It is known, cf. [11], that the ordinary Galerkin method for the problem (3.3) with S^{μ} a space of smoothest splines and $\mu \geq 3$ gives optimal order error in L_2, but the best that can be said for general spaces S^{μ} seems to be the estimate, cf. [8],

$$\|U(t)(\cdot) - u(\cdot, t)\|_{L_2(I)} \leq C h^{\mu - 3} .$$

Hence it may be of interest to generalize the method (1.6) to the problem (3.3).

We shall consider spaces S^{μ} with $k \geq 2$, and we define the Galerkin approximation by

$$(3.4) \quad \begin{aligned} &(U_t + U_{xxx} + UU_x, \chi + h^3 \chi_{xxx}) = 0, \ \chi \in S^{\mu}, \ 0 \leq t \leq T, \\ &U(0) \in S^{\mu}. \end{aligned}$$

We shall prove the following result.

Theorem 3.2.

Assume that $k \geq 2$ and that the initial data $U(0)$ satisfy

$$(3.5) \qquad \| U(0) - v_0 \|_{L_2(I)} \leq C_1 h^{\mu}.$$

Then there exist constants C and h_0 depending on T, v_0, and C_1 such that for $0 \leq t \leq T$ and $0 < h \leq h_0$, the Galerkin solution given by (3.4) exists, and

$$\| U(t)(\cdot) - u(\cdot, t) \|_{L_2(I)} \leq C h^{\mu}.$$

In the proof of Theorem 3.2 the following two results will be used.

Lemma 3.1. (Inverse property.)

There exists a constant C such that for all $\chi \in S^{\mu}$

$$\| \chi_x \|_{L_2(I)} \leq C h^{-1} \| \chi \|_{L_2(I)}.$$

Lemma 3.2.

Let $k \geq 2$. There exist constants K and h_0, depending on u, such that $W \in S^{\mu}$ is well defined for $h \leq h_0$ by

$$\rho = u - W,$$

$$(\rho_{xxx} + u\rho_x, \chi + h^3\chi_{xxx}) + K(\rho, \chi) = 0, \quad \chi \in S^\mu. \tag{3.6}$$

Furthermore there exists a constant C such that

$$\sum_{j=0}^{3} h^h \|(\tfrac{\partial}{\partial t})^\ell \rho\|_{H^j(I)}$$

$$\leq C h^\mu (\|u\|_{H^\mu(I)} + \|u_t\|_{H^\mu(I)}), \quad \ell = 0, 1. \tag{3.7}$$

Here we use the notation $\|u\|_{H^r(I)} = (\sum_{i=0}^{r} \|(\frac{\partial}{\partial x})^i U\|^2_{L_2(I)})^{\frac{1}{2}}$.

Lemma 3.1 is well known and easy to prove. We prove Lemma 3.2.

Proof of Lemma 3.2.

Let $\|\cdot\| = \|\cdot\|_{L_2(I)}$ and let C denote a generic constant. For

$$K \geq \sup (\tfrac{1}{2}\|u_x\|_{L_\infty(I)} + \tfrac{1}{2}h(C')^2\|u\|^2_{L_\infty(I)} + 1)$$

where C' is the constant of Lemma 3.1, the form $\varphi, \psi \mapsto (\varphi_{xxx} + u\varphi_x, \psi + h^3\psi_{xxx}) + K(\varphi, \psi)$ is positive definite. Hence, W exists. Then for any $\chi \in S^\mu$, and h small,

$$\begin{aligned} & h^3\|\rho_{xxx}\|^2 + \|\rho\|^2 \\ & \leq C((\rho_{xxx} + u\rho_x, \rho + h^3\rho_{xxx}) + K(\rho, \rho)) \\ & = C((\rho_{xxx} + u\rho_x, (u-\chi) + h^3(u-\chi)_{xxx}) + K(\rho, u-\chi)). \end{aligned} \tag{3.8}$$

By well known approximation properties of the space S^μ, $\mu \geq 3$, $k \geq 2$, we can find χ such that

$$\|u-\chi\| + h^3\|(u-\chi)_{xxx}\| \leq C h^\mu \|u\|_{H^\mu(I)}.$$

Hence we obtain from (3.8)

(3.9.a) $$\|\rho_{xxx}\| \leq C\, h^{\mu-3}\, \|u\|_{H^{\mu}(I)} ,$$

(3.9.b) $$\|\rho_{x}\| \leq C\, h^{\mu-2}\, \|u\|_{H^{\mu}(I)} ,$$

(3.9.c) $$\|\rho\| \leq C\, h^{\mu-3/2}\, \|u\|_{H^{\mu}(I)} .$$

Next apply duality, [6]. Solve the equation

(3.10) $$Kf - (uf)_x - f_{xxx} - h^3 (uf_{xxx})_x - h^3 f_{xxxxxx} = \rho .$$

Note that, for h small enough,

(3.11) $$\|f\|_{H^3(I)} \leq C\|\rho\| .$$

Then, for any $\chi \in S^{\mu}$,

(3.12) $$\|\rho\|^2 = (\rho_{xxx} + u\rho_x,\ (f-\chi) + h^3 (f-\chi)_{xxx}) + K(\rho, f-\chi).$$

Choose χ such that

(3.13) $$\|f-\chi\| + h^3 \|(f-\chi)_{xxx}\| \leq Ch^3 \|f\|_{H^3(I)} \leq Ch^3 \|\rho\|$$

where we used (3.11) in the last step. By (3.9), (3.12), and (3.13) we find that

$$\|\rho\|^2 \leq Ch^{\mu} \|u\|_{H^{\mu}(I)} \|\rho\| .$$

By interpolation between this result and (3.9.a), the result (3.7) follows in the case $\ell = 0$.

To prove (3.7) for $\ell = 1$, differentiate (3.6) with respect to time to obtain

$$(3.14)\qquad (\rho_{txxx} + u\rho_{tx}, \chi + h^3\chi_{xxx}) + K(\rho_t, \chi) + (u_t\rho_x, \chi + h^3\chi_{xxx}) = 0 .$$

We have now, using (3.7) for $\ell = 0$,

$$\begin{aligned} h^3\|\rho_{txxx}\|^2 + \|\rho_t\|^2 &\le C((\rho_{txxx} + u\rho_{tx}, \rho_t + h^3\rho_{txxx}) + K(\rho_t, \rho_t)) \\ &\le C((\rho_{txxx} + u\rho_{tx}, \rho_t + h^3\rho_{txxx}) + K(\rho_t, \rho_t) \\ &\qquad + (u_t\rho_x, \rho_t + h^3\rho_{txxx})) \\ &\qquad + C(h^{\mu-1}\|\rho_t\| + h^{\mu+\frac{1}{2}}h^{3/2}\|\rho_{txxx}\|) \end{aligned}$$

and it follows by use of (3.14) with the proper χ that the analogue of (3.9) is valid. Next solve (3.10) with the right hand side replaced by ρ_t. We have then, corresponding to (3.12),

$$(3.15)\qquad \|\rho_t\|^2 = (\rho_{txxx} + u\rho_{tx}, (f - \chi) + h^3(f - \chi)_{xxx}) + K(\rho_t, f - \chi) - (u_t\rho_x, \chi + h^3\chi_{xxx}) .$$

By the analogues of (3.11) and (3.13) we find that, for the proper choice of χ,

$$\begin{aligned} |(u_t\rho_x, \chi + h^3\chi_{xxx})| &\le C(\|\rho\|\,\|\chi\|_{H^1(I)} + \|\rho_x\|\,h^3\|\chi_{xxx}\|) \\ &\le C\,h^{\mu}\,\|\rho_t\| . \end{aligned}$$

Hence (3.15) yields

$$\|\rho_t\| \le C\,h^{\mu}$$

and (3.7) follows in the case $\ell = 1$ as well.

Proof of Theorem 3.2.

Let $\|\cdot\| = \|\cdot\|_{L_2(I)}$ and let C denote a generic constant. Let

$$\vartheta = U - W$$

where W was defined by (3.6). We assume a priori that the solution U to (3.4) exists for $0 \leq t \leq T_1 \leq T$, $T_1 > 0$, and that

$$\|\vartheta\|_{L_\infty(I)} + \|\vartheta_x\|_{L_\infty(I)} \leq h, \quad 0 \leq t \leq T_1 \tag{3.16}$$

for $h \leq h_0$. Using these a priori assumptions we shall derive a better estimate for ϑ and in the end we can then remove the a priori constraints by a standard continuity argument.

Note that since

$$\|\chi\|_{L_\infty(I)} \leq C h^{-\frac{1}{2}} \|\chi\|_{L_2(I)}, \quad \chi \in S^\mu, \tag{3.17}$$

(3.16) implies via Lemmas 3.1 and 3.2 that

$$\|U\|_{L_\infty(I)} + \|U_x\|_{L_\infty(I)} \leq C, \qquad 0 \leq t \leq T_1 . \tag{3.18}$$

We now start to derive an $O(h^\mu)$ estimate for $\|\vartheta\|$. We have

$$\begin{aligned} (\vartheta_t + \vartheta_{xxx} + U\vartheta_x + \vartheta W_x, \chi + h^3 \chi_{xxx}) \\ = (\rho_t + \rho W_x, \chi + h^3 \chi_{xxx}) + K(\rho, \chi) . \end{aligned} \tag{3.19}$$

Take first $\chi = \vartheta$ in (3.19). By (3.18) and a similar uniform bound for W_x we obtain, using also Lemma 3.2,

$$(3.20)\qquad \tfrac{1}{2}\frac{d}{dt}\|\vartheta\|^2 + h^3(\|\vartheta_{xxx}\|^2 + (\vartheta_t + U\vartheta_x + \vartheta W_x, \vartheta_{xxx})) \le Ch^{2\mu} + C(\|\vartheta\|^2 + h^6\|\vartheta_{xxx}\|^2).$$

Next let $\chi = \vartheta_t$ in (3.19). Then we get

$$(3.21)\qquad \|\vartheta_t\|^2 + (\vartheta_{xxx} + U\vartheta_x + \vartheta W_x, \vartheta_t) + \tfrac{1}{2}h^3\frac{d}{dt}\|\vartheta_{xxx}\|^2 = -h^3(U\vartheta_x + \vartheta W_x, \vartheta_{txxx}) + (\rho_t + \rho w_x, \vartheta_t + h^3\vartheta_{txxx}) + K(\rho, \vartheta_t).$$

We note that (3.21) and Lemmas 3.1 and 3.2 imply that

$$(3.22)\qquad \|\vartheta_t\|^2 + h^3\frac{d}{dt}\|\vartheta_{xxx}\|^2 \le C h^{2\mu} + C(\|\vartheta\|^2 + \|\vartheta_{xxx}\|^2).$$

Here we also used the fact that, by the a priori assumptions,

$$|(U\vartheta_x, \vartheta_t)| \le C\|\vartheta_x\|\,\|\vartheta_t\| \le \tfrac{1}{4}\|\vartheta_t\|^2 + C(\|\vartheta\|^2 + \|\vartheta_{xxx}\|^2)$$

and the corresponding result with ϑ_t replaced by $h^3\vartheta_{txxx}$.

Now multiply (3.21) by h^3 and add to (3.20). We obtain

$$(3.23)\qquad \tfrac{1}{2}\frac{d}{dt}(\|\vartheta\|^2 + h^6\|\vartheta_{xxx}\|^2) + h^3[\|\vartheta_{xxx}\|^2 + (\vartheta_t + U\vartheta_x + \vartheta W_x, \vartheta_{xxx}) + \|\vartheta_t\|^2 + (\vartheta_{xxx} + U\vartheta_x + \vartheta W_x, \vartheta_t)]$$

$$\leq C h^{2\mu} + C(\|\vartheta\|^2 + h^6 \|\vartheta_{xxx}\|^2)$$
$$+ h^3(\rho_t + \rho W_x , \vartheta_t + h^3 \vartheta_{txxx}) + K h^3(\rho, \vartheta_t)$$
$$- h^6 (U\vartheta_x + \vartheta W_x , \vartheta_{txxx}) .$$

We next estimate various terms in (3.23). First note that the terms in square brackets equal

$$\|\vartheta_{xxx} + \vartheta_t + \tfrac{1}{2}(U\vartheta_x + \vartheta W_x)\|^2 - \tfrac{1}{4}\|U\vartheta_x + \vartheta W_x\|^2 \tag{3.24}$$

and that

$$h^3 \|U\vartheta_x + \vartheta W_x\|^2 \leq C h \|\vartheta\|^2 . \tag{3.25}$$

Next note that by Lemmas 3.1 and 3.2 and by (3.22) we have

$$|h^3(\rho_t + \rho W_x, \vartheta_t + h^3 \vartheta_{txxx}) + K h^3(\rho, \vartheta_t)| \tag{3.26}$$
$$\leq C h^{3+\mu} \|\vartheta_t\| \leq c\, h^{2\mu} + \tfrac{1}{2} h^6 \|\vartheta_t\|^2$$
$$\leq C h^{2\mu} + C(\|\vartheta\|^2 + h^6 \|\vartheta_{xxx}\|^2) - \tfrac{1}{2} h^9 \frac{d}{dt} \|\vartheta_{xxx}\|^2 .$$

Finally note that

$$h^6 (U\vartheta_x + \vartheta W_x, \vartheta_{txxx}) \tag{3.27}$$
$$= h^6 (\vartheta\vartheta_x - \rho\vartheta_x + \vartheta W_x, \vartheta_{txxx}) + h^6 (u\vartheta_x, \vartheta_{txxx})$$

and that by the a priori assumptions, by Lemmas 3.1 and 3.2, and by (3.22), (since $\|\rho\|_{L_\infty(I)} \leq \|\rho\|_{H^1(I)}$),

$$(3.28)\qquad |h^6(\vartheta\vartheta_x - \rho\vartheta_x + \vartheta W_x, \vartheta_{txxx})|$$
$$\leq C h^3\|\vartheta\|\,\|\vartheta_t\| \leq C\|\vartheta\|^2 + \tfrac{1}{2}h^6\|\vartheta_t\|^2$$
$$\leq Ch^{2\mu} + C(\|\vartheta\|^2 + h^6\|\vartheta_{xxx}\|^2 - \tfrac{1}{2}h^9\frac{d}{dt}\|\vartheta_{xxx}\|^2.$$

We now use the results (3.24) to (3.28) in (3.23). We obtain then

$$(3.29)\qquad \tfrac{1}{2}\frac{d}{dt}(\|\vartheta\|^2 + (h^6 + 2h^9)\|\vartheta_{xxx}\|^2)$$
$$\leq Ch^{2\mu} + C(\|\vartheta\|^2 + h^6\|\vartheta_{xxx}\|^2) - h^6(u\vartheta_x, \vartheta_{txxx}).$$

To handle the last term in (3.29), integrate in time from 0 to τ, $\tau \leq T_1$. Note that, by Lemma 3.2 and our assumption (3.5) on initial data,

$$(3.30)\qquad \|\vartheta\|^2(0) + (h^6 + 2h^9)\|\vartheta_{xxx}\|^2(0) \leq Ch^{2\mu}.$$

Furthermore note that

$$\int_0^\tau (u\vartheta_x, \vartheta_{txxx})dt = (u\vartheta_x, \vartheta_{xxx})\Big|_0^\tau$$
$$-\int_0^\tau ((u_t\vartheta_x, \vartheta_{xxx}) + (u\vartheta_{tx}, \vartheta_{xxx}))dt$$

and that

$$\int_0^\tau (u\vartheta_{tx}, \vartheta_{xxx})dt = -\int_0^\tau ((u_x\vartheta_{tx}, \vartheta_{xx}) + (u\,\vartheta_{txx}, \vartheta_{xx}))dt =$$
$$= -\int_0^\tau ((u_x\vartheta_{tx}, \vartheta_{xx}) - \tfrac{1}{2}(u_t\vartheta_{xx}, \vartheta_{xx}))dt - (u\vartheta_{xx}, \vartheta_{xx})\Big|_0^\tau.$$

From this it follows by Lemma 3.1, (3.22), and (3.30) that, for h small enough,

$$(3.31) \qquad \left| h^6 \int_0^\tau (u\vartheta_x, \vartheta_{txxx}) dt \right|$$
$$\leq Ch^{2\mu} + \tfrac{1}{4}\|\vartheta\|^2(\tau) + C\int_0^\tau \|\vartheta\|^2 dt .$$

Integrate now (3. 29) from 0 to τ and use (3. 30) and (3. 31). It follows that

$$\tfrac{1}{4}(\|\vartheta\|^2 + 2(h^6 + 2h^9)\|\vartheta_{xxx}\|^2)(\tau)$$
$$\leq Ch^{2\mu} + C\int_0^\tau (\|\vartheta\|^2 + h^6\|\vartheta_{xxx}\|^2) dt$$

and by Gronwall's lemma we obtain

$$(3.32) \qquad \|\vartheta\| \leq C_2 h^\mu , \quad 0 \leq t \leq T_1 .$$

We shall use this estimate to show that U exists and that the a priori assumption (3. 16) is satisfied on $[0, T]$ for h small enough. Note that the constant C_2 in (3. 32) depends on C_1, v_0, u, and T (for $T_1 \leq T$) but not on h. Fix h and note that (3. 16) holds initially if $h \leq h_0$ by the assumption on initial data and since (3. 4) is a finite system of ordinary differential equations for the coefficients of U in some basis for S^μ. Assume that (3. 16) fails for some t in $[0, T]$, and let then

$$t_0 = \inf\{t \in [0, T]: (3.16) \text{ fails}\}.$$

Then $t_0 > 0$. By the analysis above and by Lemma 3. 1 and (3. 17) we obtain

$$h = \|\vartheta\|_{L_\infty(I)}(t_0) + \|\vartheta_x\|_{L_\infty(I)}(t_0)$$
$$\leq Ch^{-3/2}\|\vartheta\|_{L_2(I)}(t_0) \leq CC_2 h^{\mu - 3/2} .$$

Since $\mu \geq 3$ we arrive at a contradiction for h small enough. For existence, note that if (for fixed h) $U(t)$ ceased to exist at t_0, then $\|U(t)\| \to \infty$ as $t \to t_0^-$, by well known results for ordinary differential equations. This would contradict the estimates above.

Hence the result (3. 32) is valid without qualifications for $0 \leq t \leq T$ and $h \leq h_0$. The theorem follows from (3. 32) and Lemma 3. 2 via the triangle inequality.

References.

1. P. Brenner and V. Thomée, Stability and convergence rates in L_p for certain difference schemes, Math. Scand. 27 (1970), 5-23.

2. J. E. Dendy, Two methods of Galerkin type achieving optimum L^2-accuracy for first order hyperbolics, to appear in SIAM J. Numer. Anal.

3. T. Dupont, Galerkin methods for first order hyperbolics: An example, SIAM J. Numer. Anal. 10 (1973), 890-899.

4. T. Dupont and L. Wahlbin, L^2 optimality of weighted-H^1 projections into piecewise polynomial spaces, to appear.

5. G. Fix and N. Nassif, On finite element approximations to time dependent problems, Numer. Math. 19(1972), 127-135.

6. J. Nitsche, Ein Kriterium für die Quasioptimalität des Ritzschen Verfahrens, Numer. Math. 11 (1968), 346-348.

7. R. D. Richtmyer and K. W. Morton, Difference methods for initial value problems, Second edition, Interscience, New York 1967.

8. B. Swartz and B. Wendroff, Generalized finite-difference schemes, Math. Comp. 23 (1969), 37-48.

9. V. Thomée, Convergence estimates for semidiscrete Galerkin methods for initial value problems, Numerische, insbesondere approximations-theoretische Behandlung von Funktionalgleichungen, Springer Lecture Notes, 333, Berlin 1973.

10. V. Thomée, Spline approximation and difference schemes for the heat equation, The Mathematical Foundations of the Finite Element Method with Applications to Partial Differential Equations, A. K. Aziz, Ed., Academic Press, 1972.

11. V. Thomée and B. Wendroff, Convergence estimates for Galerkin methods for variable coefficient initial-value problems, to appear in SIAM J. Numer. Anal.

12. L. Wahlbin, A dissipative Galerkin method applied to some quasilinear hyperbolic equations, to appear in La Revue Francaise d'Automatique, Informatique, Recherche Operationelle.

13. M. F. Wheeler, A priori L^2 error estimates for Galerkin approximations to parabolic partial differential equations, SIAM J. Numer. Anal. 10 (1973), 723-759.

Department of Mathematics
University of Chicago
Chicago, Illinois 60637

C^1 Continuity via Constraints for 4th Order Problems

RIDGWAY SCOTT

1. Introduction.

Let Π be a domain in the plane whose boundary consists of a finite number of non-intersecting polygonal arcs, and let $\mathfrak{J}$ be a triangulation of Π. We denote by $S_n(\Pi, \mathfrak{J})$ the space of functions in $C^1(\Pi)$ that are polynomials of degree $\leq n$ in each triangle in $\mathfrak{J}$. Strang [St2, St3] was the first to study the spaces S_n in general, proposing a heuristic method for calculating the dimension of S_n. Together with John Morgan, we proved in [MS] that Strang's conjectured dimension was correct (modulo some restrictions on $\mathfrak{J}$ when $n < 4$ to be described later). The advantage of S_n is that it is contained in the Sobolev space $H^2(\Pi)$, hence admissible in a 4-th order variational problem. However, the usual Ritz-Galerkin procedure using the subspace S_n requires an explicit basis for S_n, which we have not found in general. Thus we study a method for calculating the Ritz-Galerkin approximation from S_n that views S_n as a subspace of a larger space (with explicit basis) satisfying constraints. The main result of this paper is that these constraints are not redundant, so that the matrix determining the Ritz-Galerkin approximation is invertible. The proof relies on the dimension result in [MS]. This constraint method has been studied in the engineering literature [HK][†] for $n=3$, and has been found to be competitive with other techniques. (In fact, once static condensation has been performed, the number of degrees of freedom for this method is almost identical to the number for the usual Galerkin method using the Clough-Tocher element, differing only at the boundary.) We study the method for all $n \geq 3$, as analysis of the general case poses no complication. We thank Carl de Boor and Todd Dupont for their helpful comments about the results presented here.

We now describe Strang's heuristic computation of $\dim S_n$ and the results of [MS]. To calculate $\dim S_n$, we view S_n as the space of discontinuous piecewise polynomials satisfying constraints along each interior edge in $\mathfrak{T}$. Let T denote the number of triangles in $\mathfrak{T}$ and E_0 the number of interior edges. Since it takes $n+1$ constraints on the values and n constraints on the normal derivatives to make two polynomials of degee n agree in a C^1 fashion across each edge, we would expect

$$\dim S_n = (\dim \mathcal{P}_n)T - (2n+1)E_0 ,$$

where $\mathcal{P}_n$ is the space of polynomials of degree $\leq n$ in two variables. However, the constraints above are redundant: if around each interior vertex we impose the $2n+1$ edge constraints on all edges but one, then functions satisfying the constraints are already C^1 at the vertex. Thus there are three redundancies for each interior vertex. There is one further redundancy at a singular vertex.

Definition 1.

Let v be a vertex in a tringulation $\mathfrak{T}$ of Π. Then v is said to be singular if

i) v is not on $\partial\Pi$,
ii) precisely four edges meet at v, and

[†] Ciarlet has informed us of a recent paper [Z] by Zienkiewicz on this topic, and there may indeed be others in the engineering literature.

iii) <u>the edges meeting at</u> v <u>form two straight lines</u>.

As we show in the next section, there is one extra redundancy associated with a singular vertex. So if we let V_0 be the number of interior vertices in $\mathfrak{T}$ and σ be the number of singular ones, we have

$$\text{(1.1)} \qquad \dim S_n = (\dim \mathcal{P}_n)T - (2n+1)E_0 + 3V_0 + \sigma .$$

This is the dimension conjectured by Strang in [St1, St2], except that singular vertices were not explicitly mentioned. We have proved in [MS] that <u>(1.1) is correct for any</u> Π <u>and</u> $\mathfrak{T}$ <u>if</u> $n \geq 4$. Further, we showed by an example that it is <u>false</u> in general for $n=2$, i.e., there can be more redundancies than those already enumerated. The case $n=3$ is of importance to us here, so we now state conditions on $\mathfrak{T}$ under which we can prove that (1.1) is correct for $n=3$ (we know of no examples where it is incorrect, however, so it may be correct for all triangulations).

Our conditions on $\mathfrak{T}$ are based on how complicated it is to construct $\mathfrak{T}$. We think of $\mathfrak{T}$ as being constructed from a single triangle by adding one triangle at a time, as in Figure 1. The three types of operations are a) adding two edges to an existing edge, b) adding one edge to two existing edges and c) adding two edges to an existing edge and vertex. (The latter is required only for non-simply connected domains Π.) It is the step b) that complicates things, and we consider that case in detail now. Suppose τ is the new triangle formed in a step b), and let τ_1 be one of the triangles sharing an edge e_1 with τ in the complex of triangles constructed so far. Let e_2 be the other edge of τ shared with the current complex, let e_3 be the other edge of τ_1 that meets e_1 at the common vertex of e_1 and e_2, and let v be the other vertex of e_1. We say e_1 is <u>free</u> if e_3 and e_2 are not co-linear and the number of triangles in the current complex (including τ) meeting at v is less than five.

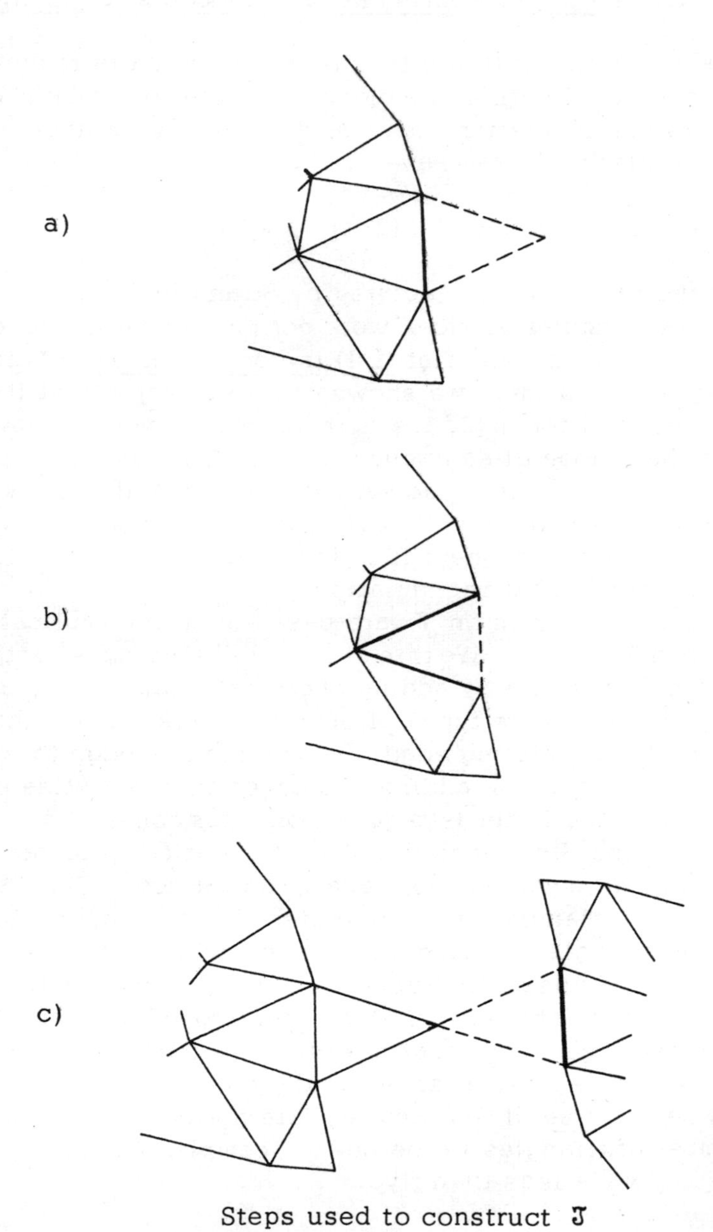

Steps used to construct $\mathfrak{J}$

Figure 1

Definition 2.

We say $\mathfrak{J}$ is unconstrained if it can be constructed via steps a), b), c), such that in each step b) one of the new interior edges is free (see above).

In [MS], we prove the following:

Proposition 1.

Formula (1.1) for $\dim S_n(\Pi, \mathfrak{J})$ is valid for $n \geq 4$ for all $\mathfrak{J}$, and for $n=3$ if $\mathfrak{J}$ is unconstrained.

We note that any triangulation in the plane can be constructed using only steps a), b), c). An example of a constrained triangulation is shown in Figure 2.

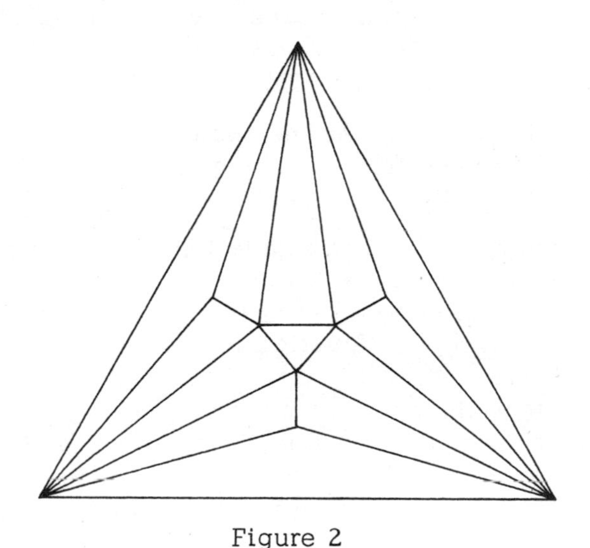

Figure 2

Constrained Triangulation

We can prove that (1.1) holds for $n=2$ under stronger restrictions on $\mathfrak{J}$ than in Proposition 1. This result indicates that the dimension of S_2 is not large enough to achieve good approximation, unless singular vertices are systematically used [P]. For example, we show in [MS] that, on a regular mesh without singular vertices, there are no C^1 piecewise quadratics with compact support. In this paper, we are interested only in the case $n \geq 3$.

Our main theorem relies on the dimension formula (1.1) being valid, so we now give two criteria for determining whether or not a triangulation is constrained. These are not intended to be definitive criteria, rather they are aimed at a situation typical for a finite element problem. In a finite element problem, it is frequently the case that one will start with a coarse mesh $\mathfrak{T}_0$, compute an approximation using $\mathfrak{T}_0$, then refine $\mathfrak{T}_0$ to get $\mathfrak{T}_1$, compute with $\mathfrak{T}_1$ and compare the solutions as a convergence test. One simple refinement is the <u>regular</u> subdivision $\mathfrak{T}_1$ of $\mathfrak{T}_0$, obtained by joining the edge midpoints of adjacent triangles in $\mathfrak{T}_0$ by a new edge, thus dividing each triangle in $\mathfrak{T}_0$ into four triangles. This process is easily programmed on a computer and has the advantage that it does not decrease any of the angles in the triangulation (each triangle is divided into four similar triangles). Our first criterion is the following.

> (1) Suppose $\mathfrak{T}_0$ is unconstrained and $\mathfrak{T}_1$ is obtained from $\mathfrak{T}_0$ by a regular subdivision. Then $\mathfrak{T}_1$ is unconstrained.

Thus checking to see if a triangulation is unconstrained can be, in some cases, reduced to checking it on a coarser mesh. To give the second criterion, we introduce the notion of a <u>quasi-singular vertex.</u> A vertex is said to be quasi-singular if

i) it is interior,
ii) more than four edges meet at it, and
iii) by ordering the edges rotationally, four consecutive edges form two straight lines.

Thus if we take a singular vertex and divide (only one) of the triangles (any number of times), we get a quasi-singular vertex. Our second criterion is as follows.

> (2) Suppose $\mathfrak{T}_0$ has no quasi-singular vertices and $\mathfrak{T}_1$ is obtained from $\mathfrak{T}_0$ via a regular subdivision. Then $\mathfrak{T}_1$ is unconstrained.

Notice that $\mathfrak{T}_0$ can be constrained without having any quasi-singular vertices (see Figure 2). Thus the subdivision "loosens" the triangulation.

The proof of the two criteria above rests on proving a slightly generalized one. If a vertex is quasi-singular, we say it is quasi-singular with respect to the triangle τ bordered by edges e_2 and e_3, where e_1, e_2, e_3, e_4 are the four consecutive edges forming two straight lines. We will prove the following:

(3) Suppose $\mathfrak{T}_0$ can be constructed by steps a), b) and c) in such a way that in a step b) the new interior vertex is not quasi-singular with respect to the new triangle (although it can be quasi-singular with respect to another triangle). Then the regular subidvision $\mathfrak{T}_1$ of $\mathfrak{T}_0$ is unconstrained.

Proof:

We construct $\mathfrak{T}_1$ four triangles at a time, based on the construction of $\mathfrak{T}_0$ given by hypothesis. When we are to add a type a) triangle in $\mathfrak{T}_0$, then we add the corresponding four triangles in $\mathfrak{T}_1$ in the order a, a, b, a. If we are to add a type b) triangle in $\mathfrak{T}_0$, we add the corresponding $\mathfrak{T}_1$ triangles in the order b, a, b, b. For type c) in $\mathfrak{T}_0$, the order is a, a, b, c in $\mathfrak{T}_1$. Because of the quasi-singularity assumption, each step b) is free, so $\mathfrak{T}_1$ is unconstrained. The proof of (2) from (3) is immediate. To prove (1), note that if a step b) adds an interior vertex that is quasi-singular with respect to the new triangle, it cannot be free. We remark that in the above criteria, it is not necessary that $\mathfrak{T}_1$ be obtained by joining edge midpoints, any point chosen on each edge would do.

2. Description of S_n as a quotient: definitions of R_n and Γ.

We denote by $R_n(\Pi, \mathfrak{T})$ the space of continuous piecewise polynomials of degree n that are C^1 at the vertices of $\mathfrak{T}$, i.e., those φ in $C^0(\Pi)$ such that $\varphi|_\tau \in \mathcal{P}_n$ for all triangles $\tau \in \mathfrak{T}$ and such that at each vertex v in $\mathfrak{T}$, the gradients of the restrictions of φ to the various triangles meeting at v all agree. It is easy to see that $R_n = S_n$ for $n \leq 2$, so it is for $n \geq 3$ that we expect to gain something by looking at R_n, and in this case we can give a (well-known) nodal description of this space. The nodal values for R_n consist of

1) the value and gradient at each vertex in $\mathfrak{T}$,

2) the value at each of $n-3$ distinct points in the interior of each edge in $\mathfrak{T}$, and

3) the value at each of $\frac{1}{2}(n-2)(n-1) = (\dim \mathcal{P}_{n-3})$ points in the interior of each triangle in $\mathfrak{T}$, chosen so that if a polynomial of degree $n-3$ vanishes at them it vanishes identically.

We now define a "nodal" basis $\{\varphi_j\}$ for R_n corresponding to these nodal values. Enumerate the nodes $z_1, z_2, \ldots$ with multiplicities in the sense that each vertex appears three times. We define $\{\varphi_j\}$ by

$$D_i \varphi_j(z_i) = \delta_{ij},$$

where the operator D_i is always multiplication by 1 if z_i is not a vertex, and if $z_i, z_{i'}, z_{i''}$ represent the same vertex, then $\{D_i, D_{i'}, D_{i''}\} = \{1, \partial_x, \partial_y\}$. We now show that S_n can be viewed as a subspace of R_n satisfying constraints.

Let us number the interior edges e_i, $i = 1, \ldots, E_0$, and let us pick $n-2$ distinct points y_i^j in the interior of each edge e_i. For each edge e_i, pick a direction $\underline{\underline{\nu}}_i$ normal to e_i and define, for φ in R_n, $\delta_i \varphi$ by

$$\delta_i \varphi = \partial_{\underline{\underline{\nu}}_i} (P_1 - P_2),$$

where P_1 and P_2 are the restrictions of φ to the triangles with common edge e_i that $\underline{\underline{\nu}}_i$ points into and out of, respectively. We define $\Gamma : R_n(\Pi, \mathfrak{T}) \to \mathbb{R}^{(n-2)E_0}$ by

$$(\Gamma\varphi)_{i,j} = \delta_i\varphi(y_i^j) \; ; \quad i = 1,\ldots,E_0; \quad j = 1,\ldots,n-2 \; .$$

Thus Γ measures the jump in the normal derivative of φ across each interior edge.

Proposition 2.

$$S_n(\Pi,\mathfrak{J}) = \ker\Gamma \; .$$

Proof:

It is clear that $\Gamma\varphi = 0$ if φ is in S_n, since $S_n \subset C^1(\Pi)$. Conversely, suppose $\Gamma\varphi = 0$ for φ in R_n. Then $\delta_i\varphi$ vanishes at n points on the edge e_i, the end-points of e_i plus the $n-2$ interior points y_i^j. Since $\delta_i\varphi$ is a polynomial of degree $n-1$, its restriction to e_i must vanish identically (one variable Lagrange interpolation). Thus φ is in $C^1(\Pi)$, hence in $S_n = C^1(\Pi) \cap R_n$. This completes the proof.

The constraints Γ as now defined will be redundant when there are singular vertices in $\mathfrak{J}$ (see the previous section, Definition 1). We now show how the redundant constraints can be removed. Suppose v is a singular vertex with edges e_1, e_2, e_3, e_4 meeting at v, where e_1, e_3 and e_2, e_4 are co-linear. Let us pick one of the edges, say e_1, and one of the constraint points, say y_1^1. We claim that if we remove the constraint

$$\delta_1\varphi(y_1^1) = 0$$

from the constraint map Γ, we still have $\ker\Gamma = S_n$. To see this, we must show that there is an added constraint associated with a singular vertex. So let $\varphi \in R_n$, and let $\Gamma\varphi = 0$, where Γ has been modified by removing the constraint above. As in the proof of Proposition 2, we know that φ is C^1 across e_2, e_3, and e_4. We want to show that it is also C^1 across e_1. Let us denote by ∂_i differentiation in the direction parallel to the edge e_i and by P_i the restriction of φ to the triangle with edges e_i and e_{i+1} (indices mod 4), $i=1,2,3,4$. (See Figure 3.) Then

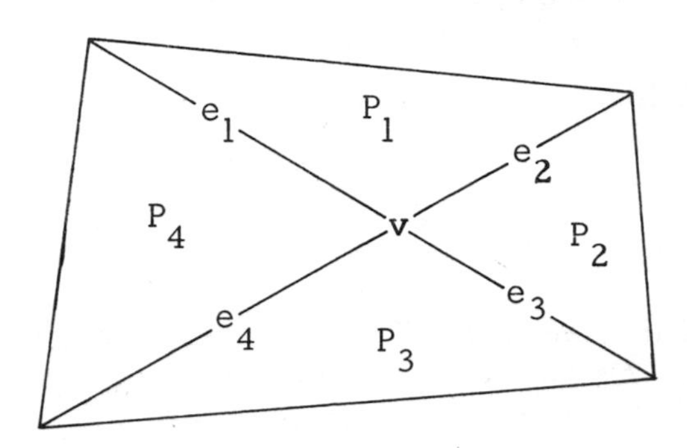

Figure 3

$$\partial_1(P_1 - P_2) = 0 \qquad \text{on } e_2 ,$$

$$\partial_2(P_2 - P_3) = 0 \qquad \text{on } e_3 , \qquad \text{and}$$

$$\partial_3(P_3 - P_4) = 0 \qquad \text{on } e_4 ,$$

because φ is C^1 across e_2, e_3, and e_4. Differentiating along each edge, we have

$$\partial_2\partial_1(P_1 - P_2) = 0 \quad \text{at } v ,$$

$$\partial_3\partial_2(P_2 - P_3) = 0 \quad \text{at } v , \text{ and}$$

$$\partial_4\partial_3(P_3 - P_4) = 0 \quad \text{at } v .$$

Using the fact that $\partial_1 = \pm\partial_3$ and $\partial_2 = \pm\partial_4$, we find that

$$\partial_1\partial_2(P_1 - P_4) = 0 \quad \text{at } v .$$

Referring to our previous notation, this means that

$$\partial_1(\delta_1\varphi) = 0 \quad \text{at } v .$$

Combined with the fact that $\delta_1\varphi = 0$ at $n-1$ points on e_1, this proves that $\delta_1\varphi \equiv 0$ on e_1, so $\varphi \in S_n$.

We now claim that we can remove one constraint from Γ for each singular vertex: for each singular vertex, pick an edge emanating from it, and remove one of the constraints on that edge. To see that this process is well-defined, we observe that both endpoints of an edge cannot be singular vertices. Thus the modified Γ maps R_n to $\mathbb{R}^{(n-2)E_0-\sigma}$ where σ is the number of singular vertices in $\mathfrak{J}$. We will show in the proof of the Theorem in the next section that Γ is onto (under the restrictions on $\mathfrak{J}$ of Proposition 1 when $n=3$), so there are no more redundancies in the constraints.

Remark:

We could have viewed S_n as the kernel of a constraint map Γ_0 defined on the space S_n^0 of C^0 piecewise polynomials of degree n, where the difference of normal derivatives is evaluated at n points on each interior edge (plus singular vertex modification). However, the constraints are very redundant:

$$\text{codim } \Gamma_0 = 2 \ V_0$$

as calculations similar to those in the proof of the Theorem in the next section show. (V_0 is the number of interior vertices in $\mathfrak{J}$.)

3. Application of S_n to a 4-th order variational problem.

Suppose we have an elliptic bilinear form $a(\ , \ )$ on $H^2(\Pi)$,

$$(3.1) \qquad a(u,v) = \int_{\Pi} \sum c_{\alpha,\beta} D^{\alpha}u \, D^{\beta}v ,$$

where α and β in the sum satisfy $|\alpha|, |\beta| \le 2$. Let us assume that $a(,)$ is continuous and coercive over $H^2(\Pi)$:

$$(3.2) \qquad |a(u,v)| \le C \|u\|_2 \|v\|_2 \quad \text{for all } u,v \text{ in } H^2(\Pi),$$

and

$$(3.3) \qquad \gamma \|v\|_2^2 \le |a(v,v)| \quad \text{for all } v \text{ in } H^2(\Pi).$$

We consider the following <u>variational problem:</u> given f in $L_2(\Pi)$, find u in $H^2(\Pi)$ such that

$$(3.4) \qquad a(u,v) = (f,v) \quad \text{for all } v \text{ in } H^2(\Pi).$$

The solution u to this variational problem may be viewed as the solution of a 4-th order elliptic boundary value problem, as is well-known. Since S_n is a subspace of $H^2(\Pi)$, the <u>Galerkin approximation</u> to u may be defined: given f in $L_2(\Pi)$, let u^* in $S_n(\Pi, \mathfrak{T})$ satisfy

$$(3.5) \qquad a(u^*,v) = (f,v) \quad \text{for all } v \text{ in } S_n(\Pi, \mathfrak{T}).$$

The following result is well-known.

<u>Proposition 3.</u>

If $a(,)$ satisfies (3.2) and (3.3), then there is a unique solution u^* to (3.5) and it satisfies

$$(3.6) \qquad \|u-u^*\|_2 \le \frac{C}{\gamma} \inf_{\varphi \in S_n} \|u-\varphi\|_2 .$$

What is unusual about the present situation is that it is difficult to compute u^*. For $n=3$, even if we were able to construct a basis for S_3, we do not expect the support of the basis functions to be very small. (We can show, for instance, that there are no C^1 piecewise cubics supported in the star of a vertex.) We now describe a way to compute u^* (using R_n and Γ from the previous section) which has been studied in the engineering literature by Harvey and

Kelsey [HK].

Let us introduce the <u>non-conforming</u> version of (3.1):

$$\tilde{a}(v,w) = \sum_{T \in \mathcal{T}} \int_T \sum_{\alpha,\beta} c_{\alpha\beta} D^\alpha v \, D^\beta w \tag{3.7}$$

for piecewise smooth functions v and w. Let $\{\varphi_j\}$ be the nodal basis for R_n defined in the previous section, and define the matrix

$$K_{ij} = \tilde{a}(\varphi_i, \varphi_j) .$$

Let $\bar{\Gamma}$ be the matrix whose i-th column is $\Gamma\varphi_i$. Thus $\bar{\Gamma}(A_i) = \Gamma(\sum A_i\varphi_i)$. We claim $u^* = \sum U_i^* \varphi_i$ can be computed by solving the equation

$$\begin{bmatrix} K & \bar{\Gamma}^t \\ \bar{\Gamma} & 0 \end{bmatrix} \begin{bmatrix} U^* \\ \lambda^* \end{bmatrix} = \begin{bmatrix} F \\ 0 \end{bmatrix} , \tag{3.8}$$

where the vector F is given by $F_i = (f, \varphi_i)$. For the proof, we translate (3.8) into its coordinate-free form, namely

$$\begin{aligned} &\tilde{a}(u^*, v) + \lambda^* \cdot \Gamma v = (f, v) \quad \text{for all } v \text{ in } R_n , \\ &\Gamma u^* = 0 . \end{aligned} \tag{3.9}$$

<u>Proposition 4.</u>

<u>If</u> (w, λ) <u>solves</u> (3.9), <u>then</u> w <u>equals the solution</u> u^* of (3.5).

<u>Proof:</u>

We have $\tilde{a}(w,v) = (f,v)$ for all v in S_n, since $\Gamma v = 0$. Since $S_n \subset H^2(\Pi)$, we find that for v in S_n

$$a(w,v) = \tilde{a}(w,v) = (f,v) ,$$

which is the defining equation for u^*.

Proposition 5.

There always exists a solution (u^*, λ^*) *to* (3.9), *where* u^* *is given by* (3.5).

Proof:

Define a linear form $\alpha: R_n \to \mathbb{R}^1$ by

$$\alpha v = (f, v) - \tilde{a}(u^*, v),$$

where $u^* \in S_n$ is defined by (3.5). Since $\ker \Gamma \subset \ker \alpha$, we can write $\alpha = \bar{\alpha} \circ \Gamma$, where $\bar{\alpha}$ is a linear form on the range of Γ, i.e., $\mathbb{R}^{(n-2)E_0 - \sigma}$. (If Γ is not onto, $\bar{\alpha}$ need not be unique.) Let us represent $\bar{\alpha}$ as the inner-product with a fixed vector:

$$\bar{\alpha}\lambda = \lambda^* \cdot \lambda \quad \text{for all } \lambda \in \mathbb{R}^{(n-2)E_0 - \sigma}.$$

Then $\alpha v = \bar{\alpha}\Gamma v = \lambda^* \cdot \Gamma v$ for all v in R_n. Thus (u^*, λ^*) satisfies (3.9).

In view of Propositions 4 and 5, a solution to (3.8) always exists, and non-uniqueness occurs only in λ^*. Rather than studying methods for solving (3.8) as a (possibly) singular system, we prove the following, our main result.

Theorem.

Suppose that $\mathfrak{T}$ *is unconstrained* (see Definition 2) *if* $n=3$. *Then the matrix in* (3.8) *is invertible.*

Proof:

The proof is based on Proposition 1 plus the following observation.

Proposition 6.

The matrix in equation (3.8) *is invertible if and only*

Γ is onto.

Proof:

Suppose Γ is not onto. If λ is chosen perpendicular to the image of Γ, then $(0, \lambda)$ is a null solution of (3.8). Thus invertibility implies that Γ is onto. Conversely, if (U, λ) is a null solution of (3.8), then $U = 0$ by Proposition 4, and so $\Gamma^t \lambda = 0$. When Γ is onto, Γ^t is one-one, so $\lambda=0$. Thus, there can be no non-trivial null solutions when Γ is onto, and this completes the proof.

We now prove that Γ is onto using (1.1). By Proposition 2, we have

$$\begin{aligned} \dim \operatorname{im} \Gamma &= \dim R_n - \dim S_n \\ &= (\dim \mathcal{P}_{n-3})T + (n-3)E + 3V \\ &\quad - \{(\dim \mathcal{P}_n)T - (2n+1)E_0 + 3V_0 + \sigma\} \\ &= -3nT + (n-3)E + (2n+1)E_0 + 3(V-V_0) - \sigma, \end{aligned}$$

where E and V denote the total number of edges and vertices (interior plus boundary), respectively, in $\mathcal{T}$. Now we claim that

$$3T = E + E_0 \quad \text{and} \quad V - V_0 = E - E_0 . \tag{3.10}$$

Substituting above, we find that

$$\dim \operatorname{im} \Gamma = (n-2)E_0 - \sigma$$

which proves that Γ is onto. Now we prove (3.10). The second equation says that the number of boundary vertices equals the number of boundary edges, and follows from the fact that $\partial\Pi$ consists of non-intersecting polygonal arcs. To prove the first equation, we associate with each interior edge the two triangles sharing that edge, and with each boundary edge we associate the triangle with that edge.

Via this association, each triangle appears exactly three times, so we have

$$3T = 2E_0 + (E-E_0) = E + E_0 .$$

This completes the proof of the theorem.

Remarks:

1) When $a(\ ,\ )$ is symmetric (and ≥ 0) we can give a variational interpretation to the equations (3.9). In this case, u minimizes the functional

$$a(v,v) - 2(f,v)$$

over $H^2(\Pi)$. The equations (3.9) arise as the first variation equations at a stationary point of the functional

$$a(v,v) - 2(f,v) + 2\lambda \cdot \Gamma v$$

defined on $R_n \times \mathbb{R}^{(n-2)E_0-\sigma}$. Thus we may view u^* as being defined by a Lagrangian multiplier technique, with the multipliers imposing the constraints on u^* exactly.

2) One can ask what happens to the matrix in (3.8) when a vertex is close to being singular. We expect that it will be numerically ill-conditioned; in fact, the solution u^* is not a continuous function of the vertex placement in the neighborhood of a singular vertex, since $\dim S_n$ jumps up by one there. If there is a reason to use nearly-singular vertices, we suggest removing one of the constraints as for a singular vertex. The result is that u^* is no longer C^1 across the edge from which the constraint is removed. Our analysis thus breaks down, but the matrix in (3.8) should remain invertible and be numerically stable.

3) We have written (3.8) in a form different from what would be used in computations. The matrix in (3.8) obviously has a banded structure similar to K, since the constraints Γ are local.

4) Notice that the analyses in this section do not depend on whether the non-conforming "stiffness" matrix K is invertible or not.

4. Plate bending example: interpretation of λ^*

Suppose we take $a(\ ,\ )$ to be the form corresponding to the plate bending problem (plus 1, to insure coerciveness over $H^2(\Pi)$):

$$(4.1)\qquad a(u,v) = \frac{D}{2}\int_{\Pi} \Delta u\,\Delta v - (1-\sigma)(u,_{11}v,_{22} + u,_{22}v,_{11} - 2\,u,_{12}v,_{12}) + (u,v)\,,$$

where σ is Poisson's ratio and D is the flexural rigidity [BS]. The solution u to the variational problem (3.4) with $a(\ ,\ )$ as specified above satisfies the differential equation $\Delta^2 u + u = f$ in Π, plus some natural boundary conditions determined by integration by parts. We now consider the inner product $\tilde{a}(u,\varphi)$ for φ in $R_n(\Pi,\mathfrak{J})$. Integrating by parts, we find that†

$$(4.2)\qquad \tilde{a}(u,\varphi) - (f,\varphi) = -\sum_{i=1}^{E_0}\int_{e_i} M_i(u)\delta_i\varphi$$

where $M_i(u)$ is the bending moment of the plate around the edge e_i given by

$$(4.3)\qquad M_i(u) = \sigma\Delta u + (1-\sigma)\sum \cos(\underline{\nu}_i, x_j)\cos(\underline{\nu}_i, x_k)u,_{jk}$$

(x_1, x_2 are Cartesian coordinates) [BS].

In our original definition of Γ, we chose any $n-2$ points y_i^j on e_i, and defined $\Gamma\varphi$ in terms of $\delta_i\varphi(y_i^j)$. We now choose specific points and some scaling factors based on the Lobatto quadrature rule.

† For notation, see Section 2.

We recall [DR, SS] that the Lobatto quadrature points for $[0,1]$ are the roots of

$$x(1-x)P'_{n-1}(x) = c_n\left(\frac{d}{dx}\right)^{n-2}[x(1-x)]^{n-1}, \tag{4.4}$$

where P_{n-1} is the (n-1)st Legendre polynomial. Let us order the roots so that the $n-2$ interior roots $\xi_i \in]0,1[$ appear first, with $\xi_{n-1} = 0$, $\xi_n = 1$. There are weights $w_1, \ldots, w_n$ such that

$$\int_0^1 P(\xi)d\xi = \sum_{i=1}^{n} w_i P(\xi_i) \tag{4.5}$$

for any polynomial P of degree $< 2n-2$. For $n=3$, this is just Simpson's rule. Our modification of Γ is given by

$$(\Gamma\varphi)_{i,j} = w_j |e_i| \delta_i\varphi(y_i^j), \tag{4.6}$$

where $|e_i|$ is the length of e_i and y_i^j are the points on e_i corresponding to ξ_j, $j = 1, \ldots, n-2$, under the affine identification of e_i with $[0,1]$. (We shall assume for simplicity that there are no singular vertices.) Now define

$$\lambda(u)_{i,j} = M_i(u)(y_i^j), \tag{4.7}$$

where $M_i(u)$ is the bending moment (4.3). The key point is that since $\delta_i\varphi$ is zero at the vertices, $\lambda(u)\cdot\Gamma\varphi$ is the Lobatto quadrature for

$$\sum_{i=1}^{E_0} \int_{e_i} M_i(u)\,\delta_i\varphi .$$

For any φ in R_n, we have

$$(\lambda^* - \lambda(u))\cdot\Gamma\varphi = \tilde{a}(u-u^*, \varphi) + \sum_{i=1}^{E_0}\int_{e_i} M_i(u)\delta_i\varphi - \lambda(u)\cdot\Gamma\varphi. \tag{4.8}$$

Thus if Γ is onto and u^* is close to u, then λ^* should be close to $\lambda(u)$. We conjecture that as the mesh size h of $\mathfrak{T}$ converges to zero, λ^* converges to the values $\lambda(u)$ of the bending moments of the plate around each edge with rate (in ℓ_∞) $O(h^{n-1})$, the accuracy of the quadrature rule. Toward proving this conjecture, we note that

$$(4.9)\qquad \left| \sum_{i=1}^{E_0} \int_{e_i} M_i(u)\,\delta_i\varphi - \lambda(u)\cdot\Gamma\varphi \right| \le c(u,n)h^{n-1}\|\Gamma\varphi\|_{\ell_1}$$

where h is the maximum length of the interior edges. (Proof: The quadrature rule is exact for polynomials of degree $< 2n-2$, and the polynomial $\delta_i\varphi$ is determined on e_i by its values at y_i^j, so its derivatives can be estimated in terms of $\Gamma\varphi$.) Let κ be the smallest constant such that

$$(4.10)\qquad \inf_{v\in S_n} \sqrt{\tilde{a}(\varphi-v,\varphi-v)} \le \kappa\|\Gamma\varphi\|_{\ell_1}$$

for all φ in R_n ($\kappa<\infty$ exists since $\ker\Gamma = S_n$, but it may depend on mesh size). From (4.8) and the Schwarz inequality, we have

$$(4.11)\qquad \|\lambda^*-\lambda(u)\|_{\ell_\infty} \le C'\kappa\|u-u^*\|_2 + c(u,n)h^{n-1}.$$

Thus, convergence of λ^* to $\lambda(u)$ depends on bounding κ plus proving convergence of u^* to u. In the next section, we prove the latter for $n \ge 5$. We feel that for $n \ge 5$, it is also possible to bound κ (at least by $O(h^{-1})$) using local arguments that, in particular, give a direct proof that Γ is onto. This belief is based on our knowledge [MS] about a local basis for S_n, $n \ge 5$. The case $n=4$ (both convergence of u^* and bounding of κ) may also be amenable to the ideas in [MS], but we feel that the important case $n=3$ will require some new technique.

5. Solution of the approximation problem for $n \geq 5$

Proposition 3 reduces the question of convergence of u^* to u with mesh refinement to an approximation problem for S_n (see estimate (3.6)). We can solve this for $n \geq 5$ by noting that in this case S_n has a well-known subspace with good approximation properties. This space, denoted by $R_n^1(\Pi, \mathfrak{T})$, is determined by the following nodal values:

1) value, first and second derivatives at each vertex,
2) value at $n-5$ points on each edge, normal derivative at $n-4$ points on each edge, and
3) value at $\frac{1}{2}(n-5)(n-4)$ (= dim $\mathcal{P}_{n-6}$) points in the interior of each triangle chosen so that if a polynomial of degree $n-6$ vanishes at them, it vanishes identically.

With this nodal description of $R_n^1(\Pi, \mathfrak{T})$, we can define an interpolating basis by

$$D_j \varphi_i(z_j) = \delta_{ij} ,$$

where D_j is either multiplication by one or a first or second derivative, depending on the nature of the j-th nodal value. Notice that $\{\varphi_i\}$ is a local basis: $\varphi_i = 0$ except in the triangles containing z_i. We can thus apply the theory in [St1], once we make the following definition.

Definition 3.

Let $\mathfrak{T}$ be a triangulation of Π. We define the condition number of $\mathfrak{T}$, denoted by $\chi(\mathfrak{T})$, as follows:

$$\chi(\mathfrak{T}) = \sup_{\tau \in \mathfrak{T}} r_1(\tau)/r_2(\tau) ,$$

where, for a triangle τ, $r_1(\tau)$ (resp. $r_2(\tau)$) is the radius of the circumscribed (resp. inscribed) circle of τ.

Proposition 7.

Let $n \geq 5$. There is a constant $c_0 < \infty$ depending only on Π, $\chi(\mathfrak{T})$ (see Definition 3), and n such that, for any function u,

$$\inf_{v \in R_n^1(\Pi, \mathfrak{T})} \|u-v\|_2 \leq c_0 h^{n-1} \|u\|_{n+1} ,$$

where h is the maximum diameter of the triangles in $\mathfrak{T}$.

Proof:

We apply Theorem 3 of [Stl], and we only need to check that the nodal basis $\{\varphi_i\}$ for R_n^1 is uniform in the sense of [Stl]. This is done by mapping each triangle τ in $\mathfrak{T}$ to an equilateral triangle with sides of length 1. The Jacobians of this map and its inverse are bounded in terms of h and $\chi(\mathfrak{T})$, proving the uniformity. This completes the proposition.

Putting together Propositions 3 and 7, we find that u^*, determined by solving (3.8), satisfies

$$\|u-u^*\|_2 \leq c_1 h^{n-1} \|u\|_{n+1}$$

if $n \geq 5$, where $c_1 = c_0\, C/\gamma$. We conjecture that a similar result holds for $n = 3$ and 4.

References.

[BS] Bergman, S., and M. Schiffer, Kernel Functions and Elliptic Differential Equations in Mathematical Physics, Academic Press (1953).

[BH] Bramble, J. H., and S. R. Hilbert, "Bounds for a certain class of linear functionals with applications to Hermite interpolation," Num. Math. 16 (1971), pp. 362-369.

[C] Ciarlet, P. G., "Sur l'element de Clough et Toucher," to appear.

[DR] Davis, P. J., and P. Rabinowitz, Numerical Integration, Blaisdell (1967).

[HK] Harvey, J. W., and S. Kelsey, "Triangular plate bending elements with enforced compatability," AIAA J. 9 (1971), pp. 1023-1026.

[MS] Morgan, John, and Ridgway Scott, "The dimension of the space of C^1 piecewise polynomials," to appear.

[P] Powell, M. J. D., "Piecewise quadratic surface fitting for contour plotting," Proc. Conf. Software for Numerical Mathematics, University of Technology, Loughborough (1973).

[St 1] Strang, G., "Approximation in the finite element method," Num. Math. 19 (1972), pp. 81-98.

[St 2] Strang, G., "Piecewise polynomials and the finite element method," Bull. AMS 79 (1973), pp. 1128-1137.

[St 3] Strang, G., "The dimension of piecewise polynomial spaces, and one-sided approximation," Proc. Conf. Numerical Solution of Diff. Equations, Dundee (1973), Lecture notes in Mathematics, #365, Springer Verlag, 1974, pp. 144-152

[SS] Stroud, A. H., and Don Secrest, Gaussian Quadrature Formulae, Prentice-Hall (1966).

[Z] Zienkiewicz, O. C., "Constrained variational principles and penalty functions in finite element analysis," Proc. Conf. Numerical Solution of Diff. Equations, Dundee (1973), Lecture Notes in Mathematics #363, Springer Verlag, 1974, pp. 207-214.

Department of Mathematics
University of Chicago
Chicago, Illinois 60637

Finite Element and Finite Difference Methods for Hyperbolic Partial Differential Equations

H.-O. KREISS AND G. SCHERER

1. Introduction.

Consider a real system of partial differential equations

(1.1) $$\partial u/\partial t = P(x, t, \partial/\partial x)u$$

in a cylindrical domain $\{\Omega \times 0 \leq t < \infty\}$ with boundary $\{\partial\Omega \times 0 \leq t < \infty\}$. Here, $x = (x^{(1)}, \ldots, x^{(s)})$ denotes a point in the real Euclidean space $\mathbb{R}^s$, $u = (u^{(1)}, \ldots, u^{(n)})'$ is a vector function and P a differential operator with matrix coefficients. For $t = 0$, initial values are given:

(1.2) $$u(x, 0) = f(x), \quad x \in \Omega,$$

and on the boundary, homogeneous boundary conditions

(1.3) $$B(x, \partial/\partial x)u = 0, \quad x \in \partial\Omega, \quad t \geq 0$$

are prescribed. Here, B denotes a differential operator whose coefficients depend on x but not on t. Let

(1.4) $$(u, v) = \int_\Omega u^* v \; dx, \qquad (u, u) = \|u\|^2$$

denote the usual L_2-scalar product and norm. We assume that the operator P is semibounded, i.e., there is a dense set $S \subset L_2(\Omega)$ of functions w satisfying the boundary conditions (1.3) such that

$$(w, Pw) + (Pw, w) \leq 0 , \qquad w \in S . \tag{1.5}$$

As is well known, (1.5) implies an energy estimate for those solutions of (1.1)-(1.3) which, for every fixed t, belong to S because

$$\begin{aligned} \frac{\partial}{\partial t} \|u\|^2 &= (u, \partial u/\partial t) + (\partial u/\partial t, u) \\ &= (u, Pu) + (Pu, u) \leq 0 . \end{aligned} \tag{1.6}$$

A Galerkin procedure to solve the above problem numerically is easily constructed. Let $\{\varphi_j\}_1^N$ denote a set of basis functions which belong to S. Make an "ansatz"

$$v(x,t) = \sum_{j=1}^{N} \alpha_j(t)\varphi_j(x)$$

and determine the $\alpha_j(t)$ from the Galerkin equations

$$(\partial v/\partial t - Pv, \varphi_i) = 0 . \tag{1.7}$$

The procedure is always stable because (1.7) implies

$$(\partial v/\partial t - Pv, v) + (v, \partial v/\partial t - Pv) = 0 ,$$

i.e.,

$$\partial \|v\|^2/\partial t = (v, Pv) + (Pv, v) \leq 0 . \tag{1.8}$$

We can write (1.7) as a system of ordinary differential equations

$$A\, d\underline{\underline{\alpha}}/dt = B\underline{\underline{\alpha}} \tag{1.9}$$

where $\underline{\underline{\alpha}} = (\alpha_i)_1^N$, $A = ((\varphi_i, \varphi_j))$, $B = ((P\varphi_i, \varphi_j))$. Here, $A = A^*$ is positive definite and $B + B^*$ is negative semidefinite.

It is also quite easy to derive error estimates as long as one does not care to get the optimal order. All this can be found in a paper by B. Swartz and B. Wendroff [1].

The advantage of the finite element method is that the resulting procedures are automatically stable, and that one has extreme flexibility in choosing the basis functions. Therefore, in very complicated domains or for problems with complicated interfaces, the method might be the only feasible one. There are two disadvantages:

1) Whatever method one uses to compute the solution of (1.9), one has to invert the matrix A. This can be very costly.

2) For hyperbolic partial differential equations it is essential to control the dispersion, dissipation and the propagation of discontinuities. This is easily done by using suitable difference approximations. It has still to be shown that this can be done easily also for finite elements techniques.

The main disadvantage of finite difference methods is that it may be difficult to handle boundaries properly. Consider for example a region with a corresponding net as shown in Figure 1.

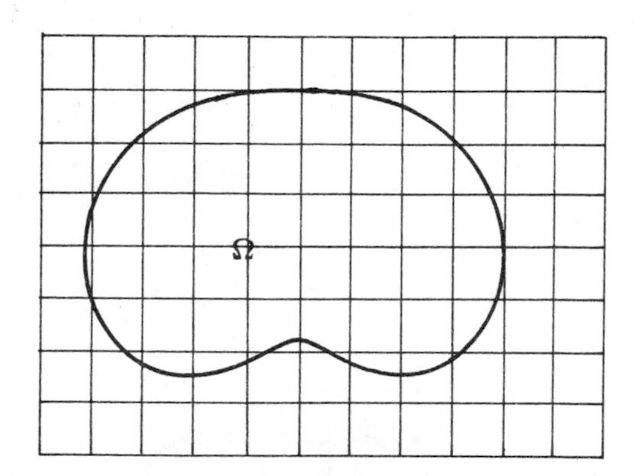

Figure 1

Assume that the boundary conditions for the differential equations are given by $u = 0$ and that we want to construct a second order accurate procedure. For that purpose, we replace $u(P)$ by a linear combination, $u(P) \approx \alpha u(P_1) + (1-\alpha)u(P_2) = 0$, corresponding to Figure 2.

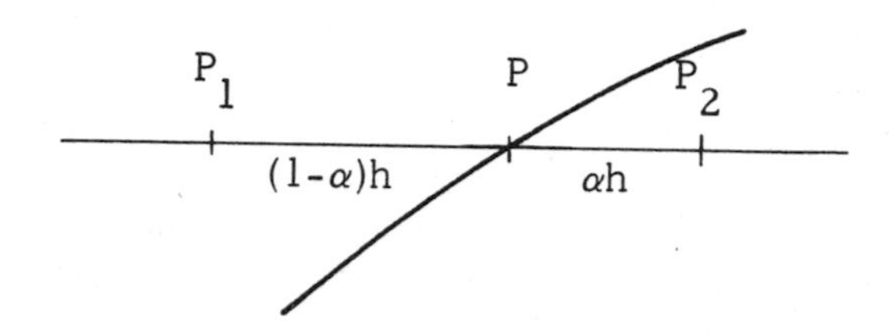

Figure 2

In general, α varies discontinuously from point to point. There is no easy method to prove stability except by some a priori energy estimate. The sad fact is that these estimates do only exist for methods of low order of accuracy. The difficulty is real because the grid introduces, in a somewhat random fashion, "local parasite" solutions which can cause instability.

Things become more tractable if one uses two overlapping nets

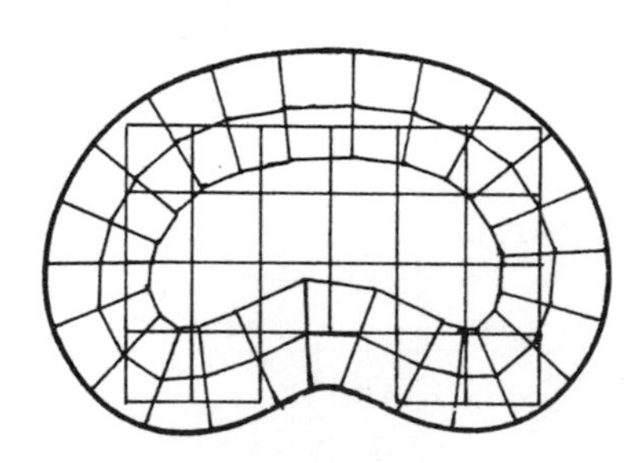

Figure 3

and interpolates between these nets. Here one net follows the boundary and is overlapped by an ordinary rectangular net in the interior. One can construct the interpolation in such a way that no stability problem arises. Therefore we need only to consider regions of the following kind:

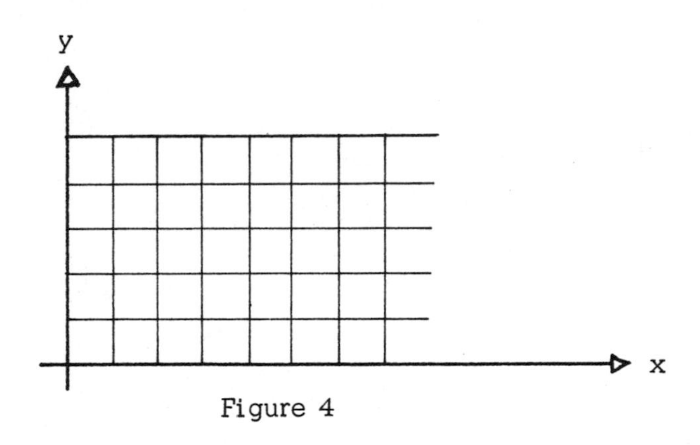

Figure 4

Here $0 \leq x < \infty$, $0 \leq y \leq 1$. For $x = 0$, we prescribe the same boundary conditions as for the original region. For $y = 0, 1$ we have periodic boundary conditions. Furthermore, the solutions are required to belong to L_2. We shall show in the next two sections how to construct approximations of high order of accuracy for regions of the above kind.

2. Difference approximations for $\partial/\partial x$.

Consider the operator $\partial/\partial x$ for $0 \leq x < \infty$. Let

$$(2.1) \qquad (u, v) = \int_0^\infty u^* v \, dx , \qquad \|u\|^2 = (u, u)$$

denote the L_2-norm. Integration by parts gives us

$$(2.2) \qquad (u, \partial u/\partial x) + (\partial u/\partial x, u) = -|u(0)|^2$$

for all smooth u which decay sufficiently fast for $x \to \infty$.

We shall now consider difference approximations for $\partial/\partial x$. Let $h > 0$ denote the grid size, $x_\nu = \nu h$, $\nu = 0, 1, \ldots$ the grid points and $v_\nu = v(x_\nu)$ the grid functions. Let Q be a difference approximation for $\partial/\partial x$. We assume that Q, written as a matrix, has the form

$$Q = \frac{1}{h}\begin{pmatrix} B & C \\ -C^* & D \end{pmatrix},$$

where $B = (b_{ij})$ is an $r \times r$ matrix,

$$D = \begin{pmatrix} 0 & \alpha_1 & \alpha_2 & & \cdots & \alpha_s 0 \cdots & & \\ -\alpha_1 & 0 & \alpha_1 & \alpha_2 & \cdots & & \alpha_s 0 \cdots & \\ -\alpha_2 & -\alpha_1 & 0 & \alpha_1 & \alpha_2 & \cdots & & \alpha_s 0 \cdots \end{pmatrix}$$

an antisymmetric band matrix, and

$$C = \left(\begin{array}{ccc:c} 0 & \cdots & 0 & 0 \\ 0 & \cdots & 0 & \\ \hdashline & C_s & & 0 \end{array}\right), \quad C_s = \begin{pmatrix} \alpha_s & & & 0 \\ \alpha_{s-1} & \alpha_s & & \\ \vdots & & \ddots & \\ \alpha_1 & & & \alpha_s \end{pmatrix}.$$

This form for Q is natural. It arises from any reasonable difference approximation of the Cauchy problem when making modifications near the boundary.

Introduce now the discrete norm

$$(2.4) \qquad (u, v)_h = \sum_{\nu=0}^{\infty} u_\nu v_\nu h, \quad \|u\|_h^2 = (u, u)_h .$$

It would be ideal if we could construct Q in such a way that the discrete version of (2.2) holds, i.e.,

$$(2.5) \qquad (u, Qu)_h + (Qu, u)_h = -|u_0|^2 .$$

However we are going to show

Lemma 2.1.

Let the norm be defined by (2.4). There are no difference operators for which (2.5) holds.

Proof.

In this case, B has to be of the form

(2.6) $$B = B_1 + B_2$$

where

$$B_1 = \begin{pmatrix} b_{11} & 0 & \dots & 0 \\ 0 & 0 & \dots & 0 \\ \vdots & \vdots & & \\ 0 & 0 & \dots & 0 \end{pmatrix}, \quad B_2 = \begin{pmatrix} 0 & b_{12} & \dots & b_{1r} \\ -b_{12} & 0 & b_{23} \cdots & b_{2r} \\ \vdots & & & \vdots \\ -b_{1r} & & & 0 \end{pmatrix} = -B_2^* .$$

Furthermore, consistency implies

(2.7) $$\sum_{j=1}^{s} 2j\alpha_j = 1 .$$

Let $h = 1$ and define the vectors $\underline{\underline{e}}_j, \underline{\underline{f}}_j$ by

$$\underline{\underline{e}}_j = (-1)^j(r^j, (r-1)^j, \dots, 1^j)' , \quad \underline{\underline{f}}_j = (0^j, 1^j, \dots (s-1)^j)',$$

$$j = 0, 1, \dots .$$

$\underline{\underline{e}}_j, \underline{\underline{f}}_j$ are nothing else but the discretization of $(x-r)^j$. Near the boundary, Q is therefore an approximation to $\partial/\partial x$ of order τ if and only if

(2.8) $$j\,\underline{\underline{e}}_{j-1} = B\,\underline{\underline{e}}_j + C_s \underline{\underline{f}}_j , \quad j = 0, 1, 2, \dots, \tau.$$

Here, $\underline{\underline{e}}_{-1}$ is defined as $\underline{\underline{e}}_{-1} = \underline{\underline{0}}$. (2.8) can also be written as

(2.9) $$B_2 \underline{\underline{e}}_j = \underline{\underline{g}}_j, \quad \underline{\underline{g}}_j = j\,\underline{\underline{e}}_{j-1} - B_1 \underline{\underline{e}}_j - C_s \underline{\underline{f}}_j .$$

By assumption, B_2 is an antisymmetric matrix. Therefore the following compatibility conditions must hold:

$$(\underline{\underline{e}}_i, \underline{\underline{g}}_j) + (\underline{\underline{e}}_j, \underline{\underline{g}}_i) = 0, \quad 0 \le i,j \le \tau . \tag{2.10}$$

Here, $(\underline{\underline{f}}, \underline{\underline{g}})$ denotes the usual Euclidean vector product. If the approximation is consistent, then $\tau \ge 1$. Therefore

$$\begin{aligned} -(\underline{\underline{e}}_0, B_2\underline{\underline{e}}_0) &= (\underline{\underline{e}}_0, B_1\underline{\underline{e}}_0) + (\underline{\underline{e}}_0, C_s\underline{\underline{f}}_0) \\ &= b_{11} + \sum_{j=1}^{s} j\alpha_j = 0, \end{aligned} \tag{2.11}$$

i.e., by (2.7),

$$b_{11} = -\tfrac{1}{2} . \tag{2.12}$$

However,

$$\begin{aligned} (\underline{\underline{e}}_0, \underline{\underline{g}}_1)+(\underline{\underline{e}}_1, \underline{\underline{g}}_0) &= (\underline{\underline{e}}_0, \underline{\underline{e}}_0) - 2(\underline{\underline{e}}_0, B_1\underline{\underline{e}}_1) - (\underline{\underline{e}}_0, C_s\underline{\underline{f}}_1) - (\underline{\underline{e}}_1, C_s\underline{\underline{f}}_0) \\ &= r - r - \sum_{\nu=1}^{s} \alpha_\nu \sum_{j=1}^{\nu}(j-1) + \sum_{\nu=1}^{s} \alpha_\nu \sum_{j=1}^{\nu} j \\ &= \sum_{\nu=1}^{s} \alpha_\nu \nu = \tfrac{1}{2} \ne 0 . \end{aligned}$$

Thus (2.10) cannot be satisfied. This proves the lemma.

If we want to satisfy the compatibility conditions (2.10), then we have to change the norm. The simplest way is to use

$$(u, v)_h = \sum_{\nu=0}^{\infty} \lambda_\nu u_\nu v_\nu h, \quad \lambda_\nu > 0, \quad \lambda_\nu = 1 \text{ for } \nu \ge r \tag{2.12'}$$

instead of (2.4). In this case, the relation (2.5) holds if Q is of the form

$$hQ = \begin{pmatrix} \Lambda^{-1} & 0 \\ 0 & I \end{pmatrix} \cdot \begin{pmatrix} B & C \\ -C^{*} & D \end{pmatrix}, \quad \Lambda = \begin{pmatrix} \lambda_0 & 0 & \dots & 0 \\ 0 & \lambda_1 & \dots & 0 \\ \vdots & & & \\ 0 & & \dots & \lambda_{r-1} \end{pmatrix}.$$

Here, B is given by (2.6) with $b_{11} = -\frac{1}{2}$. There are consistent approximations of this form. For example, the approximation

$$h\,\partial v_0/\partial x \approx (v_1 - v_0), \quad 2h\,\partial v_\nu/\partial x \approx (v_{\nu+1} - v_{\nu-1}) \qquad \nu \geq 1$$

can be written as

$$hQ = \begin{pmatrix} -1 & 1 & 0 & \dots & \\ -\frac{1}{2} & 0 & \frac{1}{2} & 0 & \dots \\ & -\frac{1}{2} & 0 & \frac{1}{2} & \\ & & \ddots & \ddots & \ddots \end{pmatrix} = \begin{pmatrix} 2 & 0 & & & \\ 0 & 1 & 0 & \dots & \\ 0 & 0 & 1 & 0 & \dots \\ & & & \ddots & \end{pmatrix} \begin{pmatrix} -\frac{1}{2} & \frac{1}{2} & 0 & \dots & \\ -\frac{1}{2} & 0 & \frac{1}{2} & & \\ 0 & -\frac{1}{2} & 0 & \frac{1}{2} & \\ & & \ddots & & \ddots \end{pmatrix}.$$

Now the relations (2.8) become

$$j\Lambda \underline{\underline{e}}_{j-1} = B\underline{\underline{e}}_j + C_s \underline{\underline{f}}_j, \quad j = 0, 1, 2, \dots, \tau$$

and therefore we get instead of (2.9) the relations

$$B_2\, \underline{\underline{e}}_j = \underline{\underline{g}}_j, \quad \underline{\underline{g}}_j = j\Lambda \underline{\underline{e}}_{j-1} - B_1 \underline{\underline{e}}_j - C_s \underline{\underline{f}}_j. \tag{2.13}$$

The compatibility conditions (2.10) now are

$$j(\underline{\underline{e}}_i, \Lambda \underline{\underline{e}}_{j-1}) + i(\underline{\underline{e}}_j, \Lambda \underline{\underline{e}}_{i-1}) - \{(\underline{\underline{e}}_i, B_1 \underline{\underline{e}}_j) + (\underline{\underline{e}}_j, B_1 \underline{\underline{e}}_i)\}$$

$$= (\underline{\underline{e}}_i, C_s \underline{\underline{f}}_j) + (\underline{\underline{e}}_j, C_s \underline{\underline{f}}_i).$$

For $i = j = 0$, we get again (2.12), i.e., $b_{11} = -\frac{1}{2}$. Let $i + j \geq 1$. Then

$$j(\underline{\underline{e}}_i, \Lambda \underline{\underline{e}}_{j-1}) = j \sum_{\nu=0}^{r-1} \lambda_\nu (r-\nu)^{i+j-1}(-1)^{i+j-1},$$

$$(\underline{\underline{e}}_i, B_1 \underline{\underline{e}}_j) = \tfrac{1}{2}(-1)^{i+j-1} r^{i+j},$$

$$(\underline{\underline{e}}_i, C_s \underline{\underline{f}}_j) = (-1)^i \sum_{\nu=1}^{s} \alpha_\nu \sum_{\mu=0}^{\nu-1} \mu^j (\nu-\mu)^i .$$

Thus

$$(2.14) \quad (j+i)(-1)^{i+j-1} \sum_{\nu=0}^{r-1} \lambda_\nu (r-\nu)^{i+j-1} - (-1)^{i+j-1} r^{i+j}$$

$$= \sum_{\nu=1}^{s} \alpha_\nu \Big(\sum_{\mu=0}^{\nu-1} \mu^j (\mu-\nu)^i + \mu^i (\mu-\nu)^j \Big)$$

$$= \sum_{\nu=1}^{s} \alpha_\nu \Big(\sum_{\mu=0}^{\nu-1} \mu^{\sigma-i} (\mu-\nu)^i + \mu^i (\mu-\nu)^{\sigma-i} \Big)$$

$$= J_{i,\sigma} \qquad 1 \le \sigma = i+j \le 2\tau .$$

The left hand side of the last relation depends only on $\sigma = i+j$. Therefore the same must be true for the right hand side, i. e.,

$$(2.15) \qquad J_{i+1,\sigma} = J_{i,\sigma} \qquad 0 < i+1 \le \sigma, \quad 0 < i+1 \le \tau .$$

We shall now prove

Lemma 2.2.

A necessary condition that there is a difference approximation which is accurate of order τ near the boundary is that

$$(2.16) \qquad \sum_{\nu=1}^{s} \alpha_\nu \nu^\sigma = 0 \qquad \text{for } \sigma = 3, 5, \ldots, 2\tau-1 .$$

Proof.

Let $N_{i,\sigma}(\nu)$ be defined by

$$N_{i,\sigma}(\nu) = \sum_{\mu=0}^{\nu-1} \mu^{\sigma-i}(\mu-\nu)^i + \sum_{\mu=1}^{\nu} (\mu-\nu)^i \mu^{\sigma-i}, \quad 0 \leq i \leq \sigma.$$

Then

$$(2.17) \qquad N_{i+1,\sigma}(\nu) = \sum_{\mu=0}^{\nu-1} \mu^{\sigma-i-1}(\mu-\nu)^{i+1} + \sum_{\mu=1}^{\nu} (\mu-\nu)^{i+1} \mu^{\sigma-i-1}$$

$$= N_{i,\sigma}(\nu) - \nu\, N_{i,\sigma-1}(\nu).$$

The well-known explicit formulas for $\sum \mu^\sigma$ give us

$$N_{0,\sigma} = (-1)^\sigma N_{\sigma,\sigma} = \sum_{\mu=0}^{\nu-1} \mu^\sigma + \sum_{\mu=1}^{\nu} \mu^\sigma = 2\sum_{\mu=1}^{\nu} \mu^\sigma - \nu^\sigma$$

$$(2.18) \qquad = 2 \begin{cases} (\sigma+1)^{-1}\nu^{\sigma+1} + \frac{1}{2}B_2\binom{\sigma}{1}\nu^{\sigma-1} + \frac{1}{4}B_4\binom{\sigma}{3}\nu^{\sigma-3} + \ldots + B_\sigma \nu, & \text{for } \sigma \text{ even}, \\ (\sigma+1)^{-1}\nu^{\sigma+1} + \frac{1}{2}B_2\binom{\sigma}{1}\nu^{\sigma-1} + \frac{1}{4}B_4\binom{\sigma}{3}\nu^{\sigma-3} + \ldots + \frac{\sigma}{2} B_{\sigma-1}\nu^2, & \text{for } \sigma \text{ odd}. \end{cases}$$

Therefore by induction using (2.17) we get expressions

$$(2.19) \qquad N_{i,\sigma}(\nu) = \gamma_1^{(i,\sigma)}\nu^{\sigma+1} + \gamma_3^{(i,\sigma)}\nu^{\sigma-1} + \ldots .$$

Here the last term is proportional to ν or ν^2 depending on whether σ is even or odd, respectively. Furthermore, $\gamma_1^{(i,\sigma)} \neq 0$, because in first approximation

$$N_{i,\sigma} \approx 2\int_0^\nu x^{\sigma-i}(x-\nu)^i dx = 2(-1)^i \nu^{\sigma+1}\int_0^1 x^{\sigma-i}(1-x)^i dx.$$

Assume now that σ is even. The relation

$$(2.20) \qquad \sum_{\mu=0}^{\nu-1} \mu^i(\mu-\nu)^{\sigma-i} = (-1)^\sigma \sum_{\mu=1}^{\nu} (\mu-\nu)^i \mu^{\sigma-i}$$

implies

$$J_{i,\sigma} = \sum_{\nu=1}^{s} \alpha_\nu N_{i,\sigma}(\nu) .$$

Thus by (2.17)

$$J_{i+1,\sigma} = J_{i,\sigma} - \sum_{\nu=1}^{s} \alpha_\nu \, \nu \, N_{i,\sigma-1}(\nu) , \tag{2.21}$$

and therefore by (2.15)

$$0 = \sum_{\nu=1}^{s} \alpha_\nu \, \nu \, N_{i,\sigma-1}(\nu) = \gamma_1^{(i,\sigma-1)} \sum_{\nu=1}^{s} \alpha_\nu \, \nu^{\sigma+1} + \dots \tag{2.22}$$

for all i, j with $0 < i+1 \leq \tau$, $0 < j \leq \tau$, i.e., all σ with $0 \leq \sigma \leq 2\tau - 2$. The lemma follows easily from (2.20) using induction.

Assume now that the conditions (2.16) are fulfilled. We want to show that the relations (2.15) are valid. For even σ this follows from (2.21), (2.22) and (2.16). $\sigma = 1$ gives $i = 0$ and the relation (2.15) is no condition. We have

$$J_{0,1} = \sum_{\nu=1}^{s} \alpha_\nu \Big(\sum_{\mu=0}^{\nu-1} \mu + (\mu-\nu) \Big) = \sum_{\nu=1}^{s} \alpha_\nu (\nu^2 - \nu - \nu^2) = -\tfrac{1}{2}. \tag{2.23}$$

For odd $\sigma > 1$, we get from (2.20) and (2.16)

$$\begin{aligned} J_{i,\sigma} &= \sum_{\nu=1}^{s} \alpha_\nu \Big(\sum_{\mu=0}^{\nu-1} \mu^{\sigma-i} (\mu-\nu)^i - \sum_{\mu=1}^{\nu} (\mu-\nu)^i \mu^{\sigma-i} \Big) \\ &= \sum_{\nu=1}^{s} \alpha_\nu \big(0^{\sigma-i} (-\nu)^i - 0^i \nu^{\sigma-i} \big) = 0 . \end{aligned}$$

Therefore (2.15) also holds for odd σ. Thus the equations (2.14) are equivalent to

$$\sigma \cdot \sum_{\nu=0}^{r-1} \lambda_\nu (r-\nu)^{\sigma-1} = r^\sigma - (-1)^\sigma g_\sigma , \qquad \sigma = 1, 2, \dots, 2\tau \tag{2.24}$$

where by (2.18) and (2.23)

$$g_\sigma = \begin{cases} \sum_{\nu=1}^{s} \alpha_\nu N_{0,\sigma}(\nu) = 2\, B_\sigma \sum_{\nu=1}^{s} \alpha_\nu\, \nu = B_\sigma , & \text{for } \sigma = 2,4,\ldots,2\tau-2, \\ 2 \sum_{\nu=1}^{s} \alpha_\nu \sum_{\mu=0}^{\nu-1} \mu^\tau (\mu-\nu)^\tau , & \text{for } \sigma = 2\tau , \\ -\frac{1}{2} , & \text{for } \sigma = 1 , \\ 0 , & \text{for } \sigma = 3,5,\ldots,2\tau-1 . \end{cases}$$

If we set $r = 2\tau$, then (2.24) defines a linear system of equations which has a unique solution.

We have

Theorem 2.1.

Assume that $r = 2\tau$ and that the system of equations (2.24) has a positive solution. Then there is a scalar product (2.12') and a difference approximation Q which is accurate of order τ near the boundary and accurate of order 2τ in the interior such that (2.15) holds.

Proof.

We have to construct B_2 from (2.9). B_2 is of order $2\tau \times 2\tau$ and (2.9) represents a mapping of $\tau+1$ vectors. All the compatibility conditions are fulfilled. Therefore, there are antisymmetric matrices which fulfill the requirements.

By direct calculations we have computed the λ_j's for $\tau = 1,2,3,4$. The results are given in Table 1. All λ_j's are positive.

It is very doubtful whether these formulas have any practical interest for $\tau > 4$, because the boundary matrix B and the difference formula in the interior become rather complicated. It is therefore of interest to generalize the scalar product (2.5) even more. Replace (2.5) by

$\tau = 1$

$\lambda_0 = 0.5$ $\qquad$ $\lambda_1 = 1.0$

$\tau = 2$

$\lambda_0 = 0.3541666667$ $\qquad$ $\lambda_2 = 1.0208333333$

$\lambda_1 = 1.2291666667$ $\qquad$ $\lambda_3 = 0.8958333333$

$\tau = 3$

$\lambda_0 = 0.3159490741$ $\qquad$ $\lambda_3 = 1.240509259$

$\lambda_1 = 1.390393519$ $\qquad$ $\lambda_4 = 0.9116898148$

$\lambda_2 = 0.6275462963$ $\qquad$ $\lambda_5 = 1.013912037$

$\tau = 4$

$\lambda_0 = 0.2948906762$ $\qquad$ $\lambda_4 = 0.4127080578$

$\lambda_1 = 1.525720624$ $\qquad$ $\lambda_5 = 1.278484623$

$\lambda_2 = 0.2574528770$ $\qquad$ $\lambda_6 = 0.9232955798$

$\lambda_3 = 1.798113701$ $\qquad$ $\lambda_7 = 1.009333861$

Table 1

$$(u,v)_h = (u,Av) + \sum_{\nu=r}^{\infty} u_\nu v_\nu h .$$

Here

$$(u,Av) = \sum_{0\le i,j<r} a_{ij} u_i v_j h$$

is a positive definite form defined by a positive definite symmetric matrix A. In this case, the relation (2.5) holds if Q is of the form

$$hQ = \begin{pmatrix} A^{-1} & 0 \\ 0 & I \end{pmatrix} \cdot \begin{pmatrix} B & C \\ -C^* & D \end{pmatrix} .$$

Now the compatibility conditions become

$$(2.25) \qquad j(\underline{\underline{e}}_i, A\underline{\underline{e}}_{j-1}) + i(\underline{\underline{e}}_j, A\underline{\underline{e}}_{i-1}) = (-1)^{\sigma-1} r^\sigma + J_{i,\sigma} .$$

Furthermore, A has to be symmetric and therefore also

$$(2.26) \qquad (\underline{\underline{e}}_\nu, A\underline{\underline{e}}_\mu) = (\underline{\underline{e}}_\mu, A\underline{\underline{e}}_\nu)$$

must hold.

By explicit calculations one can show:

Lemma 2.3.

Let $\tau \le 8$. *The compatibility conditions* (2.25), (2.26) *can be satisfied if and only if*

$$\sum_{\nu=1}^{s} \alpha_\nu \nu^\sigma = 0 \quad \text{for all odd } \sigma \text{ in } [3,\tau+1].$$

Thus the approximation is, in the interior, of order $\tau+2$ $\{\tau+1\}$ *for even* $\{$*odd*$\}$ τ.

Therefore, one can construct matrices A, B such that the conditions (2.9) are fulfilled. If in addition $A > 0$ then we have constructed an approximation of the kind desired. We have done this for $\tau = 3$ and $\tau = 5$. These are given in Table 2.

Matrix A =

0.2251157407	0.1990740728	-0.1255787034	0.0
	1.039930552	0.125	0.03182870336
		0.6788194403	0.05092592584
			0.9936342537

Matrix B =

-0.5	0.7013888889	-0.2013888889	0.0
	0	0.7083333333	-0.006944444444
		0	0.5902777778
			0

Eigenvalues of A (single precision only)

1.121775746
0.9856095612
0.6935795248
0.1365351193

for $\tau = 3$

Matrix A =

0.1934552379	0.2380016800	-0.1739483122	0.1073399670	-0.01872541150	-0.01267408940
	1.253478542	-0.2470744140	0.0	0.01813904103	0.04034866253
		1.299999997	0.04968584888	-0.0722222164	-0.05389460037
			0.6932098716	0.1973700300	0.01790353889
				0.8729229867	0.001705384784
					1.003023133

Matrix B =

-0.5	0.8136493826	-0.5171084777	0.2427611687	-0.02224350860	-0.01705856494
	0	1.282950328	-0.5417487949	0.0	0.07244784942
		0	0.8587800939	0.02615243964	-0.1190906832
			0	0.5743083528	-0.03118255181
				0	0.7115506172
					0

Eigenvalues of A (single precision only)

1.596797839
1.042205215
1.000163600
0.9921823218
0.5828833729
0.1018573642

for $\tau = 5$

Table 2

3. Difference approximations for hyperbolic systems.

It is now very easy to write down stable approximations for systems. For simplicity we do not discretize the time variable. Consider a symmetric system,

$$\partial u/\partial t = H_1 \partial u/\partial x + H_2 \partial u/\partial y , \tag{3.1}$$

in the domain Ω of Figure 4. Assume that the boundary conditions at $x = 0$ are given by

$$u^{(1)}(0, y) = u^{(2)}(0, y) = \ldots = u^{(r)}(0, y) = 0 \tag{3.2}$$

where r denotes the number of ingoing characteristics. Assume furthermore that, for $x = 0$,

$$(y, H_1 y) \geq 0 \tag{3.3}$$

for all vectors y with $y^{(1)} = y^{(2)} = \ldots = y^{(r)} = 0$.

Let Q_y denote a centered difference operator in the y-direction and Q_x a difference operator of our kind in the x-direction. Then we approximate (3.1) by

$$d\, v_h/dt = H_1 Q_x v_h + H_2 Q_y v_h . \tag{3.4}$$

Here v_h is a grid function which satisfies the boundary conditions. There is no difficulty in deriving an energy estimate for this equation.

References.

1. Swartz, B. and Wendroff B., Generalized finite difference schemes, Math. Comp. 23, (1969).

Professor H. -O. Kreiss
Department of Computer Science
Uppsala University
Sturegatan 4B 2tr
Uppsala, Sweden

Godela Scherer
Department of Computer Science
Uppsala University
Sturegatan 4B 2tr
Uppsala, Sweden

Solution of Problems with Interfaces and Singularities

IVO BABUŠKA

Chapter 1: Introduction.

During recent years, considerable progress has been made in theoretical investigations of the finite element method. Let us mention here, e.g., the references [1], [2], [3], [4]. We refer also to the proceedings of various conferences, e.g., [5], [6], [7] where additional information may be found, etc. Also significant progress in the research of the computational aspects of the finite element method has been made during the last few years. (See, e.g., [7], [8].) The finite element technique has been studied in many books from different points of view (see, e.g., [9], etc.).

There is an area in the finite element method where neither theoretical nor practical progress has been made, which would be comparable with the progress in the general treatment of this method. This area is connected with the question of how to handle the singularities of the solution arising especially because of interfaces and unsmoothness of the domain. There are two kinds of problems which have to be investigated separately: the global and local problems.

1.1. The global problem.

Let us introduce a few problems which motivate our question.

1.1.1. Power reactor problem.

A power reactor is in essence constructed by repeating the core cells. In Figure 1.1 we see a typical core cell as used at the Duane Arnold Energy Center, Iowa. A core cell is a control rod and the four fuel assemblies which immediately surround it. A fuel assembly consists of a fuel bundle and the channel which surrounds it. The fuel assemblies are arranged in the reactor core to approximate a circular cylinder inside the core shroud. There are 1472 bundles in this reactor.

Mathematically, the reactor is described by either the multi-group diffusion equation or the transport equation with the material properties different in and outside rods, etc. This structure creates an interface problem with many interfaces. Every interface creates a singularity in the solution. Obviously it is impossible to take account of all these singularities individually.

In practice, the computation is made by the use of so-called homogenization where the cells are "smeared up" to an average behavior (which may differ from cell to cell). Homogenization is computed by some techniques based on intuition and there is no uniquely accepted approach to this problem. No strict mathematical analysis has been found in the literature. Because of extreme difficulties related to the nonhomogenized solution, there are very likely no bench mark computations which would experimentally show the magnitude of the error caused by the homogenization approach. References [10] through [15] inclusive deal with the different aspects of this problem.

1.1.2. Problems of composite materials.

Composite materials are used today more and more in many technical areas, e.g., the composite aircraft structures, and many others. There is even a special journal devoted to such problems (Journal of Composite Materials). Especially important here are the dynamic problems of the composite materials. There is a significant effort to clarify experimentally and theoretically questions related to the

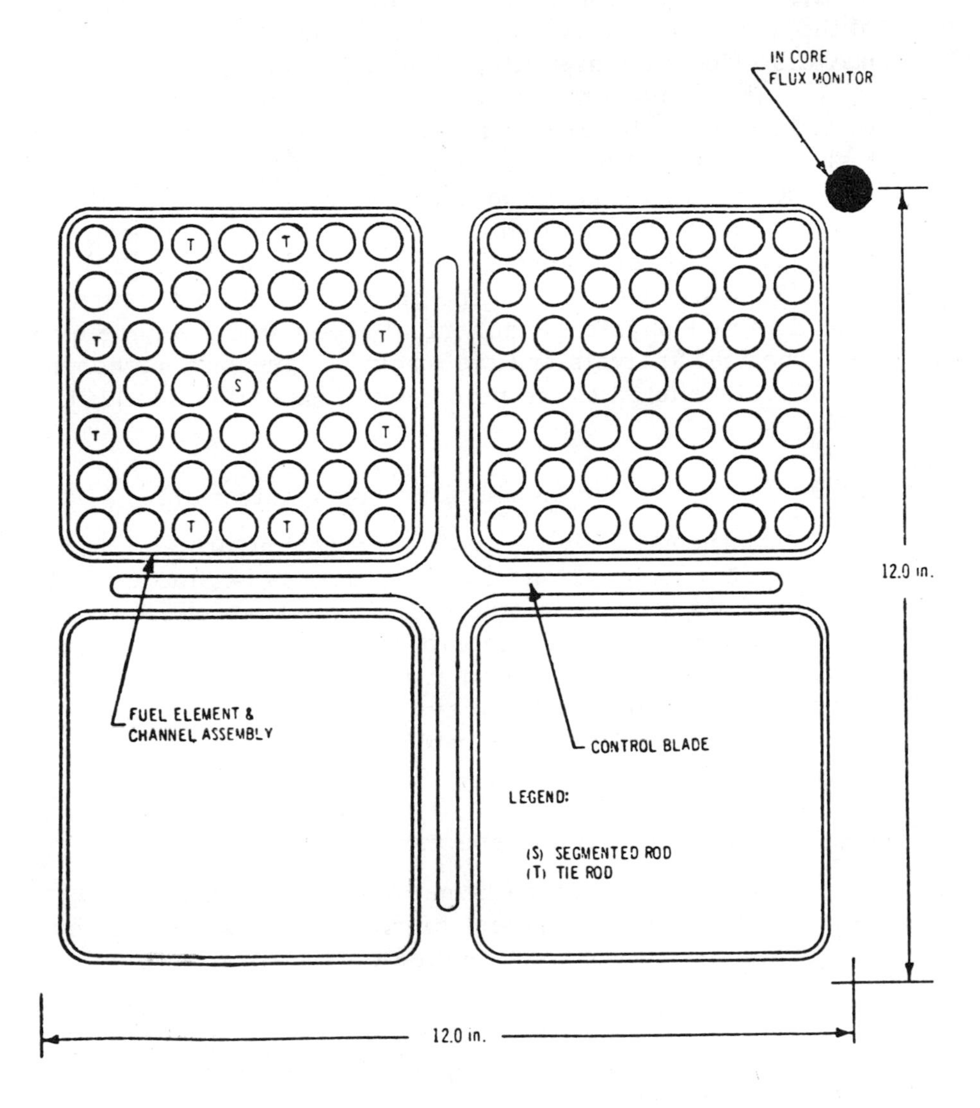
IN CORE
FLUX MONITOR
T
S
FUEL ELEMENT &
CHANNEL ASSEMBLY
CONTROL BLADE
LEGEND:
(S) SEGMENTED ROD
(T) TIE ROD
12.0 in.
12.0 in.

Figure 1.1

behavior of such composite materials. A good description of the situation in this field may be found in [16] where one may also find extensive literature citations.

The problem here is analogous to the reactor problem (it too has many interfaces). The question is how to handle them in a "global" way (see, e. g., [17], [18], etc.). With respect to the dynamic behavior, the initial study has likely been made in [19] and then further study has been made in [20], [21], [22], and others. For surveys see, e. g., [23], [29], [25]. The analysis of the problem in the papers mentioned is of a physical-mechanical type without any detailed mathematical analysis and statements. For some mathematical aspects related to the problem of composites see [26], [27].

1. 1. 3. Some other problems of the homogenization.

Mathematically very similar problems arise in many other areas. We mention here the problems occurring in biology such as studies related to the flow of electrical current in the brain, the diffusion of metabolics in the tissues, etc. For some results related to these topics see, e. g., [28].

The homogenization plays an important role in physics and in mechanics. As an example we mention here a series of papers by Eringen [see, e. g., [29], [30],) etc.

1. 1. 4. The mathematical aspect.

Let us show the mathematical aspects of the problem mentioned in its simplest form. We will be interested in the boundary value problem on a domain Ω for the self-adjoint elliptic differential equation

$$-\sum_{i,j=1}^{2} \frac{\partial}{\partial x_i} a^H_{i,j}(x_1, x_2) \frac{\partial u^H}{\partial x_j} + q^H(x_1, x_2) u^H = f \tag{1.1}$$

where the functions $a^H_{i,j}$ and q^H are measurable and the associated operator is uniformly elliptic (with resp. to H). The question is how does the solution of our problem behave for small H and how to find the solution numerically. An example is when $a^H_{i,j}$ and q^H are "nearly" periodic functions with the period H and f is slowly changing with respect to the size of H.

If $a^H_{i,j} \to a_{i,j}$ and $q^H \to q$ strongly in L_2 then the solution converges to the solution of the differential equation with coefficients $a_{i,j}$ and q. A quite different situation occurs when $a^H_{i,j} \to a_{i,j}$ only weakly. Then the main theoretical questions are still unresolved. Let us mention here references [31], [32], and [33] which are related to this problem.

Another question arises in connection with the numerical solution, namely, whether it is possible to use ideas of a small parameter H. These problems are in general unsolved. Some particular results are contained in [15], [34], [35], and [36].

We mentioned the problem related to the equation (1.1). There is also a problem which will be called the problem with the <u>grained boundary</u>. Let us show a simple model example. Let a continuous periodic function $\varphi^H(x_1, x_2) \geq 0$ with the period H be given. Then denote by

$$\Omega_H = \{(x_1, x_2) \mid \varphi(x_1, x_2) > 0\}$$

and solve the problem

$$-\Delta u^H = f \tag{1.2}$$

on Ω_H, with zero Dirichlet boundary condition on $\partial\Omega_H$. We will assume that the function f is slowly changing with respect to the size of H. This type of problem also has many applications. For some mathematical analysis of this type of problem see [37], [38], [39].

1.1.5. Outline of the results for the global problem.

Chapter 2 will deal with the problems described above. The first part deals with the one dimensional problem as a simple model which shows in a simple way some of the essence of the problem. We will prove a few theorems and show some numerical results. A simple example will show that the straight forward use of the finite element method may lead to a very significant error. The second part of chapter 2 deals with two dimensional problems. These results are based on references [34], [35], and [39]. An illustrative example is introduced to show some of the aspects of the theory.

1.2. The local problem.

The local problem has a quite different character. The difficulties lie in the problem of determining how much the convergence of the finite element method will be slowed down when the solution of the problem has some singularities. Let us name a few areas in which this problem arises.

1.2.1. Mechanics of fracture.

The study of the stresses around a crack in the theory of elasticity, e.g., the study of the stress concentration factor, has to deal with singularities of the solution. These singularities are local and are located in some isolated points (ends of the cracks). For the study of these problems see, e.g., [44] and especially the survey [45].

1.2.2. Reactor computations.

After the homogenization has been made, the reactor problem leads to the interface problem. Here singularities are present which are of a type similar to the singularities created by the cracks. The critical areas are where the interface has a corner. These singularities will slow down the asymptotic rate of convergence and the use of higher order elements cannot improve this rate. Nevertheless, the numerical experiments show that for practical accuracy range it is worthwhile to use higher order elements. See, e.g., the references [46], [47], and also [15], [48].

1.2.3. Mathematical problems.

The influence of the singularities caused by corners of the domain and the interfaces on the rate of convergence has been studied, e.g., in [49], [50], [51], and [52]. Two different approaches showing how to deal with such problems have been studied in the finite element method. First is the addition of "special elements", see, e.g., [48], [53], [54], and [2] chapter 8. The other possibility is to use proper refinement of elements, see, e.g., [55]. Many important computational problems here remain open, especially in the 3 dimensional case. In addition we will analyze the problem of nonuniform convergence.

Chapter 2: Global Solution.

We will analyze in this chapter the question which arises when many singularities are present as explained in Chapter 1. We call it the problem of the global solution. To explain the main ideas, we will first study (in section 1) the one dimensional problem. In section 2 we will analyze the two dimensional problem.

2.1. Global solution for the one dimensional problem.

In this section the one dimensional case will be analyzed. We will analyze the source and eigenvalue problem and bring some illustrational numerical examples.

2.1.1. Formulation of the model-source problem.

We will study as model problem the boundary value problem

$$-\frac{d}{dx}\, p^H(x)\, \frac{du}{dx} + q^H(x)u = f^H(x), \quad 0 < x < 1 \tag{1.1}$$

with boundary condition

$$p^H(0)\, \frac{du}{dx}(0) = u(1) = 0 . \tag{1.2}$$

Here $p^H(x)$, $q^H(x)$ and $f^H(x)$ are one parameter sets of measurable functions such that

$$0 < \alpha_1 \leq p^H(x) \leq \alpha_2 < \infty \tag{1.3}$$

$$0 \leq q^H(x) \leq \alpha_2 . \tag{1.4}$$

The numbers α_1, α_2 are independent of H. For the sake of brevity we will denote

$$u' = \frac{du}{dx}, \quad u^{[j]} = \frac{d^j u}{dx^j}, \quad j = 0,1,\ldots .$$

Denote as usual by $\mathcal{D}$ the set of all infinitely differentiable functions on $(0,1)$ with compact support, and by L_2 the space of square integrable functions u on $(0,1)$ with the norm

$$\|u\|^2_{L_2} = \int_0^1 |u|^2\,dx . \tag{1.5}$$

The scalar product will be written $(\cdot,\cdot)_{L_2}$. Sometimes the notation $L_2 = H^0$ will be used.

Suppose now that $\ell \geq 1$ is an integer. We will denote by H^ℓ the Sobolev space with the norm

$$\|u\|^2_{H^\ell} = \sum_{j=0}^{\ell} \|u^{[j]}\|^2_{L_2} \tag{1.6}$$

and associated scalar product $(\cdot,\cdot)_{H^\ell}$.

Further, let $H^{1;0} \subset H^1$ be the subspace of all functions u with $u(1) = 0$.

For ℓ not integral, $\ell = \lfloor \ell \rfloor + \alpha$, $0 < \alpha < 1$, $\lfloor \ell \rfloor$ integral part of ℓ, H^ℓ will be understood as the interpolated space (Hilbert scale) between $H^{\lfloor \ell \rfloor}$ and $H^{\lfloor \ell+1 \rfloor}$. The space $H^{\ell;0}$ $0 < \ell < 1$ is defined analogously. For more about Hilbert scales and interpolation see, e.g., [1] Chapter 2, and [56].

Finally, let us introduce the norm $\|\cdot\|_C$ for continuous functions on $[0,1]$

$$\|u\|_C = \sup_{x \in [0,1]} |u(x)| .$$

Using the well known Sobolev imbedding theorem, we have for any $u \in H^{1-\varepsilon}$, $0 \leq \varepsilon < \frac{1}{2}$,

$$\|u\|_C \leq C\|u\|_{H^{1-\varepsilon}} , \tag{1.7}$$

where C does not depend on u, and u is continuous on $[0,1]$ (possibly only after changing it on a set of measure zero) (see, e.g., [59].)

The solution of our problem (1.1), (1.2) will be understood in the weak sense in $H^{1;0}$; i.e., $u^H \in H^{1;0}$ will be the solution of the problem (1.1), (1.2) if, and only if,

$$B^H(u^H, v) = (f^H, v)_{L_2} \tag{1.8}$$

for any $v \in H^{1;0}$ where

$$B^H(u, v) = \int_0^1 (p^H u'v' + q^H uv)dx \, . \tag{1.9}$$

Assuming $f \in L_2$ and using the well known Lax-Milgram theorem (see also [1] Theorem 5.2.1) the (weak) solution u^H exists, is unique and satisfies

$$\|u^H\|_{H^1} \leq C\|f^H\|_{L_2} \tag{1.10}$$

where C in (1.10) is independent of f and H (depends on α_1, α_2).

2.1.2. The convergence problem.

In this section we will analyze the question of convergence of u^H for $H \to 0$ when $p^H(x) \to p_0(x)$, $^H(x) \to q_0(x)$, $f^H(x) \to f_0(x)$. We will discuss the problem when p^H, q^H, f^H converge in L_2 (Theorem 2.1) and will analyze the case when p^H, q^H, f^H converge only weakly in L_2 (Theorem 2.2). We will need the following well known lemma.

<u>Lemma 2.1.</u>

Let $u_i \to u$ *strongly in* H, $v_i \rightharpoonup v$ *weakly in* H, *then* $(u_i, v_i) \to (u, v)$.

<u>Theorem 2.1.</u>

Let $p^H(x)$ *and* $q^H(x)$ *satisfy conditions* (1.3) *and*

(1.4). Further let $p^H(x) \to p_0(x)$, $q^H(x) \to q_0(x)$, $f^H(x) \to f_0(x)$ strongly in L_2. Denote by u^H the solution of (1.1) and (1.2). Then $u^H \to u_0$ in H^1 (strongly), where u_0 is the solution of (1.1) and (1.2) replacing p^H, q^H, f^H with p_0, q_0, f_0.

Proof.

1) For any $v \in H^{1;0}$ we have

$$(2.1) \qquad \int_0^1 (p^H u^{H'} v' + q^H u^H v)dx = \int_0^1 f^H v\, dx .$$

Because $f^H \to f_0$ in L_2, $\|f^H\|_{L_2}$ is uniformly bounded. Using (1.10) we see that $\|u^H\|_{H^1}$ is uniformly bounded. Therefore u^H form a precompact set in $H^{1-\varepsilon}$, $0 < \varepsilon \leq 1$, and we may select a sequence $H_i \to 0$ such that

a) $\qquad u^{H_i} \to \bar{u} \qquad$ (strongly in $H^{1-\varepsilon}$)

b) $\qquad u^{H_i} \rightharpoonup \bar{u} \qquad$ (weakly in H^1).

Because the functions p^{H_i} and q^{H_i} converge strongly in L_2 and using Lemma 2.1 we get

$$(2.2) \qquad B_0(\bar{u}, v) = (f_0, v)_{L_2}$$

for any $v \in H^{1,0}$ where

$$(2.3) \qquad B_0(u, v) = \int_0^1 (p_0 u'v' + q_0 uv)dx .$$

Because of (1.7), $\bar{u}(1) = 0$, and so $\bar{u} \in H^{1;0}$. Equality (2.2) proves that $\bar{u}$ is the solution of (1.1), (1.2) with p_0, q_0, f_0. This solution is unique. Therefore $u^H \rightharpoonup u_0$ weakly in H^1 and $u^H \to w_0$ strongly in $H^{1-\varepsilon}$, for any $\varepsilon > 0$.

2) Let us prove now that $u^H \to u_0$ (strongly in H^1). Denote $w^H = p^H(u^H)'$, $w_0 = p_0 u_0'$. Using (2.1), (2.2), (2.3) we get for any $v \in H^{1;0}$

$$(2.4) \quad \int_0^1 (w^H - w_0)v'dx = \int_0^1 (f^{H} - f_0)v\,dx - \int_0^1 (q^H u^H - q_0 u_0)v\,dx .$$

Denote

$$\eta^H(x) = \int_x^1 (w^H(t) - w_0(t))dt .$$

Because $w^H - w_0$ is uniformly bounded in L_2 we may select the subsequence H_i so that η^{H_i} converges in C. Taking η^{H_i} for v in (2.4) we get $\|w^{H_i} - w_0\|_{L_2} \to 0$. So $(u^{H_i})' \to u_0'$ in L_2 and our theorem is proven.

The situation changes when p^H, q^H and f^H converge only weakly in L_2.

Theorem 2.2.[†]

Let $p^H(x), q^H(x)$ satisfy conditions (1.3) and (1.4). Further let $1/p^H \rightharpoonup 1/p_0$, $q^H \rightharpoonup q_0$, $f^H \rightharpoonup f_0$ in L_2 (weakly). Then $u^H \rightharpoonup u_0$ weakly in H^1 (not strongly in general) and strongly in $H^{1-\varepsilon;0}$, for any $0 < \varepsilon \leq 1$. In addition, $p^H(u^H)' \to p_0 u_0'$ strongly in L_2, where we denote by u_0 the solution of (1.1), (1.2) with p_0, q_0, f_0.

Proof.

Because $f^H \rightharpoonup f_0$ in L_2, $\|f^H\|$ is uniformly bounded. Using (1.10) we see that $\|u^H\|_{H^1}$ is uniformly bounded too, i.e., $\|u^H\|_{H^1} \leq C$ with C independent of H. Denote $w^H = p^H(u^H)'$. Because of (1.3), $w^H \in L_2$. Let us show that $w^H \in H^1$ when $(w^H)'$ is understood in the weak sense. In fact, we get

$$(2.5) \quad \int_0^1 w^H \varphi' dx = -\int_0^1 \varphi[q^H u^H - f^H]dx$$

[†] Proof of this theorem is joint with R. B. Kellogg.

for any $\varphi \in H^{1;0}$ and so $(w^H)' = q^H u^H - f^H \in L_2$ and therefore $\|w^H\|_{H^1} \le C$ with C independent of H.

This makes it possible for us to select $H_i \to 0$ so that

a) $(w^{H_i})' \rightharpoonup w_0'$ (weakly in L_2)

b) $w^{H_i} \to w_0$ (strongly in $H^{1-\varepsilon}$, $0 < \varepsilon \le 1$)

c) $(u^{H_i})' \rightharpoonup u_0'$ (weakly in L_2)

d) $u^{H_i} \to u_0$ (strongly in $H^{1-\varepsilon}$, $0 < \varepsilon \le 1$).

Because $w^{H_i} \to w_0$ strongly in $H^{1-\varepsilon}$ and $1/p^{H_i} \to 1/p_0$ weakly, Lemma 2.1 is applicable and we see that

$$\frac{1}{p^{H_i}} w^{H_i} \rightharpoonup \frac{w_0}{p_0} \text{ (weakly in } L_2). \tag{2.6}$$

But $\frac{1}{p^{H_i}} w^{H_i} = (u^{H_i})'$. Because $(u^{H_i})' \rightharpoonup u_0'$ (weakly in L_2) we have $\frac{w_0}{p_0} = u_0'$. Therefore $w_0 = u_0' p_0$.

On the other hand, applying Lemma 2.1 once more, we see that

$$\int_0^1 \varphi(q^{H_i} u^{H_i} - f^{H_i})dx \to \int_0^1 \varphi(q_0 u_0 - f_0)dx \tag{2.7}$$

for any $\varphi \in H^{1;0}$. Therefore by (2.5) we have for any $\varphi \in H^{1;0}$

$$\int_0^1 (p_0 u_0' \varphi' + q_0 u_0 \varphi)dx = \int_0^1 f_0 \varphi dx . \tag{2.8}$$

Because $\|u^{H_i}\|_{H^1} \le C$ and $u^{H_i}(1) = 0$, we have also $u_0(1) = 0$. Therefore u_0 is the solution of (1.1) and (1.2) with p_0, q_0, f_0. We will show that the conditions (1.3) and (1.4) are satisfied for p_0 and q_0 (after modification on a set of measure zero if necessary). Let us prove that $1/p_0 \leq 1/\alpha_1$. Assuming to the contrary that $1/\alpha_1 - 1/p_0 < 0$ on a set of positive measure, we get a contradiction. In fact, let χ_0 be the characteristic function of this set. Then

$$0 > \int_0^1 (1/\alpha_1 - 1/p_0)\chi_0 dx = \lim_{H \to 0} \int_0^1 (1/\alpha_1 - 1/p_H)\chi_0 dx \geq 0 .$$

In a similar way we get that $1/p_0 \geq 1/\alpha_2$ and so condition (1.3) is satisfied for the function p_0. Similarly (1.4) may be proved for q_0. Therefore, condition (2.8) determines u_0 uniquely. This proves that u^{H_i} converges to u_0 and not just the subsequence.

We have shown that u^H converges weakly in H^1 and strongly in $H^{1-\varepsilon}$.

To show that u^H does not converge strongly in H^1 (in general), it is sufficient to study problem (1.1), (1.2) with $1/p^H = 1 + \frac{1}{2} \sin \frac{1}{H} x$ and $q^H = 0, f^H = 1$. Its solution may be found in closed form. Our proof is now complete.

Let us remark that our theorem together with (1.7) yields convergence in the C-norm (i.e. uniformly pointwise).

2.1.3. Some applications to numerical solution.

In the spirit of the introduction where the importance of a cell problem has been shown, we will analyze now the question of the numerical solution of a special simple problem.

Let $p_N^H(x) = p_N(x/H)$ with $H = 1/R$, N, R integers, be a periodic function with period H, so that $p_N(x)$ has period 1 and

$$(3.1) \quad p_N(x) = \begin{cases} p_1 & \text{for } -\frac{1}{2} + \frac{2i}{2N} \leq x < -\frac{1}{2} + \frac{2i+1}{2N} , \\ p_2 & \text{for } -\frac{1}{2} + \frac{2i+1}{2N} < x < -\frac{1}{2} + \frac{2(i+1)}{2N} , \end{cases}$$

$$i = 0, \ldots, N-1 ; \; p_1, p_2 \text{ constants.}$$

Similarly let $q_N^H(x) = q_N(x/H)$ with $q_N(x)$ the periodic function with period 1 so that

$$(3.2)\quad q_n(x) = \begin{cases} q_1 & \text{for } -\tfrac{1}{2} + \dfrac{2i}{2N} < x < -\tfrac{1}{2} + \dfrac{2i+1}{2N}, \\ q_2 & \text{for } -\tfrac{1}{2} + \dfrac{2i+1}{2N} < x < -\tfrac{1}{2} + \dfrac{2(i+1)}{2N}, \end{cases}$$

$$i = 0, \ldots, N-1, \quad q_1, q_2 \text{ constants.}$$

We assume that (1.3) and (1.4) are satisfied, i.e.,

$$0 < \alpha_1 \leq p_i \leq \alpha_2 < \infty \qquad i = 1, 2,$$

$$0 \leq q_i \leq \alpha_2 < \infty \qquad i = 1, 2.$$

Further let $f^H = 1$. We have $2NR$ points at which the coefficients p^H and q^H have singularities.

Let us assume that we are not able for computational reasons to take account of (in detail) all these singularities. In other words, let us assume that we will be able to use finite elements with step H only. As as example, let us use piecewise linear elements and denote by $\tilde{u}_H$ this approximate solution. Obviously, for the parameters $\{y_{H,i}\}$ of this solution, we are led to a tridiagonal system of linear equations.

$$\sum_{j=0}^{R-1} a_{ij} y_{H,j} = b_{H,i}, \qquad i = 0, \ldots, R-1$$

where

$$(3.3)\qquad a_{ii} = \frac{2}{H} A(N) + H[2D(N) + \tfrac{1}{2}B(N)]$$

$$i = 1, 2, \ldots, R-1$$

$$a_{0,0} = \frac{1}{H} A(N) + H[D(N) - C(N) + \tfrac{1}{4}B(N)]$$

$$a_{i-1,i} = a_{i,i+1} = -\frac{1}{H} A(N) + H[D(N) - \tfrac{1}{4}B(N)]$$

$$(3.4)\qquad i = 0, 1, \ldots, R-1$$

$$a_{ij} = 0 \quad \text{for} \quad |i-j| > 1$$

$$b_{H,i} = H \qquad b_{H,0} = H/2$$

and

$$A(N) = \tfrac{1}{2}(p_1+p_2) = \bar{p} \tag{3.5}$$

$$B(N) = \tfrac{1}{2}(q_1+q_2) = \bar{q} \tag{3.6}$$

and for large N, $C(N) \approx 0$, $D(N) \approx \frac{1}{12}\bar{q}$.

Using standard results of the theory of the finite element method we see that $\tilde{u}_H \approx \tilde{U}$ (with error $O(H^2)$ in L_2) where $\tilde{U}$ is the solution of the problem (1.1), (1.2) with the constant coefficients $\bar{p}$ and $\bar{q}$.

On the other hand, we see that for $N \to \infty$ and H fixed, $1/p^H$ is weakly convergent to the constant function $1/p_0 = \frac{1}{2}[1/p_1 + 1/p_2]^N$ and q_N^H converges weakly to $q_0 = \frac{1}{2}[q_1+q_2] = q$. Denoting the exact solution of (1.1), (1.2) by u_N^H, we see from Theorem 2.2 that for large N we have $u_N^H \approx U$ (in L_2) where U is the solution of (1.1), (1.2) with coefficients $p_0, \bar{q}$. Because $p_0 \neq \bar{p}$, the solution $\tilde{u}_H$ gives very incorrect results.

To get a rough idea about the error let us define for $z = p_1/p_2$

$$\frac{\bar{p}}{p_0} = \lambda(z) = \frac{1}{4}\frac{(z+1)^2}{z} \tag{3.7}$$

with values shown in Table 3.1.

TABLE 3.1

z	λ(z)
0.01	25.5025
0.1	3.025
0.5	1.125
1.0	1.0
2.0	1.125
10.0	3.025
100.	25.5025

Table 1 shows that when $z > 2$ then the error may be significant, e.g., for the ratio $z = 10$ we get an error of 70%. A simple computation shows that $\lambda(z) \geq 1$.

2.1.4. The asymptotic analysis.

We saw that the numerical solution of a "cell" problem causes difficulties. It is obvious that it is not possible to avoid a structural analysis of a single cell. For, since all the cells are assumed to be identical, an approach based on single cell analysis is numerically acceptable. To show what kind of results one may expect (especially in higher dimensional cases), let us study the problem when $p^H(x) = p(\frac{y}{H})$ with $p(x)$ a periodic function with period 1 and satisfying the condition (1.3). Let us further assume that $q = 0$; so we obtain the problem

(4.1) $$-(p^H(u^H)')' = f$$

(4.2) $$p^H(0)(u^H)'(0) = 0 = u(1) .$$

We will assume that $H = \frac{1}{R}$ where R is an integer.

Let us define

(4.3) $$\chi_i(x) = \int_0^x \frac{t^{i-1}}{p(t)} dt, \quad i = 1, 2, 3; \ -\tfrac{1}{2} < x < \tfrac{1}{2},$$

and

(4.4) $$A_i = \chi_i(\tfrac{1}{2}) - \chi_i(-\tfrac{1}{2}) .$$

Further let

(4.4) $$\chi_i^H(x) = H^i \chi_i(\frac{x}{H}); \quad -\tfrac{1}{2}H < x < \tfrac{1}{2}H .$$

Denoting

(4.5) $$F(x) = \int_0^x f(t)dt ,$$

we get

$$-p^H(u^H)' = F \tag{4.6}$$

and

$$u^H = -\int_0^x \frac{F(t)}{p^H(t)}\,dt + C^H \tag{4.7}$$

with

$$C^H = \int_0^1 \frac{F(t)}{p^H(t)}\,dt\ . \tag{4.8}$$

Let us study now

$$v(jH) = \int_0^{jH} \frac{F(t)}{p^H(t)}\,dt\ , \quad j \text{ integral}\ . \tag{4.9}$$

Assuming that $F(x)$ is smooth and using the periodicity of p^H, we get

$$\begin{aligned} v(jH) &= \sum_{k=0}^{j-1} \int_{kH}^{(k+1)H} \frac{F(t)}{p^H(t)}\,dt \\ &= \sum_{i=0}^{2} \frac{1}{i!} \sum_{k=0}^{j-1} F^{(i)}((k+\tfrac{1}{2})H)\cdot A_{i+1}\cdot H^{i+1} + O(H^3) \\ &= A_1 \int_0^{jH} F(t)dt - A_1 \frac{H^2}{24} \int_0^{jH} F''(t)dt \\ &\quad + A_2 H \int_0^{jH} F'(t)dt + \tfrac{1}{2} A_3 H^2 \int_0^{jH} F''(t)dt \\ &\quad + O(H^3)\ . \end{aligned} \tag{4.10}$$

On the other hand, it is easy to see that

$$(p^H)^{-1} \rightharpoonup A_1 \quad \text{(weakly in } L_2)\ . \tag{4.11}$$

Using Theorem 2.2 we have

$$\lim_{H \to 0} u^H = U\ , \tag{4.12}$$

with

$$-A_2^{-1} U'' = f , \tag{4.13}$$

$$U'(0) = U(1) = 0 . \tag{4.14}$$

So

$$U = C - A_1 \int_0^x F(t)dt \tag{4.15}$$

with

$$C = A_1 \int_0^1 F(t)dt . \tag{4.16}$$

Using (4.15), (4.16), (4.9), (4.10) we get

$$\begin{aligned} C^H &= C + A_2 H F(1) \\ &\quad + H^2[(-\frac{1}{24} A_1 + \tfrac{1}{2}A_3)F'(1)] + O(H^3) \end{aligned} \tag{4.17}$$

and

$$\begin{aligned} u^H(jH) &= C^H - A_1 \int_0^{jH} F(t)dt \\ &\quad -A_2 H\int_0^{jH} F'(t)dt - H^2[-\frac{1}{24}A_1 + \tfrac{1}{2}A_3]\int_0^{jH} F''(t)dt \\ &= U(jH) + A_2 H[F(1) - F(jH)] \\ &\quad + H^2[(-\frac{1}{24}A_1 + \tfrac{1}{2}A_3)(F'(1) - F'(jH))] \\ &\qquad + O(H^3) . \end{aligned} \tag{4.18}$$

Let us define the auxiliary problem

$$\begin{aligned} -A_1^{-1} V'' &= A_2 A_1^{-2} U''' = -A_2 A_1^{-1} F'' , \\ A_1^{-1} V'(0) &= -A_1^{-1} A_2 F'(0) = A_1^{-2} A_2 U''(0) , \\ V(1) &= 0 . \end{aligned} \tag{4.19}$$

Then

$$V(jH) = (-F(jH) + F(1))A_2 \tag{4.20}$$

and so we see that

$$u^H(jH) = U(jH) - HV(jH) + O(H^2). \tag{4.21}$$

In addition, we have for $|x - jH| < \frac{1}{2}H$ that

$$\begin{aligned} u^H(x) &= u^H(jH) - F(jH)\chi_1^H(x-jH) \\ &\quad -F'(jH)\chi_2^H(x-jH) + O(H^3) \\ &= U(jH) + U'(jH)A_1^{-1}\chi_1^H(x-jH) \\ &\quad + U''(jH)[A_1^{-1}\chi_2^H(x-jH) + A_2A_1^{-1}H\chi_1^H(x-jH)] \\ &\quad - H[V(jH)+V'(jH)A_1^{-1}\chi_1^H(x-jH)] + O(H^2). \end{aligned} \tag{4.22}$$

Introducing now the norm $\|\cdot\|_{H_{\hat{H}}^1}$ so that

$$\|u\|^2_{H_{\hat{H}}^1} = \sum_{k=0}^{R-1} \int_k^{(k+1)H} (u^2 + (u')^2)dx \tag{4.23}$$

we see from (4.22), (4.21) and (4.18) that

$$\|u^H - (W^H - H\,{}^1\tilde{W}^H)\|_{H_{\hat{H}}^1} \leq CH^2 \tag{4.24}$$

where, for $(j-\frac{1}{2})H < x < (j + \frac{1}{2})H$,

$$\begin{aligned} W^H(x) &= U(jH) + U'(jH)u^{[1];H}(x-jH) \\ &\quad + U''(jH)u^{[2];H}(x-jH) \end{aligned} \tag{4.25}$$

$${}^1W^H = U(jH) + U'(jH)u^{[1];H}(x-jH) \tag{4.26}$$

with

$$u^{[1];H} = A_1^{-1} \chi_1^H (x - jH), \tag{4.27}$$

$$u^{[2];H} = A_1^{-1} \chi_2^H (x - jH) + H A_2 A_1^{-1} \chi_1^H (x - jH). \tag{4.28}$$

${}^{1}\tilde{W}^H$ is given by (4.26) using V defined by (4.19) instead of U.

Summarizing, we have proved the following theorem.

Theorem 4.1.

Let $p^H = p(\frac{x}{H})$ *satisfy* (1.2) *and be of period one and let* $q^H = 0$. *Further let* f *be sufficiently smooth. Denote the solution of* (1.1) *and* (1.2) *by* u^H. *Let* U *be given by* (4.13) *and* (4.14), V *by* (4.19). *Then we have*

$$\|u^H - U\|_{L_2} \leq CH \tag{4.29}$$

$$\|u^H - {}^{1}W^H\|_{H_{\hat{H}}^1} \leq CH \tag{4.30}$$

$$\|u^H - (W^H - H\,{}^{1}\tilde{W}^H)\|_{H_{\hat{H}}^1} \leq CH^2 \tag{4.31}$$

where W^H *resp.* ${}^{1}W^H$ *is given by* (4.25) *resp.* (4.26), *and* ${}^{1}\tilde{W}^H$ *is given by* (4.26) *using* V *instead of* U.

An analogous theorem may be proved in the general case. We will not do it. We presented this theorem as an introduction to the theorem in the two dimensional case discussed in Section 2.

Let us make a few additional comments. The solution of the equation using the limiting coefficients p_0 and q_0 is often called the *homogenized solution.* Using this equation as a basis for numerical computation, we see that the lowest order elements serve well when elements with diameter of the magnitude of the cell diameter are used, because the order (in L_2) of the error of the homogenized solution is $O(H)$ only. Any improvement of the numerical approach of the homogenized equation has to be related to the analysis of the homogenization approach.

2.1.5. Formulation of the model-eigenvalue problem.

In section 1.1 the source problem has been analyzed. Let us study here the model problem

$$(5.1) \qquad -\frac{d}{dx} p^H(x) \frac{du}{dx} = \lambda q^H(x) u$$

with the boundary conditions

$$(5.2) \qquad p^H(0) \frac{du}{dx}(0) = 0 = u(1).$$

We will assume here

$$(5.3) \qquad 0 < \alpha_1 \leq p^H(x) \leq \alpha_2 < \infty$$

$$(5.4) \qquad 0 < \alpha_1 \leq q^H(x) \leq \alpha_2 < \infty$$

and, as in section 1.1, we assume that the numbers α_i are independent of H.

Under these assumptions, the eigenvalues of the model problem can be ordered for every H, i.e., $\lambda_1^H \leq \lambda_2^H \leq \lambda_3^H \leq \ldots$ with $u_1^H, u_2^H, \ldots$ being the corresponding eigenfunctions. We will be interested in the analysis of the behavior of λ_i^H and u_i^H when $p^H \to p_0$, $q^H \to q_0$ for $H \to 0$.

2.1.6. Convergence for the eigenvalue problem.

It it possible to prove theorems analogous to Theorem 2.1 and 2.2. Let us show it for the more interesting case, namely, the case of weak convergence.

Theorem 6.1.

Let $p^H(x)$, $q^H(x)$ *satisfy conditions* (5.3) *and* (5.4). *Let* $1/p^H \to 1/p_0$, $q^H \to q_0$ *converge weakly in* L_2. *Let* $\lambda_1^H \leq \lambda_2^H \leq \ldots$, *resp.* $\lambda_1 \leq \lambda_2 \leq \ldots$, *be the eigenvalues for* (5.1) *and* (5.2) *with* p^H, q^H, *resp.* p_0, q_0.

Further, let u_i^H, $\|u_i^H\|_{L_2} = 1$, resp. u_i, $\|u_i\|_{L_2} = 1$, be the associated eigenfunctions. Then $\lambda_i^H \to \lambda_i$ and $u_i^H \to u_i$ (weakly in L_2) and $u_i^H \to u_i$ strongly in $H^{1-\varepsilon}$, $0 < \varepsilon \leq 1$.

Proof.

1) We will give the proof for the first eigenvalue only, and give a hint of the general proof. First, observe that the Rayleigh quotient may be applied here, i.e.,

$$\lambda_1^H = \inf_{u \in H^{1;0}} \frac{\int_0^1 p^H (u')^2 dx}{\int_0^1 q^H u^2 dx} \tag{6.1}$$

and

$$\lambda_1 = \inf_{u \in H^{1;0}} \frac{\int_0^1 p_0 (u')^2 dx}{\int_0^1 q_0 u^2 dx} \tag{6.2}$$

so λ_1^H is uniformly bounded. On the other hand, for any $v \in H^{1;0}$,

$$\int_0^1 p^H (u_1^H)' v' dx = \lambda_1^H \int_0^1 q^H u_1^H v \, dx . \tag{6.3}$$

Using the inequality (1.10), we see that

$$\|u_1^H\|_{H^1} \leq C \tag{6.4}$$

because

$$\|u_1^H\|_{L_2} \leq 1 .$$

Therefore there exists a sequence H_i such that

a) $$\lambda_1^{H_i} \to \lambda_1^0$$

b) $$u_1^{H_i} \to u_1^0 \quad \text{(strongly in } L_2) .$$

Because $q^H \rightharpoonup q_0$ (weakly in L_2) we may apply Theorem 2.2. Therefore

$$\int_0^1 p^{H_i}\left(u_1^{H_i}\right)' v'dx \to \int_0^1 p_0 u_1^{0\prime} v'dx \tag{6.5}$$

$$\lambda_1^{H_i} \int q^{H_i} u_1^{H_i} v\, dx \to \lambda_1^0 \int q_0 u_1^0 v dx \tag{6.6}$$

for any $v \in H^{1;0}$. So

$$\int_0^1 p_0 u_1^{0\prime} v'dx = \lambda_1^0 \int_0^1 q_0 u_1^0 vdx \tag{6.7}$$

holds for any $v \in H^{1\,0}$ and therefore λ_1^0, resp. u_1^0, is an eigenvalue, resp. an (associated) eigenfunction, of the problem (5.1) and (5.2) when one uses p_0 and q_0 instead of p and q.

2) Let us show that $\lambda_1^0 = \lambda_1$. Let u_1 be the eigenfunction of the homogenized problem (i.e., problem (5.1), (5.2) with the coefficients p_0 and q_0). Let $\bar{u}^H \in H^{1;0}$ be such that

$$\int_0^1 p^H \bar{u}^{H\prime} v'dx = \lambda_1 \int_0^1 q^H u_1 vdx$$

for any $v \in H^{1;0}$. Using Theorem 2.2 we see that for $H \to 0$

$$p^H(\bar{u}^H)' \to p_0 \bar{u}' \qquad \text{strongly in } L_2 \tag{6.8}$$

$$(\bar{u}^H)' \rightharpoonup \bar{u}' \qquad \text{weakly in } L_2 \tag{6.9}$$

$$\bar{u}^H \to \bar{u} \qquad \text{strongly in } H^{1-\varepsilon}, \quad 0 \leq \varepsilon \leq 1, \tag{6.10}$$

and therefore, using inequality (1.7), $\bar{u}^H \to \bar{u}$ uniformly i.e., in $\|\cdot\|_C$. So we get

$$\int_0^1 p_0 \bar{u}' v'dx = \lambda_1 \int_0^1 q_0 u_1 vdx \tag{6.11}$$

and therefore

$$\bar{u} = u_1 . \tag{6.12}$$

Let H_i be the sequence mentioned in the first part of our proof. We have

$$\lambda_1^{H_i} \leq \frac{\int_0^1 p^H (\bar{u}^H)'^2 dx}{\int_0^1 q^H (\bar{u}^H)^2 dx} . \tag{6.13}$$

Using (6.9) and (6.10) and (6.12), and applying Lemma 2.1, we get

$$\int_0^1 p^H (\bar{u}^H)'^2 dx = \int_0^1 p^H (\bar{u}^H)'(\bar{u}^H)' dx \tag{6.14}$$

$$\to \int_0^1 p_0 \bar{u}'^2 dx = \int_0^1 p_0 u_1'^2 dx.$$

By virtue of (6.10), (1.7) and (6.12) we have also

$$\int_0^1 q^H (u^H)^2 dx \to \int_0^1 q_0 u_1^2 dx. \tag{6.15}$$

So we get

$$\lambda_1^0 \leq \frac{\int_0^1 p_0 (u_1')^2}{\int_0^1 q_0 u_1^2 dx} = \lambda_1 . \tag{6.16}$$

λ_1^0 is, as proved above, an eigenvalue of the homogenized problem. Because λ_1 is the smallest eigenvalue, we get $\lambda_1^0 = \lambda_1$ which was to be proven.

The proof of theorem 6.1 in its full generality follows along similar lines but one must also use the minimax principle for the eigenvalue problem.

2.1.7. Applications to the numerical solution.

In the spirit of the introduction, let us solve numerically the eigenvalue problem for a cell problem. We will take $p^H(x) = p(x/H)$. $q^H(x) = q(x/H)$ where $p(x)$, resp. $q(x)$, are the periodic functions with period 1 so that

$$(7.1) \qquad p(x) = p_1 \quad \text{for} \quad -\tfrac{1}{2} < x < 0$$
$$= p_2 \quad \text{for} \quad 0 < x < \tfrac{1}{2}$$

$$(7.2) \qquad q(x) = q_1 \quad \text{for} \quad -\tfrac{1}{2} < x < 0$$
$$= q_2 \quad \text{for} \quad 0 < x < \tfrac{1}{2} .$$

Let $N = H^{-1}$.

Let us be interested in the smallest eigenvalue of the problem (5.1) with the boundary conditions

$$(7.3) \qquad u(0) = p^H(1)u(1) = 0$$

and compare this true eigenvalue with the eigenvalue of the homogenized problem.

Denote by $\lambda_1^{true}(H)$ the first eigenvalue of the problem (depending on p_i, q_i, $i = 1,2$) as a function of the number of cells $N(N = 1/H)$†. As proved in the previous section, $\lim_{H\to 0} \lambda_1^{true}(H)$ converges to the eigenvalue λ_1 of the homogenized problem which is easy to compute analytically. Further more let $\lambda_1^{Homog}(H)$ be the approximate value of the first eigenvalue when finite piecewise linear elements are used with the mesh size H. In Figure 7.1a,b we see the error

$$\varepsilon^{true}(H) = |\lambda_1^{true}(H) - \lambda_1|$$

and

$$\varepsilon^{Homog}(H) = |\lambda_1^{Homog}(H) - \lambda_1|$$

---

†computed numerically.

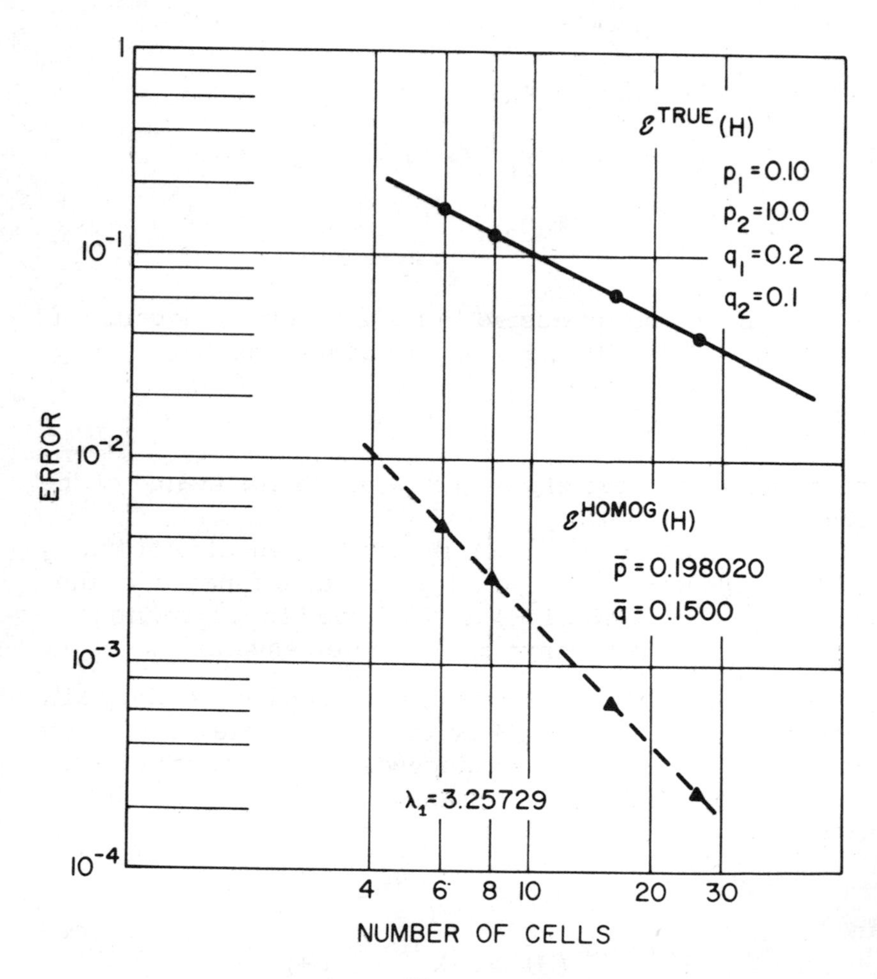

Figure 7.1.a

Error of the homogenization and finite element method

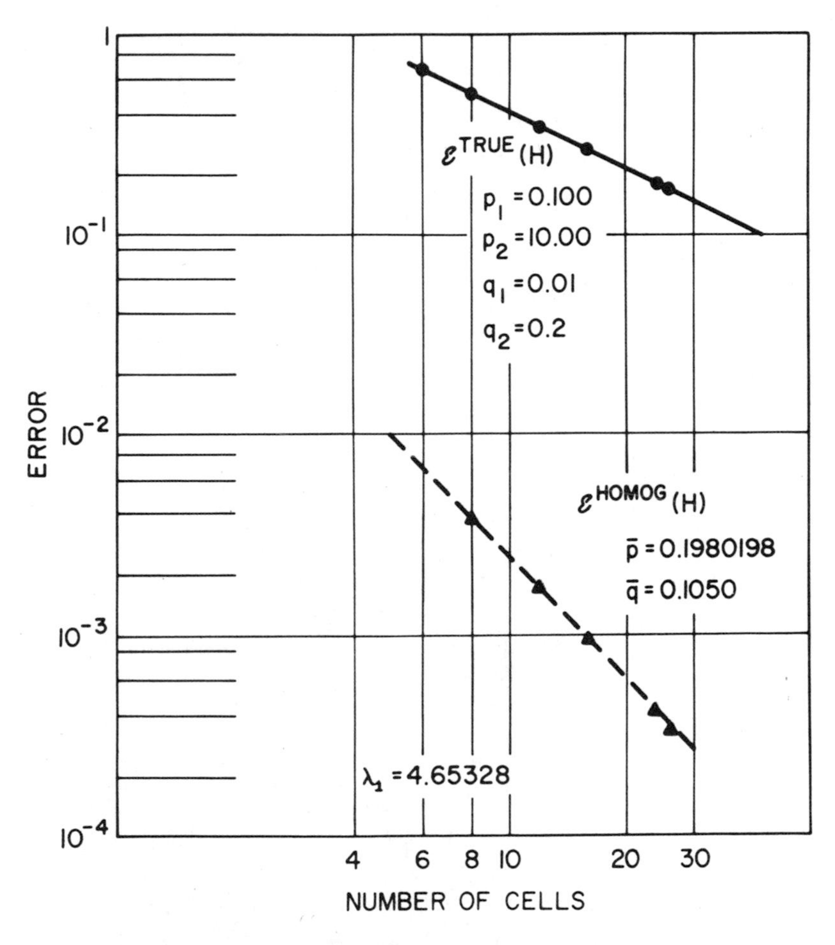

Figure 7.1.b.

Error of the homogenization and finite element method

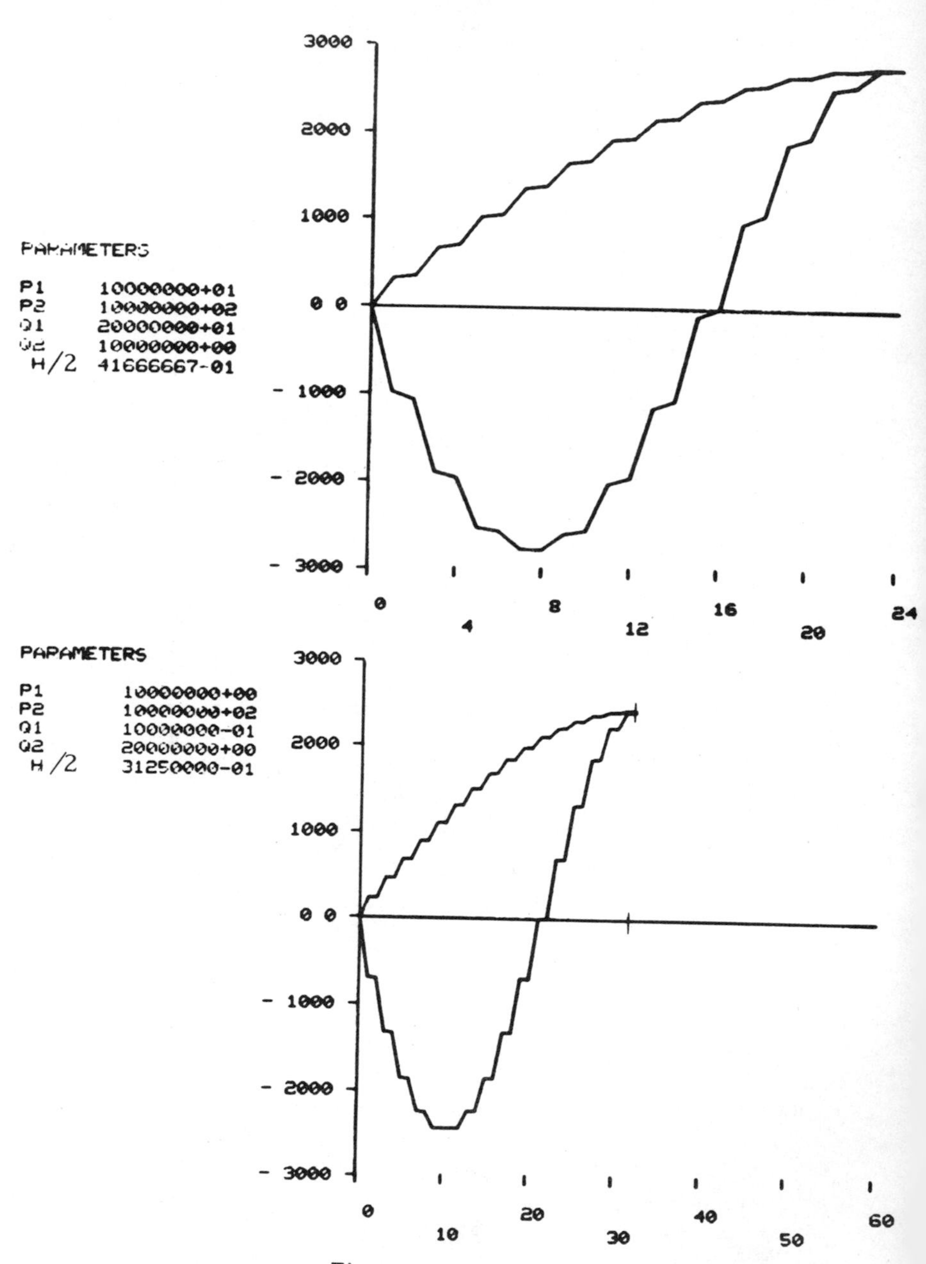

Figure 7.2
The first two eigenfunctions for $H = \frac{1}{12}$ and $H = \frac{1}{16}$

for some p and q. We see that the rate of convergence is of order H^2 for $\lambda_1^{Homog}(H)$ and is of the order H for convergence of $\lambda^{true}(H)$. The coefficients $\bar{p}$ and $\bar{q}$ in Figure 7.1a,b represents the coefficients of the homogenized equation.

The figures show very clearly that the error of the homogenized solution is a significant one and much larger than the error of the approximate eigenvalue of the homogenized equation. In other words, there is no reason to compute the homogenized solution too accurately if the error of the homogenization is the principal one. [Let us remark that the reactor computations in practice are performed via the homogenized solution. It is not clear that the elementary example we gave characterizes complicated multigroup analysis, but it is also not clear that it does not. No bench mark computations have been found to clarify this question.]

In Figure 7.2, we show the first two eigenfunctions for different H.

Let us now mention an interesting observation. Instead of the boundary conditions (7.3), let us be interested in the condition

$$u(0) = u(1) = 0 . \tag{7.4}$$

In this case we see that the rate of convergence is of order H^2 in contrast with the rate H for the boundary conditions (7.3). We see this difference in Figure 7.3.

In this very special case, it is possible to explain this behavior. The theoretical background can be obtained from a theorem of J. Osborn [57]. The essential part of this theorem for our case is the estimate of the value σ^H,

$$\sigma^H = \int_0^1 (v^H - u)u\, dx, \tag{7.5}$$

where u is the eigenfunction of the homogenized problem and v^H is the solution of the problem

$$-\frac{d}{dx} p^H(x) \frac{d}{dx} v^H = \lambda_1 q^H(x)u$$

$$v^H(0) = v^H(1) = 0 .$$

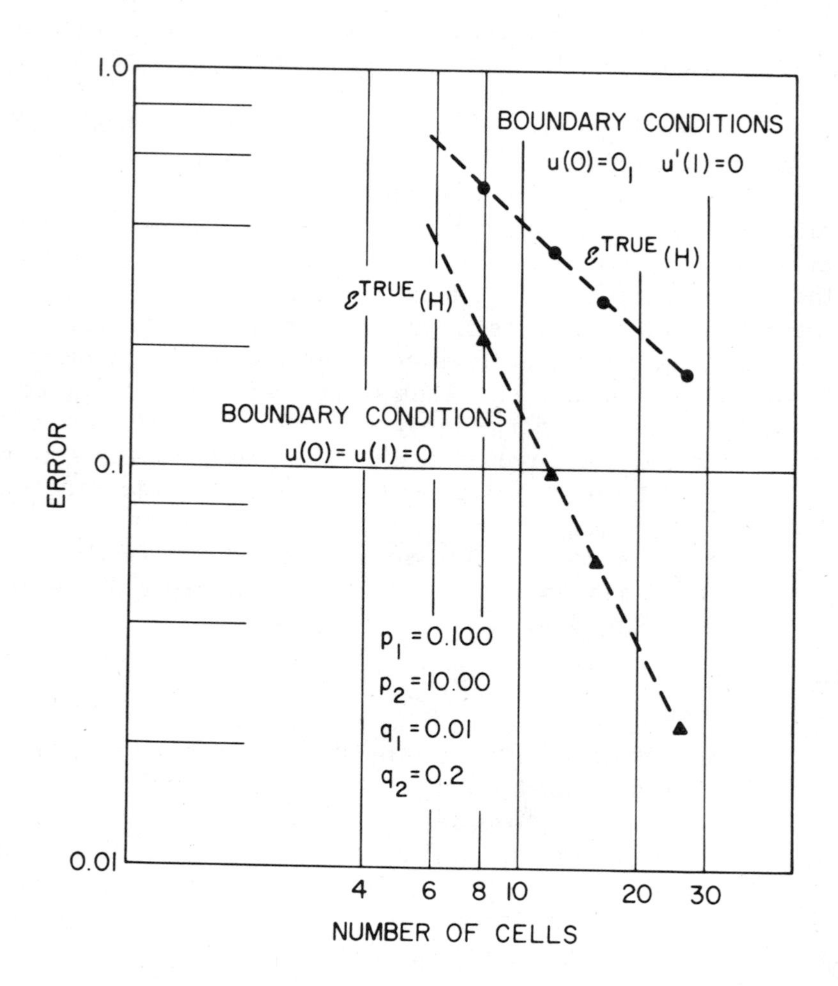

Figure 7.3

The error of the homogenization for different boundary conditions

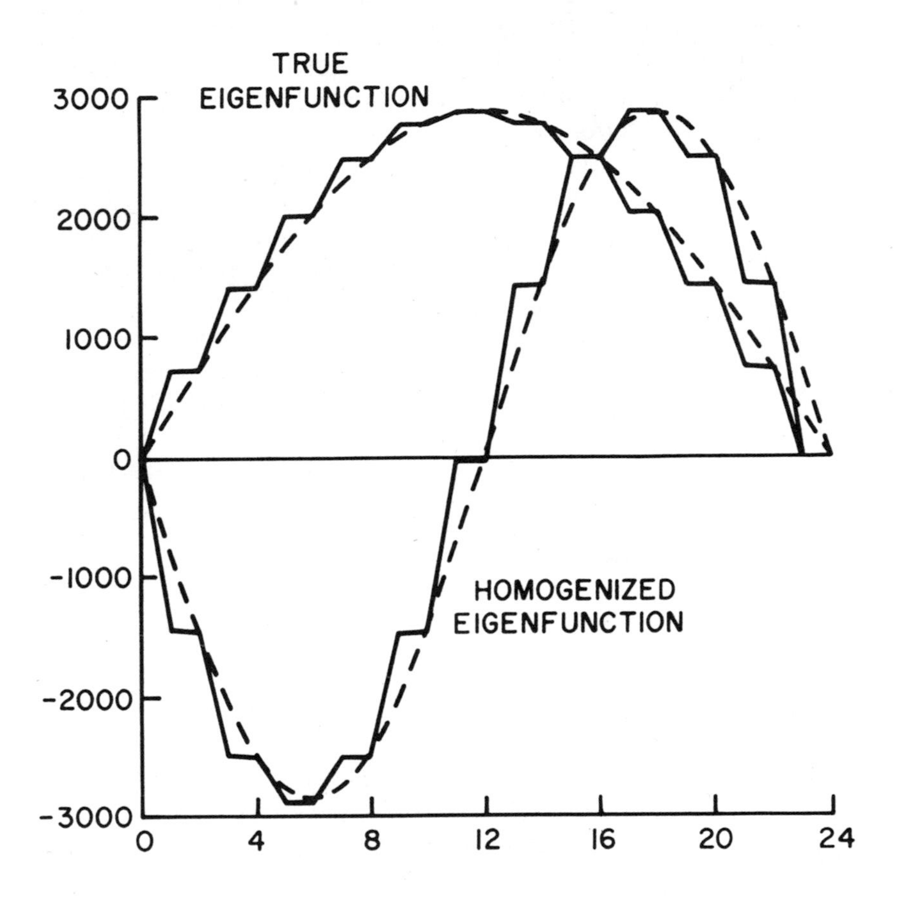

Figure 7.4

First two eigenfunctions for $H = \frac{1}{12}$ and of the homogenized problem

In our case, $|\sigma^H| \leq CH^2$ because u is symmetric and the term in $v^H - u$ of order H is antisymmetric with respect to the point $x = \frac{1}{2}$. This feature could also be seen on the relation between the eigenfunction of the homogenized and the original equation. In Figure 7.4, we see the eigenfunction for $\frac{2}{H} = 24$ and the eigenfunction for the homoggenized case (broken line).

2.2. The global solution for two dimensional problems.

In contrast to the one dimensional case, we will not present the proofs of the theorems stated. They are too complicated. We refer the reader to [34], [35], [39].

2.2.1. The formulation of the problem for the entire plane $\mathbb{R}^2$.

The function $a(x_1, x_2)$, $(x_1, x_2) = x \in \mathbb{R}^2$, will be called an H-periodic function if

$$a(x_1 + k_1 H,\ x_2 + k_2 H) = a(x_1, x_2) \tag{1.1}$$

for all integers k_1 and k_2. Given a function $a(x_1, x_2)$ which is 1-periodic, we set

$$a^H(x_1, x_2) = a(x_1/H,\ x_2/H)\ . \tag{1.2}$$

Suppose the measurable functions $a_{ij}(x_1, x_2)$ are given and that they are 1-periodic functions and that

$$|a_{i,j}(x_1, x_2)| < \beta < \infty \tag{1.3}$$

$$\sum_{i,j=1}^{2} a_{i,j}(x_1, x_2)\xi_i \xi_j \geq \alpha(\xi_1^2 + \xi_2^2), \quad \alpha > 0\ . \tag{1.4}$$

In addition, we will impose a regularity condition which will be explained later in this paragraph.

As usual, let $L_2(\mathbb{R}^2)$ be the space of square integrable functions on $\mathbb{R}^2$, i.e., let

$$\|u\|^2_{L_2(\mathbb{R}^2)} = \int_{\mathbb{R}^2} |u|^2 dx$$

where $dx = dx_1 dx_2$. In addition, let $L_{2,\alpha}(\mathbb{R}^2)$, $\alpha \geq 0$, be the weighted space with the norm

(1.6) $$\|u\|^2_{L_{2,\alpha}(\mathbb{R}^2)} = \int_{\mathbb{R}^2} (1+\|x\|^2)^\alpha |u|^2 dx$$

where we used $\|x\|^2 = x_1^2 + x_2^2$. Further, let us introduce the space $L^1(\mathbb{R}^2)$ as the space of classes modulo the constant functions, with the norm

(1.7) $$\|u\|^2_{L^1(\mathbb{R}^2)} = \int_{\mathbb{R}^2} \left(\sum_{i=1}^{2} \left| \frac{\partial u}{\partial x_i} \right|^2 \right) dx .$$

As usual, we define the space H^1 so that

(1.8) $$\|u\|^2_{H^1(\mathbb{R}^2)} = \|u\|^2_{L^1(\mathbb{R}^2)} + \|u\|^2_{L_2(\mathbb{R}^2)} .$$

We will also use the notation $H^0(\mathbb{R}^2) = L_2(\mathbb{R}^2)$. When the domain in question is a bounded domain Q, we will use $L_2(Q)$, $L^1(Q)$, $H^1(Q)$ in place of $L_2(\mathbb{R}^2)$, $L^1(\mathbb{R}^2)$, $H^1(\mathbb{R}^2)$, etc.

Let $f \in L_2(\mathbb{R}^2)$ with compact support ($\operatorname{supp} f = \Omega_0$) and

(1.9) $$\int_{\mathbb{R}^2} f dx = 0$$

be given. Then, using the Lax Milgram Lemma, we see that there is a unique $u_0 \in L^1(\mathbb{R}^2)$ such that

(1.10) $$\int_{\mathbb{R}^2} \left(\sum_{i,j=1}^{2} a_{i,j} \frac{\partial u_0}{\partial x_i} \frac{\partial v}{\partial x_j} \right) dx = \int_{\mathbb{R}^2} f\, v\, dx$$

for any $v \in L^1(\mathbb{R}^2)$, and we get

(1.11) $$\|u_0\|_{L^1(\mathbb{R}^2)} \leq C \|f\|_{L_2(\mathbb{R}^2)}$$

where C is independent of f but depends on Ω_0 (i.e. the support of f).

Now we are able to define the regularity condition which we need to impose on $a_{i,j}$. Let $u \in L^1(\mathbb{R}^2)$ be defined on Q_3 with $Q_\ell = \{x \mid |x_1| < \ell/2, /x_2| < \ell/2\}$ (ℓ integral), so that for any $v \in \mathcal{D}(Q_3)$ we have

$$(1.12) \qquad \int_{Q_3} \left(\sum_{i,j=1}^{2} a_{i,j} \frac{\partial u}{\partial x_i} \frac{\partial v}{\partial x_j} \right) dx = \int_{Q_3} f\, v\, dx$$

with $f \in L_2(Q_3)$. As usual, $\mathcal{D}(Q_3)$ is the space of all infinitely differentiable functions on Q_3 with compact support. Defining

$$(1.13) \qquad F(v) = \int_{Q_1} \left(\sum_{i,j=1} a_{i,j} \frac{\partial u}{\partial x_i} \frac{\partial v}{\partial x_j} \right) dx - \int_{Q_1} fv\, dx\,,$$

we easily see that $F(v_1) = F(v_2)$ when $v_1 = v_2$ on ∂Q_1 (the boundary of Q_1). We will call the coefficients $a_{i,j}$ <u>regular</u> if there exists a function ξ defined on ∂Q_1 so that

$$(1.14) \qquad F(v) = \oint_{\partial Q_1} \xi\, v\, ds$$

and

$$(1.15) \qquad \left(\oint_{\partial Q_1} |\xi|^p ds \right)^{1/p} \leq C[\, \|u\|_{L^1(Q_3)} + \|f\|_{L_2(Q_2)}]$$

with $p > 1$.

Let us make some remarks about this notion.

1) If the coefficients $a_{i,j}$ are smooth, then they are regular. This follows from the general theory of elleptic equations.

2) If the coefficients are piecewise smooth with interfaces which do not have turning points, then these coefficients are also regular. This conclusion follows from [40], [41], [42], [43].

Let the 1-periodic function $a_{i,j}$, $i,j = 1,2$ be given so that (1.3) and (1.4) are satisfied. In addition, we will assume that these coefficients are regular as explained above.

Given $a^H_{i,j}(x_1, x_2) = a(x_1/H, x_2/H)$, we will be interested in the weak solution u^H of the problem

$$-\sum_{i,j=1}^{2} \frac{\partial}{\partial x_j} a^H_{i,j} \frac{\partial u^H}{\partial x_i} = f \tag{1.16}$$

where f is smooth, has compact support and

$$\int_{\mathbb{R}^2} f\,dx = 0 . \tag{1.17}$$

$u^H \in L^1(\mathbb{R}^2)$ is called the weak solution of (1.16) if for every $v \in L^1(\mathbb{R}^2)$

$$\int_{\mathbb{R}^2} \left(\sum_{i,j=1}^{2} a^H_{i,j} \frac{\partial u^H}{\partial x_i} \frac{\partial v}{\partial x_j} \right) dx = \int_{\mathbb{R}^2} fv\,dx . \tag{1.18}$$

The weak solution u^H exists and is uniquely determined (in $L^1(\mathbb{R}^2)$). The question is how the solution behaves when $H \to 0$.

Further let $q(x_1, x_2)$ be a 1-periodic function with $0 \le q(x_1, x_2) \le \beta$, $\int_{Q_1} q(x_1, x_2) dx > 0$, and $q^H(x_1, x_2) = q(x_1/H, x_2/H)$.

We will also be interested in the solution of the problem

$$-\sum_{i,j=1}^{2} \frac{\partial}{\partial x_j} a^H_{i,j} \frac{\partial u^H}{\partial x_i} + q^H u^H = f . \tag{1.19}$$

$u^H \subset H^1(\mathbb{R}^2)$ is a weak solution of the problem given by (1.19) if for every $v \in H^1(\mathbb{R}^2)$

$$\int_{\mathbb{R}^2} \left(\sum_{i,j=1}^{2} a_{i,j} \frac{\partial u^H}{\partial x_j} \frac{\partial v}{\partial x_j} + q^H u\,v \right) dx = \int_{\mathbb{R}^2} f\,v\,dx; \tag{1.20}$$

in contrast to the problem with $q(x_1, x_2) = 0$, we need not assume (1.17).

2.2.2. The homogenization problem in $\mathbb{R}^2$.

We are interested here in the study of the behavior of u^H which satisfies (1.16), resp. (1.18). For the proofs of the theorems we refer the reader to [34], [35].

Theorem 2.1.

Let $u^H \in L^1(\mathbb{R}^2)$, $\int_{Q_1} u^H dx = 0$, and let it satisfy (1.18), with $f \in \mathcal{D}(\mathbb{R}^2)$ satisfying (1.17). Then, for any $\alpha > 1$,

$$\|u^H - U\|_{L_{2,\alpha}(\mathbb{R}^2)} \leq CH \tag{2.1}$$

where U is the solution of

$$-\sum_{i,j} \frac{\partial}{\partial x_j} b_{i,j} \frac{\partial U}{\partial x_i} = f, \tag{2.2}$$

[i.e., $U \in L^1(\mathbb{R}^2)$, $\int_{Q_1} U\, dx = 0$, and

$$\int_{\mathbb{R}^2} \left(\sum_{i,j=1}^{2} b_{i,j} \frac{\partial U}{\partial x_i} \frac{\partial v}{\partial x_j} \right) dx = \int_{\mathbb{R}^2} f\, v\, dx \tag{2.3}$$

for any $v \in L^1(\mathbb{R}^2)$] where $b_{i,j}$ are constants such that for any choice of (ξ_1, ξ_2)

$$\sum_{i,j=1}^{2} b_{i,j}\, \xi_i \xi_j \geq \gamma(\xi_1^2 + \xi_2^2), \quad \gamma > 0 \tag{2.4}$$

and C in (2.1) is independent of H (but depends on f and α).

The coefficients $b_{i,j}$ are computable. For a proof see [34]. Next we show how to compute $b_{i,j}$.

Theorem 2.2.

Let $\chi_\ell \in L^1(Q_1)$, $\ell = 1, 2$, be a 1-periodic function and such that

$$\int_{Q_1} \left(\sum_{i,j=1}^{2} a_{i,j} \frac{\partial \chi_\ell}{\partial x_i} \frac{\partial v}{\partial x_j} \right) dx = \int_{Q_1} \left(\sum_{j=1}^{2} a_{\ell,j} \frac{\partial v}{\partial x_j} \right) dx \tag{2.5}$$

for any 1-periodic function $v \in L^1(Q_1)$. Then

(2.6) $$b_{i,j} = B_{0,0}(u_j, \omega_i)$$

where

(2.7) $$B_{0,0}(u,v) = \int_{Q_1} \left(\sum_{i,j=1}^{2} a_{i,j} \frac{\partial u}{\partial x_i} \frac{\partial v}{\partial x_j} \right) dx$$

and

(2.8) $$u_j \equiv x_j - \chi_j \qquad j = 1,2$$

(2.9) $$\omega_i = \chi_i + \tfrac{1}{2} \qquad i = 1,2 .$$

We see that we need to find functions χ_j prior to computing the coefficients $b_{i,j}$. Knowing these functions χ_j, we may then construct a better approximation to the function u^H simply by using the function U.

For this purpose, define first

(2.10) $$Q^H_{k_1,k_2} = \{x_1, x_2 \mid |x_1 - k_1 H| < \tfrac{1}{2}H, |x_2 - k_2 H| < \tfrac{1}{2}H\} ,$$

k_1, k_2 integers. Obviously

$$\mathbb{R}^2 = \bigcup_{k_1,k_2} \bar{Q}^H_{k_1,k_2}$$

where $\bar{Q}^H_{k_1,k_2}$ is the closure of $Q^H_{k_1,k_2}$.

Now we define the function ${}^1W^H$ separately in every $Q^H_{k_1,k_2}$ with the help of U. We define in $Q^H_{k_1,k_2}$

(2.11) $$\begin{aligned} {}^1W^H(x_1,x_2) &= U(k_1H, k_2H) \\ &+ \sum_{i=1}^{2} \frac{\partial U}{\partial x_i}(k_1H, k_2H)\left[(x_i - k_iH) - H\chi_i\left(\frac{x_1-k_1H}{H}, \frac{x_2-k_2H}{H}\right)\right]. \end{aligned}$$

Then we have

Theorem 2.3.

Let u^H and U be the functions introduced in Theorems 2.1 and 2.2 and let ${}^1W^H$ be defined by (2.11). Then, for $\alpha > 1$,

$$[\sum_{k_1,k_2} \|u^H - {}^1W^H\|^2_{L^1(Q^H_{k_1,k_2})} + \|u^H - {}^1W^H\|^2_{L_{2,\alpha}(\mathbb{R}^2)}]^{\frac{1}{2}} \tag{2.12}$$

$$\leq CH$$

where C depends on f and α but is independent of H.

We stress that u^H does not converge in the space $L^1(\mathbb{R}^2)$.

Theorem 2.3 shows very well an analogy to the one dimensional case discussed in the previous section. The function ${}^1W^H$ is discontinuous on the boundary of $Q^H_{k_1,k_2}$. If we are interested in the behavior of u^H on these boundaries, then we may obviously shift the center into the desired location. Theorem 2.3 shows how to find the approximate solution with an error of order H. It is also possible to find the approximation with the error of order H^2.

Theorem 2.4.

Let u^H and U be the functions introduced in Theorems 2.1 and 2.2. Further let $V \in L^1(\mathbb{R}^2)$ be such that

$$-\sum_{i,j=1}^{2} \frac{\partial}{\partial x_j} b_{i,j} \frac{\partial V}{\partial x_i} = \Lambda U \tag{2.13}$$

where

$$\Lambda = \sum_{i+j=3} r_{i,j} \frac{\partial^3}{\partial x_1^i \partial x_2^j} \tag{2.14}$$

is a differential operator (homogenous) of third order with constant coefficients. There exist functions $u^{[1;i]}(x_1,x_2)$, $i=1,2$,

$u^{[2;i,j]}(x_1, x_2)$, $i \geq 0$, $j \geq 0$, $i+j = 2$, $|x_i| < \frac{1}{2}$, $i = 1, 2$, <u>such that, by defining</u> $W^H(x_1, x_2)$ <u>in</u> $Q^H_{k_1,k_2}$ <u>by</u>

$$W^H(x_1, x_2) = U(k_1H, k_2H)$$
$$+ H \sum_{i=1}^{2} \frac{\partial U}{\partial x_i}(k_1H, k_2H) u^{[1;i]} \left(\frac{x_1 - k_1H}{H}, \frac{x_2 - k_2H}{H}\right)$$
$$+ H^2 \sum_{\substack{i+j=2 \\ i \geq 0, j \geq 0}} \frac{\partial^2 U}{\partial x_1^i \partial x_2^j}(k_1H, k_2H) u^{[2;i,j]}\left(\frac{x_1 - k_1H}{H}, \frac{x_2 - k_2H}{H}\right)$$
$$- H\left[V(k_1H, k_2H) + H \sum_{i=1}^{2} \frac{\partial V}{\partial x_i}(k_1H, k_2H) u^{[1;i]}\left(\frac{x_1 - k_1H}{H}, \frac{x_2 - k_2H}{H}\right)\right]$$

<u>we get</u>

$$\left[\sum_{k_1,k_2} \|u^H - W^H\|^2_{L^1(Q^H_{k_1,k_2})} + \|u^H - W^H\|^2_{L_{2,\alpha}(\mathbb{R}^2)}\right] \leq CH^2 . \tag{2.16}$$

Let us make some comments about the theorem.

1) We have

$$u^{[1;i]} = x_i - \chi_i(x_1, x_2), \qquad i = 1, 2 \tag{2.17}$$

where the functions χ_i, $i = 1, 2$ were introduced in Theorem 2.2 and it was possible to determine them by the analysis of a single cell. The functions $u^{[2;i,j]}$ can also be determined by the analysis of a single cell.

2) The coefficients $b_{i,j}$ have been determined by functions $u^{[1;i]}$. Knowing the functions $u^{[2;i,j]}$, we may determine the coefficients $r_{i,j}$ in (2.14).

Let us now be interested in the solution of (1.19). The situation here is essentially similar to the previous case. Given the function $q(x_1, x_2)$, we compute the coefficient b_0 by

(2.18) $$b_0 = \int_{Q_1} q(x_1, x_2)dx > 0$$

and the homogenized equation (the analog to (2.2)) is

(2.19) $$-\sum_{i,j=1}^{2} \frac{\partial}{\partial x_j} b_{i,j} \frac{\partial U}{\partial x_i} + b_0 U = f$$

where $b_{i,j}$ are the same coefficients as before. Using this function, we may get an approximate solution in a similar manner as before. Let us state the analogs to the Theorems 2.1 and 2.3.

Theorem 2.5.

Let $u^H \in H^1(\mathbb{R}^2)$ be the weak solution of (1.19) [resp. (1.20)] for $f \in \mathcal{D}(\mathbb{R}^2)$. Further let U be a solution of (2.19); then, for any $\alpha > 1$,

$$\|u^H - U\|_{L_{2,\alpha}(\mathbb{R}^2)} \leq CH$$

where C does not depend on H, but depends on f.

Theorem 2.6.

Let u^H and U be the functions introduced in Theorem 2.5. Let ${}^1W^H$ be defined by (2.11) (with U the solution of (2.19)). Then (2.12) holds.

We see that for an error of order H we do not need to do any additional computation of the single cell (except for the computation of the coefficient b_0 - which is simple). A more complicated situation arises with the analog to Theorem 2.4. Here we have to analyze a single cell more and compute, in addition to $u^{[1;i]}$, $u^{[2;i,j]}$, some other functions $\pi^{[0]}$, $\nu^{[1;i]}$ and $\lambda^{[i]}$, $i = 1, 2$. The operator Λ_1 in (2.13) has to be changed to the operator $\Lambda_1 = \Lambda + \Lambda_0$ where Λ_0 is a first order differential operator with constant coefficients (which depend on the functions $\pi^{[0]}$, $\nu^{[1;i]}$, $\lambda^{[i]}$). We will not go into the details here but rather refer the reader to [35].

2.2.3. Some applications of the homogenization process.

Given the problem of finding the solution of the equation (1.16), we compute first the coefficients $b_{i,j}$ of the homogenized equation (2.2). Because of (2.4), the homogenized equation is elliptic. The coefficients $b_{i,j}$ have been defined in Theorem 2.2 by the expression (2.6). This expression may also be easily transformed by integration by parts to another form. Because of (2.5), we may write

$$B_{0,0}(u_j, v) = \oint_{\partial Q_1} \xi_j \, v \, dx$$

and so we get, e.g.,

$$b_{1,1} = \int_{-\frac{1}{2}}^{+\frac{1}{2}} \xi_1(\tfrac{1}{2}, x_2) dx_2 .$$

We emphasize that the differential equation

$$-\left(\frac{\partial}{\partial x_1} a^H \frac{\partial u^H}{\partial x_1} + \frac{\partial}{\partial x_2} a^H \frac{\partial u^H}{\partial x_2}\right) = f \tag{3.1}$$

(of isotropic form) may lead to the equation

$$-\left(b_{1,1} \frac{\partial^2 U}{\partial x_1^2} + (b_{1,2} + b_{2,1}) \frac{\partial^2 U}{\partial x_1 \partial x_2} + b_{2,2} \frac{\partial^2 U}{\partial x_2^2}\right) = f$$

where $b_{1,1} \neq b_{2,2}$ and $b_{1,2} + b_{2,1} \neq 0$; i.e., we have the anisotropic case.

To see that, let us give an example. Let

$$a(x_1, x_2) = p_1 > 0 \quad \text{for} \quad -\tfrac{1}{2} < x_1 < \tfrac{1}{2}, \quad 0 < x_2 < \tfrac{1}{2}$$

$$a(x_1, x_2) = p_2 > 0 \quad \text{for} \quad -\tfrac{1}{2} < x_1 < \tfrac{1}{2}, \quad -\tfrac{1}{2} < x_2 < 0 .$$

In this case, it is easy to find the functions χ_j. We have $\chi_1(x_1, x_2) = 0$. In fact, we have $(\ell = 1)$

$$\int_{Q_1} \Big(\sum_{j=1}^{2} a_{\ell,j} \frac{\partial v}{\partial x_j} \Big) dx = \int_{Q_1} a \frac{\partial v}{\partial x_1} dx = 0 \tag{3.2}$$

because v is assumed to be periodic. So we have

$$b_{1,1} = \int_{Q_1} a \, dx = \tfrac{1}{2}(p_1 + p_2) \tag{3.3}$$

$$b_{2,1} = 0 . \tag{3.4}$$

The following expression (3.5) defines the function $\chi_2(x_1, x_2)$:

$$\int_{Q_1} \Big(a \frac{\partial \chi_2}{\partial x_1} \frac{\partial v}{\partial x_1} + a \frac{\partial \chi_2}{\partial x_2} \frac{\partial v}{\partial x_2} \Big) dx = \int_{Q_1} a \frac{\partial v}{\partial x_2} dx \tag{3.5}$$

Put

$$\chi_2(x_1, x_2) = \begin{cases} (-x_2 + \frac{1}{2}) & \text{for} \quad 0 < x_2 < \frac{1}{2} , \\ (x_2 + \frac{1}{2}) & \text{for} \quad -\frac{1}{2} < x_2 < 0 . \end{cases} \tag{3.6}$$

This function is obviously periodic and belongs to $L^1(Q_1)$. Taking into consideration that v is periodic, we get

$$\begin{aligned} \int_{Q_1} \Big(a \frac{\partial \chi_2}{\partial x_1} \frac{\partial v}{\partial x_1} + a \frac{\partial \chi_2}{\partial x_2} \frac{\partial v}{\partial x_2} \Big) dx &= C(p_1 + p_2) \int_{-\frac{1}{2}}^{+\frac{1}{2}} v(x_1, 0) dx_1 \\ &\quad - C(p_1 + p_2) \int_{-\frac{1}{2}}^{\frac{1}{2}} v(x_1, \tfrac{1}{2}) dx_1 . \end{aligned} \tag{3.7}$$

On the other hand

$$\int_{Q_1} a \frac{\partial v}{\partial x_2}\, dx = (p_1 - p_2) \int_{-\frac{1}{2}}^{\frac{1}{2}} v(x_1, \tfrac{1}{2}) dx_1 \tag{3.8}$$
$$- (p_1 - p_2) \int_{-\frac{1}{2}}^{\frac{1}{2}} v(x_1, 0) dx_1 \,.$$

So, putting $C = (p_1 - p_2)/(p_1 + p_2)$, we satisfy (3. 5), and so we get

$$b_{2,2} = p_1 \left(1 - \frac{p_1 - p_2}{p_1 + p_2}\right) = p_1 \left(\frac{2p_2}{p_1 + p_2}\right) \tag{3.9}$$
$$= \left[\tfrac{1}{2}\left(\frac{1}{p_1} + \frac{1}{p_2}\right)\right]^{-1}$$

$$b_{1,2} = 0 \,. \tag{3.10}$$

The homogenized differential equation associated with

$$-\left(\frac{\partial}{\partial x_1} a^H \frac{\partial u}{\partial x_1} + \frac{\partial}{\partial x_2} a^H \frac{\partial u}{\partial x_2}\right) = f$$

with the coefficient, a, (defined by (3. 1)) is therefore

$$-\left(b_{1,1} \frac{\partial^2 U}{\partial x_1^2} + b_{2,2} \frac{\partial^2 U}{\partial x_2^2}\right) = f \tag{3.11}$$

where $b_{1,1}$ is given by (3. 3) (average of p_1 and p_2) and $b_{2,2}$ by (3. 9) (harmonic average of p_1 and p_2).

Next, we use the finite element method with linear elements, neglecting the presence of the interfaces. See Figure 3. 1. By simple computation, we get a scheme which solves the (isotropic) differential equation

$$\frac{p_1 + p_2}{2} \Delta U = f \,. \tag{3.12}$$

Comparing equation (3. 12) with (3. 11), we see that the finite element method gives results which are not correct. We see quite an analogy to the one dimensional case we

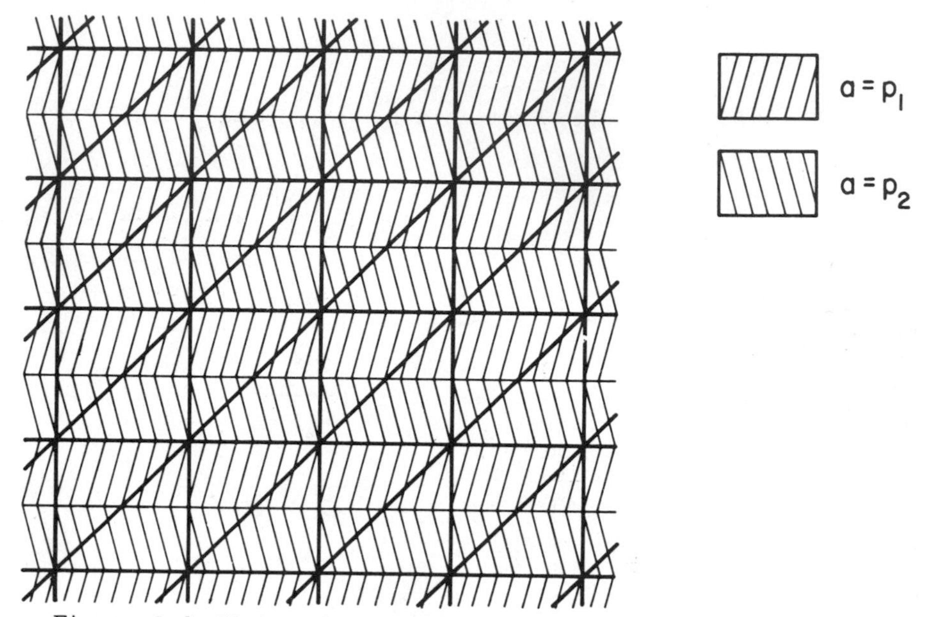

Figure 3.1 Finite element scheme with triangular elements

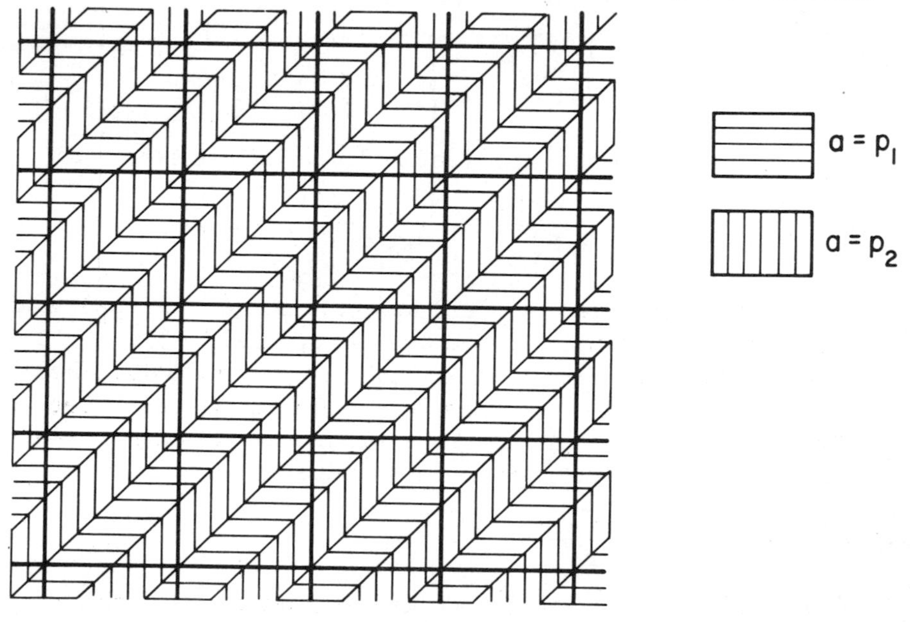

Figure 3.2 An example of the cell structure

discussed earlier.

In (3.11) we do not have mixed derivatives. A simple example shows that the mixed term may appear. For this, we turn once more to the equation (3.1), and consider the cells which are drawn in Figure 3.2. Rotating coordinates by $\frac{\pi}{2}$, we get the previous case. Because $b_{1,1} \neq b_{2,2}$, we obviously have a mixed term in the homogenized equation.

We present another example. The cell structure is shown in Figure 3.3. In this case, the homogenized

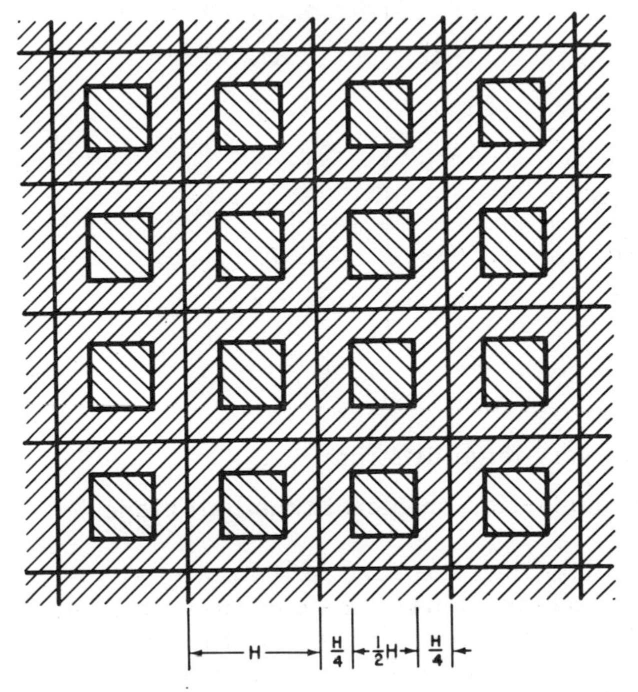

Figure 3.3 The cell structure

equation will have the form

$$b_{1,1} = b_{2,2} = b; \quad b_{1,2} = b_{2,1} = 0$$

$$b\,\Delta U = f\,.$$

The constant b cannot be computed analytically. It is necessary to compute χ_i numerically, e.g., by the finite element method. Table 3.1 shows the values of b for different Φ. Together with that, we show the average value $\bar{b}$ which will appear when finite elements (e.g. piecewise linear) are used, neglecting the presence of interfaces.

Φ	b	$\bar{b}$
1	1.	1
2	1.1822	1.25
4	1.3622	1.75
6	1.4522	2.25
10	1.5434	3.25
100	1.7084	25.75
10,000	1.7308	2500.75

Table 3.1.

We see once more very large differences.

2.2.4. The homogenization problem in the half plane.

So far we discussed the case of the entire space, i.e., the case without boundary. The question arises as to how the boundary influences the results of the homogenization. The existence of the boundary injects a boundary layer behavior.

We examine equations (1.16), resp. (1.18), on $\mathbb{R}^+_H = \{(x_1, x_2) \mid x_2 > 0\}$ with the boundary condition $u = 0$. In this case we are interested in finding $u^H \in L^1_0(\mathbb{R}^+)$ where the lower subscript zero indicates that u^H has zero trace on the boundary of $\mathbb{R}^+$ and we request that (1.18) hold for any $v \in L^1_0(\mathbb{R}^+)$. As the analog to theorems 2.1 and 2.3 we get

Theorem 4.1.

Let $u^H \in L_0^1(\mathbb{R}^+)$ *satisfy* (1.18) *with* $f \in \mathcal{D}(\mathbb{R}^2)$ (*for any* $v \in L_0^1(\mathbb{R}^+)$). *Further, let* U *be the solution of* (2.2) *with boundary condition* $U(x_1, 0) = 0$. *Then*

$$(4.1) \qquad \|u^H - U\|_{L_{2,\alpha}(\mathbb{R}^+)} \leq CH .$$

Defining ${}^1W^H$ *by* (2.11), *we get*

$$(4.2) \qquad \Big[\sum_{k_1, k_2} \|u^H - {}^1W^H\|^2_{L^1(Q^H_{k_1,k_2} \cap \mathbb{R}^+(z))} + \|u^H - U\|^2_{L_{2,\alpha}(\mathbb{R}^+)} \Big]^{\frac{1}{2}} \leq C[H^{\frac{1}{2}} e^{-\frac{\beta}{H} z} + H]$$

where $\mathbb{R}^+(z) = \{(x_1, x_2) \mid x_2 > z > 0\}$ *and* $\beta > 0$ ($\beta > 0$ *independent of* H). *The term* $H^{\frac{1}{2}}$ *cannot be removed*.

There is also an analog to Theorem 2.4. where the boundary layer term remains but the term H is replaced by H^2. The function V continues to be the solution of (2.13) but with the non-homogenous boundary condition

$$V(x_1, 0) = \rho \frac{\partial U}{\partial x_2}(x_1, 0)$$

where ρ is a constant which has to be computed. For more about this we refer the reader to [34].

So far, we have been interested in the solution of (1.16). Analogous results could be obtained for equation (1.19). We refer the reader to [35].

Let us remark that for a more general domain (i.e. not $\mathbb{R}^+$) the behavior of the solution near to the boundary (boundary layer) is not clear. It is possible to get estimates of the types appearing in (4.1) and (4.2) where the term $e^{-\frac{\beta}{H} z}$ is replaced by 1, i.e., the error is of order $H^{\frac{1}{2}}$.

2.2.5. The solution of the problem with grained boundary.

Let us describe first a model problem. Let

$$K^H_{k_1,k_2} = \{(x_1,x_2)\,|\,(x_1-k_1H)^2 + (x_2-k_2H)^2 \leq \rho^2H^2\}$$

with $\rho < \frac{1}{2}$, k_1, k_2 integral. Let

$$\Omega_H = \mathbb{R}^2 \setminus \bigcup_{k_1,k_2} K^H_{k_1,k_2}$$

(see Figure 5.1) and consider the problem of finding u^H so that

$$-\Delta u^H = f \quad \text{in} \quad \Omega_H \tag{5.1}$$

with boundary condition

$$u^H = 0 \qquad \text{on} \qquad \partial\Omega_H \,. \tag{5.2}$$

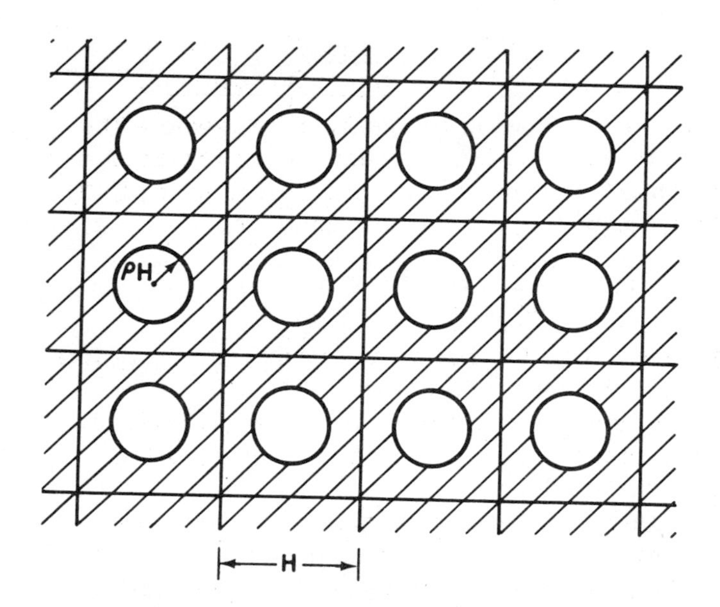

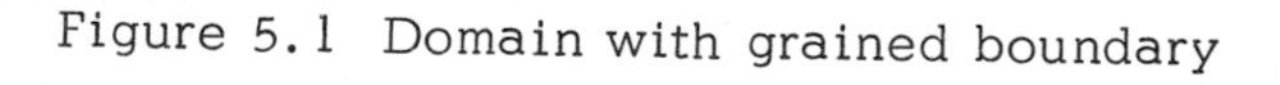

Figure 5.1 Domain with grained boundary

Assume that the derivatives of f are bounded. Then the following theorems hold. See [39].

Theorem 5.1.

Let $f \in \mathcal{D}(\mathbb{R}^2)$, *and let* u^H *be the solution of* (5.1) *and* (5.2). *Then*

$$\|u^H\|_{L_2(\Omega_H)} \leq CH^2 \tag{5.3}$$

$$\|u^H\|_{H^1(\Omega_H)} \leq CH . \tag{5.4}$$

Theorem 5.2.

Let $f \in \mathcal{D}(\mathbb{R}^2)$, *and let* u^H *be the solution of* (5.1) *and* (5.2), *as in Theorem* 2.1. *There exist functions* $u^{[0]}$, $u^{[1;i]}(x_1, x_2)$, $|x_i| < \frac{1}{2}$, $i = 1, 2$ *such that, with*

$${}^1W^H(x_1, x_2) = H^2 f(k_1 H, k_2 H) u^{[0]}\left(\frac{x_1 - k_1 H}{H}, \frac{x_2 - k_1 H}{H}\right) \tag{5.5}$$

and

$$\begin{aligned} W^H(x_1, x_2) &= H^2 f(k_1 H, k_2 H) u^{[0]}\left(\frac{x_1 - k_1 H}{H}, \frac{x_2 - k_1 H}{H}\right) \\ &+ H^3 \sum_{\ell=1}^{2} \frac{\partial f}{\partial x_1}(k_1 H, k_2 H) u^{[1;i]}\left(\frac{x_1 - k_1 H}{H}, \frac{x_2 - k_2 H}{H}\right) \end{aligned} \tag{5.6}$$

on $Q^H_{k_1, k_2}$, *we get*

$$\|u^H - {}^1W^H\|_{L_2(\Omega_H)} \leq CH^3 \tag{5.7}$$

$$\|u^H - W^H\|_{L_2(\Omega_H)} \leq CH^4 \tag{5.8}$$

(5.9)
$$\Big[\sum_{k_1,k_2} \|u^H - {}^1W^H\|^2_{H^1(Q^H_{k_1,k_2} \cap \Omega_H)}\Big]^{\frac{1}{2}} \le CH^2$$

(5.10)
$$\Big[\sum_{k_1,k_2} \|u^H - W^H\|^2_{H^1(Q^H_{k_1,k_2} \cap \Omega_H)}\Big]^{\frac{1}{2}} \le CH^3.$$

The functions $u^{[0]}$ _and_ $u^{[1;i]}$ _are computable by an analysis of a single cell._

Comparing this result with the previous one, we see a different situation. We do not have to find the function U by solving a homogenized equation.

Although we only formulated the theorem for the Laplace operator, the theorem holds for general operators. The situation here is also much simpler with respect to the boundary layer behavior. Let Ω be a bounded domain with smooth boundary. Define $S_H = \Omega_H \cap \Omega$ where Ω_H has been defined above. Further, for $z > 0$, let

$$\Omega^z = \{x \in \Omega \quad \mathrm{dist}(x, \partial\Omega) > z\}$$

and

$$S^z_H = \Omega^z \cap \Omega_H .$$

Now we turn our attention to the solution of the differential equation

(5.11)
$$-\Delta u^H = f \quad \text{in} \quad S_H$$

with the boundary condition

(5.12)
$$u^H = 0 \qquad \text{on } \partial S_H .$$

Then we may prove the following theorem.

__Theorem 5.3.__

Let $u^H, W^H, {}^1W^H$ _be defined as in Theorem_ 5.2. _Then_

$$(5.13)\qquad \Big[\sum_{k_1,k_2} \|u^H - {}^1W^H\|^2_{H^1(Q^H_{k_1,k_2} \cap S^z_H)}\Big]^{\frac{1}{2}} \le C[H^{\frac{3}{2}} e^{-\frac{\beta}{H}z} + H^2]$$

$$(5.14)\qquad \Big[\sum_{k_1,k_2} \|u^H - W^H\|_{H^1(Q^H_{k_1,k_2} \cap S^z_H)}\Big]^{\frac{1}{2}} \le C[H^{\frac{3}{2}} e^{-\frac{\beta}{H}z} + H^3]$$

with $\beta > 0$ (β independent of H).

Chapter 3: The Problem of the Singularities - The Local Problem.

3.1. The Problem of non-uniform convergence.

Let us introduce a model problem. Given $\Omega \subset \mathbb{R}_n$ with smooth boundary, we seek the weak solution of the problem

$$(1.1)\qquad -\Delta u + u = f$$

$$(1.2)\qquad \frac{\partial u}{\partial n} = 0$$

with $f \in H^r(\Omega)$, where $H^r(\Omega)$ is the usual Sobolov space.

The finite element method consists of the selection of a linear one-parameter (t,k)-system $S_h(t,k) \subset H^k(\Omega)$, $t > k$. A linear system $S_h \subset H^k(\Omega)$ is called a (t,k)-system when for any given $w \in H^s(\Omega)$ and $0 \le \ell \le k$, $\ell \le s$, there exists $\omega \in S_h(t,k)$ such that

$$(1.3)\qquad \|w-\omega\|_{H^\ell} \le C(s,\ell)h^\mu \|w\|_{H^s}$$

where

(1.4) $$\mu = \min(s-\ell, t-\ell)$$

and C is independent of w and h. For more about these systems see [1], chapter 4.

The function $u_k \in S_h(t,k)$, $k \geq 1$, will be called the finite element solution when

(1.5) $$B(u_h, v) = \int_\Omega fv$$

for any $v \in S_h(t,k)$ with

(1.6) $$B(u,v) = \int_\Omega \left(\sum_{i=1}^{n} \frac{\partial u}{\partial x_i} \frac{\partial v}{\partial x_i} + u\,v \right) dx .$$

The characteristic theorem concerning the error of u_k is

Theorem 1.1.

Let $f \in H^r(\Omega)$, $r \geq 0$, be the weak solution of (1.1) and (1.2). Then

(1.7) $$\|u_h - u\|_{H^1(\Omega)} \leq Ch^\mu \|f\|_{H^r}$$

with

(1.8) $$\mu = \min(r+1, t-1)$$

and C is independent of f and h (but depends on r).

Theorem 1.2.

The rate of convergence in (1.7) with coefficient μ given in (1.8) is best possible, i.e., there exists a subsequence $f_h \in H^r$ such that

(1.9) $$\|u_h - u\|_{H^1(\Omega)} \geq Ch^\mu \|f_h\|_{H^r(\Omega)} .$$

For the proofs of these theorems see [1], chapter 6, section 6.4.

It is essential here that C in (1.7) and (1.9) is independent of f, i.e., we choose in (1.9) function f_h depending on h. A natural question arises: What can one say about $\|u-u_h\|_{H^1(\Omega)}$ when f is fixed. This is answered in the following

Theorem 1.3.

Let u be the (weak) solution of (1.1) and (1.2) with $f \in H^r(\Omega)$, $r \geq 0$. Further let $u_h \in S_h(t,k)$, $k \geq 1$, be the finite element solution. Then, for $t-1 > r+1$,

$$\|u_h - u\|_{H^1(\Omega)} \leq O(h)h^{r+1}\|f\|_{H^r(\Omega)} \tag{1.10}$$

with $O(h) \to 0$ for $h \to 0$, and

$$\int_0^1 h^{-1}O^2(h)dh < \infty . \tag{1.11}$$

There exists a (t,k)-system $S_h(t,k)$ and $f \in H^r$ such that, for $t-1 \leq r+1$,

$$\|u_h - u\|_{H^1(\Omega)} \geq Ch^{r+1}\|f\|_{H^r(\Omega)} . \tag{1.12}$$

For the proof see [58].

3.2. Problems with local singularities.

We turn once more to the problem described by (1.1) and (1.2). Let us assume this time that the domain is L-shaped with the 270° angle at the origin. Let us assume that the system $S_h(t,k)$ is created by piecewise linear triangular elements with constant sides h. We have in this case $t=2$, $k=1$. Then the following theorem holds.

Theorem 2.1.

There exists an $f \in H^r$, $r > 0$, so that

$$\|u_h - u\|_{H^1(\Omega)} \geq Ch^{2/3}, \tag{2.1}$$

and, for any $f \in H^0$,

$$\|u_h - u\|_{H^1(\Omega)} \leq Ch^{2/3} \|f\|_{H^0(\Omega)} \tag{2.2}$$

where u is the solution of (1.1) and (1.2) on an L-shaped domain and u_h is the finite element solution using the system $S_h(t,k)$ of piecewise linear triangular elements mentioned earlier.

The proof of this theorem follows the lines of [52].

In Theorem 2.1 it is essential that the sides of the elements are equal. In that case, the number N of unknown parameters satisfies the inequality $C_1 h^{-2} \leq N \leq C_2 h^{-2}$. So we may rewrite (2.1) so that

$$\|u_h - u\|_{H^1(\Omega)} \leq C N^{-\frac{1}{3}} \|f\|_{H^0(\Omega)}, \tag{2.3}$$

resp.,

$$\|u_h - u\|_{H^1(\Omega)} \geq C N^{-\frac{1}{3}} \|f\|_{H^r(\Omega)}. \tag{2.4}$$

If the domain Ω is smooth then we may apply Theorem 1.1. to get

$$\|u_h - u\|_{H^1(\Omega)} \leq C N^{-\frac{1}{2}} \|f\|_{H^0(\Omega)}. \tag{2.5}$$

The next theorem answers the question as to whether a non-regular triangulation may lead to a better estimate than (2.5).

Theorem 2.2.

There exists a triangulation so that

$$\|u_h - u\|_{H^1(\Omega)} \leq C(\varepsilon)\, N^{\varepsilon - \frac{1}{2}} \|f\|_{H^0(\Omega)} \tag{2.6}$$

for arbitrary $\varepsilon > 0$.

For a proof, see [55]. (Let us remark that the term N^{ε} may be replaced by a logarithmic term.)

Let us analyze another model problem, namely, the solution of the problem

$$\frac{\partial}{\partial x_1} a \frac{\partial u}{\partial x_1} + \frac{\partial}{\partial x_2} a \frac{\partial u}{\partial x_2} = f \tag{2.7}$$

with the boundary condition

$$u = 0 \tag{2.8}$$

on the domain $\Omega = \{(x_1, x_2) \,|\, |x_1| < \frac{1}{2}, |x_2| < \frac{1}{2}\}$. Assume that a takes on constant values different in each of four quadrants of Ω. Let $a = a_1$ for $0 \leq x_1 < \frac{1}{2}$, $0 \leq x_2 < \frac{1}{2}$; $a = a_2$ for $-\frac{1}{2} < x_1 < 0$, $0 \leq x_2 < \frac{1}{2}$; $a = a_3$ for $-\frac{1}{2} < x_1 < 0$, $-\frac{1}{2} < x_2 < 0$, and $a = a_4$ for $0 \leq x_1 < \frac{1}{2}$, $-\frac{1}{2} < x_2 < 0$; $a_i > 0$, $i = 1, 2, 3, 4$. If we use the finite element method with piecewise triangular elements on a uniform mesh, then, for $a_1 = a_3$, $a_2 = a_4$ and $a_1/a_4 = \nu$ there exists a smooth f so that

$$C_1 h^{\mu} \leq \|u_h - u\|_{H^1(\Omega)} \tag{2.9}$$

and μ in dependence of ν is given in Table 2.1.

ν	1.37	2.66	5.85	17.3	161	16000
μ	0.9	0.7	0.5	0.3	0.1	0.01

Table 2.1

The proof is analogous to [52] using the fact that, around the origin, the solution u may be written in the form

$$(2.10) \qquad u = C_1 r^{\mu} \psi(\theta) + w$$

where $w \in H^2(\Omega)$ and (r, θ) represent polar coordinates. Expression (2.9) is valid in the C-norm also.

We next assume that $1 = a_2 = a_3 = a_4$, $a_1 = \Phi$. Then Table 2.2 shows the coefficient μ in this case.

Φ	2	4	6	10	100
μ	0.8934	0.8060	0.7675	0.7317	0.6739

Table 2.2

In practical computations, we cannot go to too small an h. We introduce an <u>effective rate of convergence</u> $\xi(h)$ for a given h by

$$\|u_h - u\|_C \approx C h^{\xi(h)} .$$

We computed numerically this effective rate $\xi(h)$ for the problem described by equations (2.7) and (2.8) in dependence on the coefficient Φ. The method was based on the analysis of the solution for multiples of h, namely, h, $2h$, $4h$. We have

$$\lim_{h \to 0} \xi(h) = \xi(0) = \mu .$$

Figure 2.1 shows this effective rate of convergence.

We clearly see the behavior of this effective rate. For Φ not too large and h not too small, the effective rate is much larger than $\xi(0)$. For Φ very large, we get $\xi(h) = \mu$ also for h not too small.

We remark that for $\Phi \to \infty$ the solution of (2.7) and (2.8) converges to the solution on an L-shaped domain. It is well known that the experimentally observed rate of

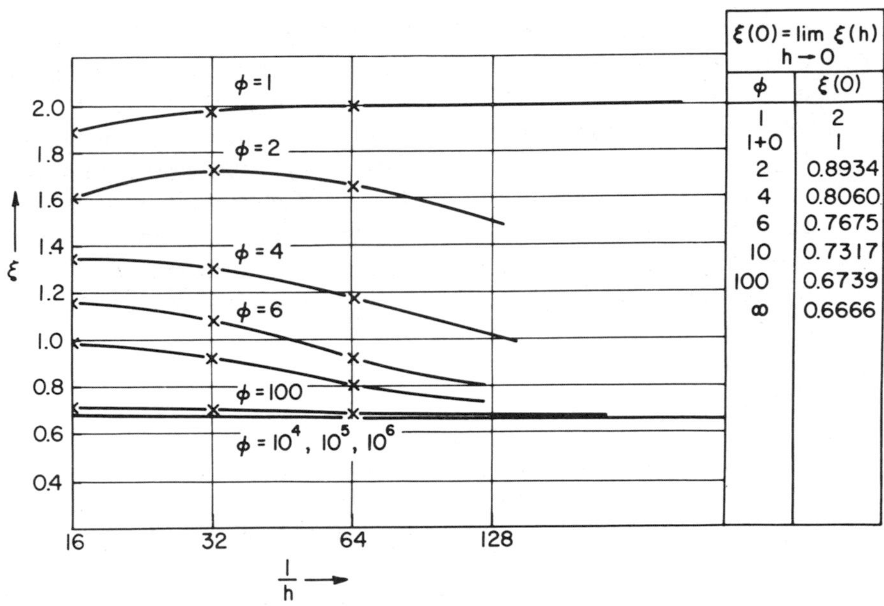

$\xi(0) = \lim_{h \to 0} \xi(h)$	
ϕ	$\xi(0)$
1	2
1+0	1
2	0.8934
4	0.8060
6	0.7675
10	0.7317
100	0.6739
∞	0.6666

Figure 2.1 The effective rate of the convergence

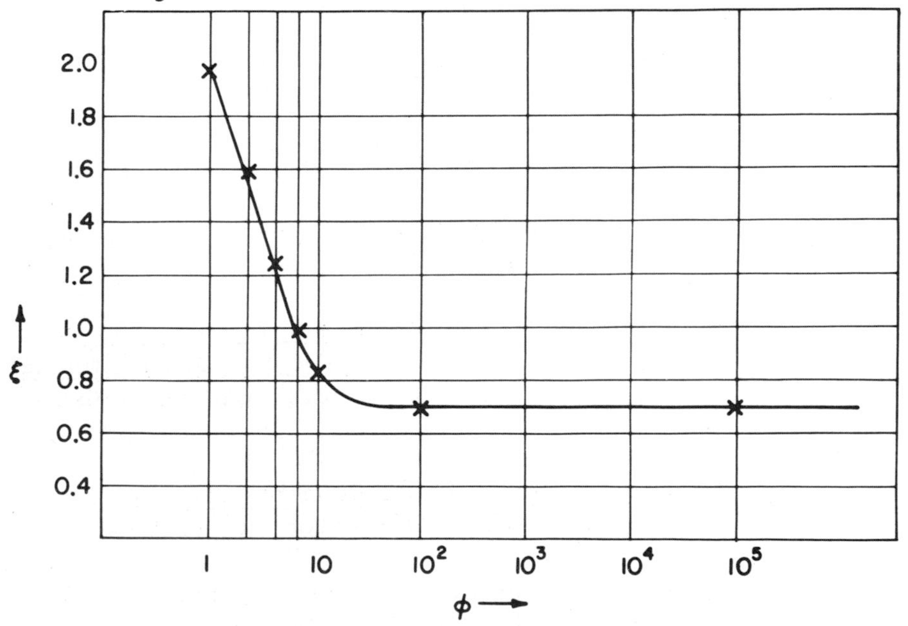

Figure 2. 2 The effective rate of the convergence for 1% accuracy

convergence in the case of an L-shaped domain is low (see e. g. [49], [50]). On the other hand, it was observed (see e. g. [46], [47]) that for Φ not too large the rate of convergence is approximately the same as in the case $\Phi = 1$ (i. e. in the case without interfaces). We see that these results are related to the size of h (which is determined by the required accuracy). Let us therefore plot in Figure 2. 2. the effective rate of convergence for 1% accuracy of the approximate solution. We see clearly what we explained above.

References.

1. I. Babuška, K. Aziz, Survey Lectures on the mathematical foundations of the finite element method, The Mathematical Foundation of the Finite Element Method with Applications to Partial Differential Equations, A. K. Aziz, Ed., Academic Press, 1972, pp. 3-363.

2. G. Strang, G. Fix, An Analysis of the Finite Element Method, Academic Press, 1973.

3. M. H. Schultz, Spline Analysis, Prentice-Hall, 1973.

4. J. T. Oden, Finite Elements of Nonlinear Continua, McGraw-Hill, 1972.

5. The Mathematical Foundations of the Finite Element Method with Applications to Partial Differential Equations, A. K. Aziz, Ed., Academic Press, 1973.

6. The Mathematics of Finite Elements and Applications, J. R. Whiteman, Ed., Academic Press, 1973.

7. Structural Mechanics Computer Programs, Ed. W. Pilkey, K. Saczalski, H. Schaeffer, University Press of Virginia, Charlottesville, 1974.

8. Finite Elemente in der Statik, Wilhelm Ernst & Sohn, Berlin, 1973.

9. O. C. Zienkiewicz, The Finite Element Method in Engineering Sciences, McGraw-Hill, 1971.

10. Naval Reactor Physics Handbook, Vol. 1, A. Radkowsky, Ed., U. S. Government Printing Office, 1964.

11. Ya V. Shevelev, Neutron diffusion in a one-dimensional uranium-water lattice, J. Nuclear Energy 6 (1957), pp. 132-141.

12. B. Davison, Effective thermal diffusion length in a sandwich reactor, J. Nuclear Energy, 7 (1958), pp. 51-68.

13. S. Ukai, A study on approximate methods for calculating the neutron flux in a heterogenous reactor, J. of Nuclear Energy, 24 (1970), pp. 479-491.

14. L. Trlifaj, A variational method for homogenization of a heterogenous medium, J. Nuclear Energy 6 (1957), pp. 142-154.

15. I. Babuška, R. B. Kellogg, Mathematical and computational problems in reactor calculations, Proceedings of the 1973 Conference, Mathematical Models and Computational Techniques for Analysis of Nuclear Systems, Ann Arbor, Michigan, April 9-11, 1973. Univ. of Michigan USAEC Tech. Inf. Center, pp. VII-67 - VII-94.

16. Dynamics of composite materials, (Proceedings of the 1972 Joint National and Western Applied Mechanics Conference, June 26-28, 1972, La Jolla, California), The American Soc. of Mech. Eng. 1972.

17. J. D. Achenbach, C. T. Sun, The directional reinforced composite as a homogeneous continuum with microstructure, Dynamics of Composite Materials, The Am. Soc. of Mech. Eng. 1972, pp. 48-70.

18. Zvi Hashin, Theory of composite materials, Mechanics of Composite Materials, Proceedings of the Fifth Symposium on Naval Structural Mechanics, Philadelphia, May 8-10, 1967, Pergamon Press, 1970, pp. 201-293.

19. S. M. Rytov, Acoustical properties on thinly laminated medium, Sov. Phys. Acoustic, 2 (1956), pp. 68-80.

20. C. T. Sun, J. D. Achenbach, G. Herrmann, Continuum theory for a laminated medium, J. Appl. Mech. 35 (1968), pp. 467-475.

21. J. C. Achenbach, G. Herrmann, Dispersion of free harmonic waves in fiber reinforced composites, A.I.A.A.J. 6 (1968), pp. 1832-1836.

22. M. Ben-Amoz, A continuum approach to the theory of elastic waves in heterogenous materials, Int. J. Eng. Sci. 6 (1968), pp. 209-218.

23. J. C. Peck, Pulse attenuation in composites, Shock Waves and the Mechanical Properties of Solids, (Proceeding of the 17th Sagamore Army Materials Research Conference, Ragnette Lake, New York, September 1970), J. J. Burke and V. Weiss, Eds., Syracuse Univ. Press, 1971, pp. 155-184.

24. J. C. Peck, Stress-wave propagation and fracture in composites, Dynamics of Composite Materials, The Am. Soc. of Mech. Engr. 1972 pp. 8-35.

25. E. H. Lee, A Survey of Variational Methods for Elastic Propagation Analysis in Composites with Periodic Structures, Dynamics of Composite Materials, The Am. Soc. of Mech. Eng. 1972, pp. 122-139.

26. L. A. Rubenfeld, J. B. Keller, Bounds on elastic moduli of composite media, SIAM J. Appl. Math. 17 (1969), pp. 495-510.

27. J. E. Flaherty, J. B. Keller, Elastic behavior of of composite media, Comm. Pure Appl. Math. 26 (1973), pp. 565-580.

28. I. Babuška, J. Evans, Flow of electrical current in the brain. To appear.

29. A. C. Eringen, E. S. Suhubi, Nonlinear theory of simple microelastic solids, Int. J. Eng. Sci. 2, (1969), pp. 189-203.

30. A. C. Eringen, D. G. B. Edelen, On nonlocal elasticity, Int. J. Eng. Sci. 10, (1972), pp. 233-248.

31. S. Spagnolo, Sul limite dille solutioni di problemi di Cauchy relativi all equazioni de colores, Ann. Scuolo Norm. Sup. Pisa, Ser. 3 , 21 (1967), pp. 657-699.

32. S. Spagnolo, Sulla convergenza di sulusioni di equazioni paraboliche ed ellipticke, Ann. Scuola Norm. Sup. Pisa, Ser. 3, 22(1968), pp. 571-597.

33. P. K. Senatorov, The stability of the solution of an ordinary differential equation of second order and of a parabolic equation with respect to the coefficients of the equation, Differencialnie Uravnenija 7 (1971), pp. 754-758.

34. I. Babuška, Solution of interface problems by homogenization I, Tech Note BN-782, January 1974, Institute for Fluid Dynamics and Applied Mathematics, University of Maryland.

35. I. Babuška, Solution of interface problems by homogenization II, Tech Note BN-787, March 1974, Institute for Fluid Dynamics and Applied Mathematics, University of Maryland.

36. I. Babuška, Numerical solution of partial differential equations, ZAMM. 54 (1974), p. T3

37. E. Ja Chruslov, Method of the orthogonal projection and Dirichlet boundary value problem in a grain domain, Mat. Sb. 88. (1972), pp. 38-60.

38. V. A. Marcenko, G. V. Suzikov, The second boundary value problem in domains with complicated boundary, Mat. Sb. 65, (1964), pp. 458-472, and Mat. Sb. 69, (1966), pp. 35-60.

39. I. Babuška, Numerical solution of problems with grained boundary, to appear.

40. R. B. Kellog, On the Poisson equation with intersecting interfaces, to appear in Applicable Analysis.

41. V. Z. Kondratjev, Boundary problems for elliptic equations with conical or angular points, Trans. Moscow Math Soc. 16 (1969), pp. 297-313.

42. P. Grisvard, Probleme de Dirichlet dans une cone, Ricerche Mat 20 (1971), pp. 175-192.

43. R. B. Kellogg, Higher order singularities for interface problems, The Math. Foundations of the Finite Element Method with Applications, A. K. Aziz, Ed., Academic Press, New York, 1972, pp. 589-602.

44. Mechanics of Fracture Methods of Analysis and Solution of Crack Problems, G. C. Sih, Ed., Noordhoff Int. Publ., Leyden, 1972.

45. P. D. Hilton, G. C. Sih, Applications of the finite element method to the calculations of stress intensity factors, Mechanics of Fracture Methods of Analysis and Solution of Crack Problems, G. C. Sih, Ed., Noordhoff Int. Publ., Leyden, 1972, pp. 426-483.

46. A. J. Lindenman, G. D. Leaf, H. G. Kaper, A computational analysis and evaluation of the finite element method for a class of nuclear reactor configurations. To appear in Computer Methods in Applied Mechanics and Engineering.

47. J. P. Hennart, Comparison of extrapolation techniques with high order finite element methods for diffusion equations with piecewise continuous material properties, Proceedings of the 1973 Conference, Mathematical Models and Computational Techniques for Analysis of Nuclear Systems, Ann Arbor, Michigan, April 9-11, 1973, pp. VII-95 - VII-119.

48. I. Babuška, R. B. Kellogg, Numerical solution of the neutron diffusion equation in the presence of corners and interfaces, Numerical Reactor Calculations, International Atomic Energy Agency, Vienna, 1972, pp. 473-485.

49. P. Laasonen, On the degree of convergence of discrete approximation for the solution of the Dirichlet problem, Ann. Acad. Sci. Finn. Ser. AI, 246 (1957), 19 pp.

50. P. Laasonen, On the truncation error of discrete approximations to solutions of Direchlet problems in domain with corners, Journal Assoc. Comp. Math. 1 (1958), pp. 32-48.

51. E. A. Volkov, Method of composite meshes for bounded and unbounded domains with piecewise smooth boundary (in Russian), Proc. Steklov Inst. Math 96 (1968), pp. 117-148.

52. I. Babuška, M. R. Rosenzweig, A finite element scheme for domains with corners, Num. Math. 20 (1972), pp. 1-21.

53. G. Fix, S. Gulati, G. I. Wakoff, On the use of singular functions with the finite element method, J. Comp. Physics.

54. L. Fox, P. Henrici, C. Moler, Approximation and bounds for eigenvalues of elliptic operators, SIAM J. Numer. Anal. 4 (1967), pp. 89-102.

55. I. Babuška, Finite element method for domains with corners, Computing 6 (1970), pp. 264-273.

56. R. S. Varga, Functional Analysis and Approximation Theory in Numerical Analysis, Conference Board of the Math. Sci. No. 3, 1971, SIAM.

57. J. Osborn, Spectral approximation for compact operators, to appear.

58. I. Babuška, R. B. Kellogg, Non-uniform error estimates for the finite element method, Tech. note BN-790 Institute for Fluid Dyn. and Appl. Math., University of Maryland, April, 1974.

59. J. L. Lions, E. Magenes, Problèmes aus limites non homogènes et applications, Vol. 1, Dunod, Paris, 1968.

Research supported in part by the Atomic Energy Commission under Grant #AEC AT(40-1) 3443. Computer time for this project was supported in part through the facilities of the Computer Science Center of the University of Maryland.

Institute for Fluid Dynamics
and Applied Mathematics
University of Maryland
College Park, Maryland 20742

The Construction and Comparison of Finite Difference Analogs of Some Finite Element Schemes

BLAIR K. SWARTZ

0. Summary.

Three subjects are considered in this paper. First, the notion of the resolving power of approximation methods, i. e. the number of intervals (or function values) per wavelength necessary to attain a preassigned error when approximating a given frequency, permits the evaluation of the relative efficiency of three types of spatial differencing used in setting up differential-difference equations for the 1-periodic parabolic problem $u_t = u_{xx}$. The three types of high-order spatial schemes considered are (explicit) centered differencing, the super-convergent smooth spline-Galerkin schemes discovered by Thomée, and the very high-order implicit schemes (Mehrstellenverfahren) which generalize Numerov's method. Seven time discretizations are introduced, namely Euler, backward-Euler, duFort-Frankel, 2nd order explicit, trapezoidal, Calahan-Zlámal, and 4th order Padé. The computational work necessary to solve each full discretization is minimized, for given error requirements, by balancing the number of intervals per wavelength against the number of time intervals per e^{th}-life. The resulting data is used to compare the relative efficiency of these finite element and finite difference methods.

Secondly, some corresponding results for the hyperbolic problem $u_t = u_x$ are briefly reviewed.

Finally, as Numerov-trapezoidal differencing turns

out to be almost optimal for the heat equation, a tridiagonal implicit difference formula which extends Numerov differencing to the general 2nd order linear differential operator in one space dimension is presented. The technique used in deriving this scheme inspires certain difference analogs of some finite element schemes. It also leads to a curious modification of the diamond difference scheme for $u_t = u_x$. This "alternating kite" scheme has $O(\Delta t^2 + \Delta x^3)$ truncation error with no more computational work than the diamond scheme itself.

1. Introduction.

There are at least three ways of attempting to compare the efficiency of difference schemes for evolutionary partial differential equations.

One way is to carefully program the competitors for a computer and compare their running times for a fixed problem. For example, Culham and Varga have done this [5] for a certain nonlinear parabolic problem.

Another, more Platonic approach is to observe with Douglas [7] that the truncation error, e, and computational work, W, for a given scheme are often approximated by

$$e = c_p(\Delta x)^p + c_q(\Delta t)^q , \quad W = c_W/(\Delta x \, \Delta t);$$

with coefficients assumed independent of Δx and Δt. For a given error e one may adjust Δx and Δt to minimize W. At the minimum we find that the balancing relationship

$$p \, c_p \, (\Delta x)^p = q \, c_q \, (\Delta t)^q$$

holds. The resulting minimal work

$$W_{min} = c_W (c_p/q)^{1/p} \, (c_q/p)^{1/q} [(p+q)/e]^{1/p+1/q}$$

could be used to compare various schemes, at least if c_p and c_q could be adequately estimated. See, e.g., Gary

[9], where the promising potential of high order differencing **for** multi-dimensional problems is also indicated.

A third method [14] [15] combines certain methods used by Kreiss and Oliger [11] for comparing spatial differencing for hyperbolic problems with Douglas' observation above. For parabolic problems this approach works out as follows.

2. Comparing differential-difference schemes for $u_t = u_{xx}$.

We consider only approximations to the simplest parabolic problem:

$$u_t = u_{xx}, \quad u(x,0) = \exp(2\pi i\omega x) \equiv u_\omega ,$$

$$u(0,t) = u(1,t), \quad u_x(0,t) = u_x(1,t) ,$$

ω integral; so that

$$u(x,t) = \exp(-\lambda t)\exp(2\pi i\omega x), \qquad \lambda = (2\pi\omega)^2 .$$

The approximation might be given by, say, the mesh function

$$v = (v_0, v_1, \dots, v_J)^T, \quad v_j = v_j(t) \approx u(x_j, t), \quad x_j \equiv j/J \equiv jh,$$

which satisfies the differential-difference scheme

$$v_t = Bv/h^2 , \quad v(0) = u_\omega ,$$

where B is the Toeplitz matrix whose generic row is

$$(\dots, 0, 1, -2, 1, 0, \dots) .$$

Since

$$Bu_\omega = -2(1-\cos 2\pi\omega h)u_\omega = -4\sin^2(\pi\omega h)u_\omega ,$$

we have, with $\theta \equiv 2\pi\omega h$,

$$v(t) = \exp(-\bar{\lambda}t)u_\omega , \quad \bar{\lambda} = 4\sin^2(\theta/2)/h^2 .$$

Suppose, somewhat more generally, that d^2/dx^2 is approximated with an implicit difference formula,

$$A\, d^2f/dx^2 - B\, f/h^2 = O(h^p)\,, \tag{2.1}$$

using Toeplitz matrices

$$A_{jk} = \alpha_{k-j}\,, \quad B_{jk} = \beta_{k-j}\,,$$

whose amplification factors (symbols) are given by

$$a(\theta) \equiv \sum_j \alpha_j \exp(ij\theta), \quad b(\theta) \equiv -\sum_j \beta_j \exp(ij\theta), \quad \theta \equiv 2\pi\omega h.$$

Then the solution of the differential-difference approximation

$$Av_t = B\, v/h^2\,, \qquad v(0) = u_\omega\,, \tag{2.2}$$

is given by

$$v(t) = \exp(-\bar{\lambda}\, t)u_\omega\,,$$

where

$$\bar{\lambda} \equiv b(\theta)/[h^2 a(\theta)]\,, \qquad \theta \equiv 2\pi\omega h\,.$$

It follows that

$$\begin{aligned}(v/u)(t) &= \exp[(\lambda-\bar{\lambda})t] \\ &= \exp\{[1-b(\theta)/(\theta^2 a(\theta))](2\pi\omega)^2 t\} \\ &= \exp(\varepsilon \cdot P)\,,\end{aligned}$$

where

$$\begin{aligned}P &\equiv (2\pi\omega)^2 t \\ &= \text{number of } e^{th}\text{-lives in } [0,t] \text{ of a wave with frequency } \omega \\ &\approx 0.7 \cdot (\text{number of half-lives in } [0,t])\,.\end{aligned} \tag{2.3}$$

To associate some meaning with the number

$$(2.4) \qquad \varepsilon \equiv 1-b(\theta)/[\theta^2 a(\theta)] \ ,$$

it may be noted that, for small εP, the relative error in the approximation v is

$$(v-u)/u = \exp(\varepsilon P) - 1 \approx \varepsilon P \ .$$

Hence we shall call ε the relative error per e^{th}-life. (We note in passing that ε, given by (2.4), is, in effect, the relative error in a rational trigonometric approximation to the polynomial θ^2.) We shall normalize things somewhat further by expressing ε in terms of

$N \equiv 1/(\omega h)$ = number of (spatial) intervals per wavelength

for the frequency ω, so that

$$(2.5) \qquad \varepsilon(N) = 1 - \frac{N^2}{4\pi^2} \frac{b(2\pi/N)}{a(2\pi/N)} \ .$$

Then we can compare two spatial approximations to d^2/dx^2 by asking, for a given ε, which scheme takes fewer intervals per wavelength to yield that error ε after one e^{th}-life.

Indeed, we shall now compare three types of high-order spatial differencing in this fashion.

The first type of spatial differencing considered is explicit (i.e., $A = I$ in (2.1) (2.2)) centered differencing. We only need $b(\theta)$; i.e., a trigonometric approximation to θ^2. A change of variable, $\tau \equiv \sin(\theta/2)$, converts this into a problem of polynomial approximation:

$$\theta^2 = 4(\arcsin \tau)^2 \approx 4\tau^2 \sum_{\nu=0}^{m-1} a_\nu(\tau^2) \equiv 4\tau^2 P_{m-1}(\tau^2) \equiv b_{ex}(m,\theta).$$

The difference scheme $(-B)$ may be recovered from $4\tau^2 P_{m-1}(\tau^2)$ by replacing τ^2 with $-h^2 D_+ D_-/4$; it will thereby have bandwidth $2m+1$. With P_{m-1} an initial segment of the MacLaurin approximation to $(\arcsin \tau)^2/\tau^2$, i.e., with

$$a_\nu = 2^{2\nu}(\nu!)^2/[(2\nu+1)!(\nu+1)] ;$$

we have, for these explicit centered differences, that (since $\theta = 2\pi/N$)

$$\varepsilon_{ex}(m, N) = 1-b(\theta)/\theta^2 = \frac{4\tau^2}{\theta^2} \sum_{\nu=m}^{\infty} a_\nu(\tau^2)^\nu \approx a_\nu(\pi/N)^{2m}, \qquad N \to \infty .$$

Since $0 \leq b_{ex}(m,\theta) \leq \theta^2$ $(0 \leq \theta \leq \pi)$, the differential-difference scheme $v_t = Bv/h^2$ is stable. Furthermore, for each m, $\varepsilon_{ex}(m, N)$ is a monotone strictly decreasing function of $N \in [2, \infty)$. These observations are useful in examining the character of time-discretizations.

Next we consider the smooth spline-Galerkin difference method due to Thomée [17] (generalized to variable coefficient periodic problems by Thomée and Wendroff [18]). The basic idea of Galerkin's method is to suppose that

$$W(x, t) = \sum_{j=1}^{J} c_j(t)\varphi_j(x)$$

is the approximate solution of $u_t = u_{xx}$ found by requiring that

$$\langle W_t - W_{xx}, \varphi_i \rangle = 0 , \qquad 1 \leq i \leq J ,$$

where $\langle \cdot , \cdot \rangle$ is the integral inner-product in space. This is equivalent to solving

$$A c_t = B c/h^2 ,$$

where

$$A_{ij} \equiv \langle \varphi_j, \varphi_i \rangle/h ; \qquad B_{ij} \equiv \langle \partial^2 \varphi_j/\partial x^2, \varphi_i \rangle h .$$

Here $\varphi_i(x) \equiv \Phi(x/h-i)$, with Φ being Schoenberg's smooth basic B-spline of degree m (order m+1) with knots at the integers {integers $+\frac{1}{2}$} for odd {even} $m > 0$ [13]. As $\varphi_i(x) \neq 0$ only for x in an interval of length $(m+1)h$ centered on x_i, both A and B have bandwidth 2m+1.

Now, $W(x_i, t) \neq c_i(t)$ since $\varphi_i(x_j) \neq \delta_{ij}$. However, Thomée observed that if one took $c_i(0) = (u_\omega)_i$, i.e., solved the differential-difference approximation

$$A\, v_t = B\, v/h^2 , \quad v(0) = u_\omega ,$$

then $\|v-u\|_{\ell_2} = O(h^{2m})$ - a superconvergence result since $\|u-w\|_{\mathbb{L}_2} \geq$ const. h^{m+1} even if w is the optimal smooth spline approximation to u of degree m. This superconvergence also follows from the fact that the matrices A and B arise from collocation with smooth splines of degree 2m+1 [16].

From Thomée [19], we may deduce that

$$\varepsilon_{sp}(m,N) = -N \frac{N \sum_{-\infty}^{\infty} j^2/(j+1/N)^{2m+2} + 2 \sum_{-\infty}^{\infty} j/(j+1/N)^{2m+2}}{\sum_{-\infty}^{\infty} (j+1/N)^{-2m-2}}$$

$$< 0 .$$

From numerical computations, $b_{sp}(m,N)/a_{sp}(m,N) > 0$ is a decreasing function of $N \geq 2$; $\varepsilon_{sp}(m,N)$ is an increasing function of $N \geq 2.5$. The first term of the numerator of ε_{sp} (when combined with the denominator's middle term) shows $O(N^{-2m})$ behavior as $N \to \infty$. The second term's contribution is in fact $O(N^{-(2m+2)})$, as is shown in [15]. This is all as it should be, since (as we observed by notation at the beginning of this section) ε is simply the relative error in the approximate eigenvalues of d^2/dx^2 associated with our implicit difference schemes (which are Rayleigh-Ritz eigenvalues in this spline-Galerkin case).

We remark that the spline amplification factors $a_{sp}(\theta)$ are also expressible as polynomials of degree m in $\cos\theta$;

as such they were generated recursively by Schoenberg [13, p. 114]. Presumably, the $b_{sp}(\theta)$ could also be so generated following his ideas there.

The third type of spatial differencing we consider is Collatz' Mehrstellenverfahren [2], high-order generalizations of Numerov's tridiagonal centered implicit difference approximation to d^2/dx^2. Their amplification factors may be found as above in the explicit centered differencing, except that now we require a Padé approximation

$$(\arcsin \tau)^2/\tau^2 = P_{m-1}(\tau^2)/Q_m(\tau^2) + O(\tau^{4m}),$$

$$\tau = \sin(\theta/2) .$$

We expect (and observe numerically) that the associated ε_{im} satisfies

$$\varepsilon_{im} = 1-N^2\tau^2 P_{m-1}/(\pi^2 Q_m) = O(N^{-4m}), \quad N\to\infty,$$

$$\tau = \sin(\pi/N).$$

Furthermore, we have observed (from the output of our Padé computer program, compiled by the physicists of Group T-9 at the Los Alamos Scientific Laboratory) that the zeroes of P_{m-1} and Q_m lie outside $[0,1]$ for $1 \leq m \leq 6$, that P_{m-1}/Q_m is positive in $[0,1]$, and that ε_{im} decreases monotonically for $N \in [2,\infty)$. Since B is recovered as before from $4\tau^2 P_{m-1}(\tau^2)$ (and A similarly from Q_m), the bandwidths of A and of B are both $2m+1$. The first two of these implicit difference approximations are given explicitly in Collatz [2, Table III of the Appendix].

To compare these three families of difference approximations we now present in Table 2.1, for $|\varepsilon| = 10^{-2}$ and $|\varepsilon| = 10^{-4}$, the number N of intervals per wavelength necessary to yield each given "relative error per e^{th}-life", ε .

Table 2.1

Intervals per wavelength for differential-difference schemes

	$\|\varepsilon\| = 10^{-2}$			$\|\varepsilon\| = 10^{-4}$		
m\Type	explicit	spline	implicit	explicit	spline	implicit
1	18.1	18.2	5.1	181	181	16.0
2	6.3	4.2	3.0	20.4	12.3	5.1
3	4.5	3.1	2.5	10.0	5.7	3.6
4	3.8	2.7	2.4	7.1	4.1	3.0
5	3.4	2.5	2.3	5.8	3.5	2.8
6	3.2	2.4	2.2	5.0	3.1	2.6
8	2.9	2.3		4.2	2.7	

We summarize this section on our three types of spatial approximations by first remarking that for all schemes $b/a \geq 0$; b/a is a monotone decreasing function of N; and that our computations indicate that ε is monotone for $N \geq 2$ for explicit and implicit differencing, for $N \geq 2.5$ for spline Galerkin differencing. Other properties are noted in Table 2.2.

Table 2.2

Type of Spatial Diffrc.	Order of acc. for fixed ω	Band-width	Sign of ε	Remarks
Explicit Centered	$O(h^{2m})$	2m+1	+	
Spline-Galerkin	$O(h^{2m})$	2m+1	-	For same order of accuracy, slightly better resolution.
Implicit Centered	$O(h^{4m})$	2m+1	+	For same band-width, slightly better resolution.

Finally, for purposes of determining stability of certain space-time discretizations to follow, we tabulate (in Table 2.3)

$$\max_{0 \le \theta \le \pi} [b(\theta)/a(\theta)]$$

which we have observed computationally to be $b(\pi)/a(\pi)$. N.B.: $\pi^2 \approx 9.9$. For general interest, we also tabulate $\max_\theta |\varepsilon(\theta)|$.

Table 2.3

	$\max[b(\theta)/a(\theta)]$			$\max|\varepsilon(\theta)|$		
m\Type	explicit	spline	implicit	explicit	spline	implicit
1	4.0	12.0	6.0	.60	.44	.39
2	5.3	10.0	7.5	.46	.13	.24
3	6.0	9.9	8.1	.39	.08	.18
4	6.5	9.9	8.5	.34	.06	.14
5	6.8	9.9	8.8	.31	.05	.11
6	7.1	9.9	8.9	.28	.04	.10
∞	π^2	π^2	π^2	0	0	0

3. Full space-time discretizations for $u_t = u_{xx}$.

Let us turn to the more important subject of describing and comparing time discretizations of our differential-difference schemes.

For example, the trapezoidal discretization of $v_t = A^{-1}B\, v/h^2$ is

$$[I - \Delta t A^{-1}B/(2h^2)]w^{n+1} = [I + \Delta t A^{-1}B/(2h^2)]w^n, \quad w^0 = u_\omega,$$

with

$$w^n = (w_0^n, w_1^n, \ldots, w_J^n)^T, \quad w_j^n \approx u(jh, n\,\Delta t).$$

Thus

$$w^n = R\,w^{n-1} = \ldots = R^n u_\omega .$$

Here the amplification factor R is given by

$$R = (1 - \xi/2)/(1 + \xi/2) \qquad \text{(trapezoidal rule)}$$

where

(3.1) $$\xi \equiv \Delta t\, b(\theta)/[h^2 a(\theta)] = b(\theta)/[M\theta^2 a(\theta)] = (1 - \varepsilon)/M ;$$

and where, again, $\theta = 2\pi\omega h = 2\pi/N$; ε is given by (2.5); and where

(3.2) $$M \equiv 1/[(2\pi\omega)^2 \Delta t] = \text{the number of time intervals per } e^{th}\text{-life.}$$

Since $b/a \geq 0$ for our spatial differences, $|R| \leq 1$; hence the trapezoidal discretization is stable for all mesh frequencies.

Now $u(\cdot, t) = \exp(-P)u_\omega$, where $P = (2\pi\omega)^2 t$ [see (2.3)] is the number of eth-lives to time t. Thus

$$w^n/u(t) = (R^M e)^P \qquad (t = n\,\Delta t)$$
$$= \exp(E \cdot P)$$

where

(3.3) $$E \equiv 1 + M \log R .$$

Like ε before, E is (for small $E\cdot P$) the relative error, $w/u-1$, in w per e^{th}-life.

We summarize the time discretizations we shall consider in Table 3.1. Each amplification factor, R, is expres sed in terms of ξ as given by (3.1).

Three schemes in Table 3.1 bear further discussion:

Table 3.1

Time discretizations expressed in terms of ξ, (3.1).

	Name	R	Time truncation type	Stability condition: $\Delta t/h^2[=N^2/(4\pi^2 M)]\leq$
1.	Euler	$1-\xi$	$-\Delta t$	$2/\max[b(\theta)/a(\theta)]$
2.	Backward-Euler	$1/(1+\xi)$	$+\Delta t$	stable for $b/a \geq 0$
3.	duFort-Frankel	$[\sqrt{1-\bar{\xi}\xi+\xi^2}+\bar{\xi}/2-\xi]/(1+\bar{\xi}/2)$, where $\bar{\xi} \equiv \frac{N^2}{4\pi^2 M} \max_\theta [\frac{b(\theta)}{a(\theta)}]$	$-(\Delta t/h)^2$	" " "
4.	2nd order explicit	$1-\xi(1-\xi/2)$	$+\Delta t^2$	$2/\max[b(\theta)/a(\theta)]$
5.	Trapezoidal	$(1-\xi/2)/(1+\xi/2)$	$-\Delta t^2$	stable for $b/a \geq 0$
6.	Calahan-Zlámal	$1-\tilde{\xi}(1+\sqrt{3}\,\tilde{\xi}/6)$, where $\tilde{\xi}\equiv \xi/[1+(3+\sqrt{3})\xi/6]$	$-\Delta t^3$	" " "
7.	4th order Padé	$(1-\xi/2+\xi^2/12)/(1+\xi/2+\xi^2/12)$	$+\Delta t^4$	" " "

The 2nd order explicit-in-time scheme is, of course, also one of the Padé schemes. It is also the method from which the Lax-Wendroff schemes are created. For multidimensional variable coefficient problems we would suggest retaining the explicit character of the three explicit-in-time schemes (schemes 1, 3,4) by approximating all unmixed spatial derivatives with explicit or implicit-by-lines spatial differences. The computing cost of implicit spatial differencing would then only be twice the cost of explicit differencing of the same band width (with a possible significant increase in storage requirements). On the other hand, it would still cost the same as explicit differencing of the same order of accuracy.

The usual duFort-Frankel scheme

$$(w_j^{n+1} - w_j^{n-1})/(2\Delta t) = [w_{j+1}^n - (w_j^{n+1}+w_j^{n-1})+ w_{j-1}^n]/h^2$$

is a stabilized version of the leap-frog (Richardson) scheme for $u_t = u_{xx}$, i.e., of the mid-point rule for $v_t = B_3 v/h^2$, with B_3/h^2 the usual 3-point difference approximation to d^2/dx^2. It may be re-written (with the parameter $c = 1$) as

$$(D_c w)^n \equiv (w_j^{n+1} - w_j^{n-1})/(2\Delta t) + (c/h^2)[w_j^{n+1} - 2w_j^n + w_j^{n-1}]$$
$$= B_3 w^n/h^2 .$$

This form is suitable for using implicit spatial approximations

(3.4a) $\qquad (D_c w)^n = A^{-1} B\, w^n/h^2 . \qquad$ (duFort Frankel -in-time)

To insure unconditional stability of this explicit 3-level scheme, c must be picked so that two characteristic roots are bounded by 1 for $N \geq 2$, $M > 0$. One of these roots is given by R in the table with $\bar{\xi}$ there replaced by

$$\bar{\xi}_c \equiv c\, N^2/(\pi^2 M) \quad (= 4c\Delta t/h^2);$$

the more troublesome root is similarly given with a sign change on the square root. It may be checked that, for $\Delta t < 1$, the magnitude of both roots is less than one when

(3.4b) $\qquad c \geq \max_{\theta} \{b(\theta)/[4\, a(\theta)]\} .$

Since the main temporal truncation error is $c(\Delta t/h)^2 u_{tt}$, we pick the smallest such c . The R exhibited then applies to initial data of the form $w^0 = u_\omega$, $w^1 = Ru_\omega$; any deviation from this must be suitably small.

The potential usefulness of the next-to-last scheme in the table was discovered by Zlámal [22], [23]. He made the nice observation that the application of a certain two-stage A-stable implicit Runge-Kutta scheme, which Gear [10, p. 223] attributes to Calahan, means that solving the same banded linear system twice to move from one time-line to the next can yield $O(\Delta t^3)$ accuracy in comparison to the $O(\Delta t^2)$ accuracy of the trapezoidal scheme.

4. Minimizing the work in order to compare full discretizations, with a caveat.

We would like to say that scheme (a) is better than scheme (b) for error level E if the computational work to attain E using scheme (a) is less than the work using scheme (b). But if E is attainable at all using a given scheme with some Δt_0 and h_0, then the computational work to attain the same E with that scheme can be made arbitrarily large by picking suitable $\Delta t \ll \Delta t_0$.

Thus, in order to compare discretizations, we follow Douglas' idea in Section 1 by first proceeding as follows for each combination of spatial and temporal differencing. We assume that the computational work per wavelength per e^{th}-life, W, is given by

$$W_S = c_S NM , \tag{4.1}$$

where

(4.2) $c_S \equiv$ the computational work per space-time mesh point

certainly depends on the scheme, S, but is independent of M and N. Then, for each of two error levels $|E| = 10^{-2}$ and $|E| = 10^{-4}$, we shall pick M and N to minimize W_S subject to the constraint that

$$|E| = |1 + M \log R_S(N, M)| . \tag{4.3}$$

From the resulting minimizing pair (N_S, M_S) we can compute W, given additionally an estimate of c_S. For fixed $|E|$ we are then able to compare schemes on the basis of their optimized work.

If we use the fast Fourier transform in our computations, then c_S would in fact be proportional to $\omega \log \omega N$, a slowly varying function of N. As we did examine this case in preliminary calculations for a hyperbolic problem [15] and found little change in the minimizing pairs, we do not intend to pursue it further.

So far our error analysis has been exact. We now must describe circumstances in which we make an approximation to ease the strain of minimizing W, given $|E|$.

For suppose, given one of our discretizations, that we define η by the relation

$$R(\xi) = e^{-\xi}(1+\eta) ,$$

with ξ as in (3.1). Then the error per e^{th}-life (3.3) is

$$E = 1-M\xi + M \log(1+\eta) \tag{4.4}$$
$$= \varepsilon + M \log(1+\eta) .$$

For small ε, $\xi \approx 1/M$, so that (for large M) $M\eta$ is primarily the truncation error in the time discretization. It may have a different sign from ε. For example, with the trapezoidal time discretization, this kind of cancellation takes place both for explicit and for implicit centered differencing; it does not take place for spline-Galerkin differencing.

Such cancellation of the spatial and temporal discretization errors has advantages and disadvantages. For example, Crandall [4] observed, in his discussion of the Numerov-trapezoidal case, that a particular ratio of $\Delta t/h^2$ would yield $O(h^6)$ accuracy instead of the usual $O(\Delta t^2+h^4)$ accuracy associated with that scheme for $u_t=u_{xx}$. On the other hand, it is certainly wise in the practical solution of parabolic problems to attempt an optimal balancing of the spatial and the temporal mesh (see, e.g., Gary [9]); a cancellation of the two truncations errors may make this task more difficult. Furthermore, our optimization of W will be difficult and unrealistic if $|E|$ is not monotone decreasing in N and M; indeed an $\hat{E}(N, M) \equiv \sup_{K \geq M, L \geq N} |E(L, K)|$ would have to be used.

Rather than do this, we now describe the only approximation in our error analysis. In our computer program for optimizing W, we redefine the sign of ε (depending on the type of spatial and temporal discretization) so that the two terms in (4.4) have the same sign for large M and small ε;

this revised (4.4) now approaches the usual truncation error expression.

This artifice has apparently succeeded in making W a unimodal function of N on the curves $|E| = \text{const.}$ we consider below. The reader can tell a priori when this approximation was made by comparing the signs of the truncation errors in Table 3.1 with the signs of ε in Table 2.2; like signs mean no cancellation takes place in (4.4); unlike signs mean cancellation actually takes place but we did not permit it. In these cases of unaccounted-for cancellation our values for W will be higher than would actually occur.

The Appendix contains pairs of the optimal values of N and M for each combination of spatial and time differencing type for a reasonable range of orders of spatial accuracy. We have there noted when we avoided spatial-temporal truncation error cancellation; the reader may wish to regard such pairs with some skepticism.

5. A comparison of the full discretizations for $u_t = u_{xx}$ based on the work involved in solving linear systems.

We have completed the major effort involved in comparing our space-time discretizations by balancing N and M, for each scheme, to minimize computational work of the form (4.1) necessary to compute to a given error level $|E|$ per e^{th}-life. We are left with the problem of estimating the computational work per space-time mesh point, c_S (4.2).

We might be wise to leave this task (and the associated final comparisons) to the reader, for there are many options open. See, e.g., [14][15]; an attempt is made to consider nonlinear problems in the latter reference.

However, as an example of the use of the tables in the Appendix, we take the case of a linear parabolic problem with coefficients varying both in x and t, yet where the main computational work is found in the matrix manipulations and not in the evaluation of the matrix elements themselves. Now, the optimal balancing of N and M will also depend on the problem (compare, e.g., Gary's balanced N and M for the Crank-Nicolson scheme involving a simple source term [9] with ours for the same scheme with no source). Nevertheless, we shall assume by necessity that all schemes degrade in the same way so that our tables are applicable.

In this context, we take the computational work per mesh point (4.2) to be the number of multiplications per mesh point in solving the linear system. It will depend on the bandwidth of the matrices (2m+1); for more specifics, see Table 5.1.

The last entry in Table 5.1 deserves a further remark. We are trying to compute

$$w^{n+1} = [Q(C^{-1}D)]^{-1}P(C^{-1}D)w^n ,$$

where C and D are banded matrices which do not commute in general, and where

$$P(z) = c\prod_{i=1}^{\mu}(z_i - z) , \quad Q(z) = \prod_{i=1}^{\nu}(w_i - z)$$

are polynomials (of the same degree in this instance). One way of solving this problem without having to compute the inverse of any band matrices is to observe that

$$\begin{aligned}[Q(C^{-1}D)]^{-1}P(C^{-1}D) &= c\prod_1^{\mu}(w_i - C^{-1}D)^{-1}\prod_1^{\mu}(z_i - C^{-1}D) \\ &= c\prod_1^{\mu}[(w_i - C^{-1}D)^{-1}(z_i - C^{-1}D)] \\ &= c\prod_1^{\mu}[(w_iC - D)^{-1}CC^{-1}(z_iC - D)].\end{aligned}$$

The roots of the two quadratics in the 4th order Padé scheme are indeed complex; hence the operation count given in the table. Dendy [6] is responsible for observing that the (real) quadratic factors may be used in the case of explicit spatial differencing (when C commutes with D). This computational work for the direct solution of 4th order Padé schemes seems huge. As Dendy [6] has further observed, perhaps iterative methods may be worthwhile in context where the FFT is inapplicable.

Table 5.1

Multiplies per mesh point to solve the 2m+1 banded systems

Type of Scheme	Mult. per Mesh point	Rationalization
Explicit-Euler	2m+1	One matrix multiply
Explicit-back. Euler	m^2+3m+1	Multiplies to solve one 2m+1 banded linear system
Explicit-duFort Frankel	2m+1	
Spline / Implicit } duFort Frankel	4m+3	Taking the view of approximating only d^2/dx^2 by $A^{-1}B/h^2$ fixed for all time. Use $(LU)^{-1}$. Double work if d/dx is also present.
Explicit-2^{nd}ord. expl.	4m+2	Two matrix multiplies.
Spline / Implicit } -2^{nd}ord.expl.	8m+6	As in D. F., but do it twice.
Trapezoidal	m^2+5m+2	One matrix mult., solve one system.
Calahan-Zlámal	m^2+9m+6	Two matrix multiplies, solve <u>same</u> system twice.
Explicit-4th ord. Padé	$8m^2+16m+4$	Form the quadratics in B; then as in trap. but with width 4m+1.
Spline / Implicit } 4^{th}ord. Padé	$8m^2+40m+16$	Two trap. with complex arithmetic.

Table 5.2

Typical Schemes in Current Practice; matrix manipulation work

$\|E\|$	Type	Mult. per mesh pt.	N	Work	$\Delta t/h^2$	Type Trunc. Error
	3 pt. Explicit-Euler	3	31	7170*	0.3	$h^2 - \Delta t$
	3 pt. Explicit-back. Euler	5	31	11,600	0.3	$h^2 + \Delta t$
	3 pt. Explicit-duFort Frankel	3	22	4100*	0.2	$h^2 - (\Delta t/h)^2$
10^{-2}	3 pt. Explicit-trap. (Crank-Nicolson)	8	26	850*	4.1	$h^2 - \Delta t^2$
	Linear spline-trap.	8	26	850	4.1	$-h^2 - \Delta t^2$
	Cubic spline-trap.	26	3.8	320	0.1	$-h^6 - \Delta t^2$
	Numerov-trap.	8	6.7	190*	0.3	$h^4 - \Delta t^2$
	3 pt. Explicit-Euler	3	315	7×10^6*	0.3	
	3 pt. Explicit-back. Euler	5	315	12×10^6	0.3	
	3 pt. Explicit-duFort Frankel	3	220	5×10^6*	0.2	as
10^{-4}	3 pt. Explicit-trap.	8	260	84,000*	41	above
	Linear spline-trap.	8	260	84,000	41	
	Cubic spline-trap.	26	7.1	6060	0.04	
	Numerov-trap.	8	21	5950*	0.3	

*Cancellation of spatial and temporal truncation errors not accounted for in this work estimate.

We first combine Table 5.1 with the Appendix to compare some of the simpler schemes in common use (Table 5.2). The first entry there (the usual 4-point explicit scheme for the heat equation) results from the pair (N, M) in the upper left corner of the first table in the Appendix, with the work taken from Table 5.1 to be $(2m+1)NM$ with $m=1$.

Finally, for each fixed time discretization, we consider a given family of spatial discretizations, e.g., explicit centered differences. Using Table 5.1 we compute the work required for each spatial order of accuracy. We then optimize over the order of accuracy of the spatial schemes; e.g., we find the best explicit centered-Euler scheme. Table 5.3 presents some of the resulting comparisons. The computational work at each minimum is a slowly

Table 5.3

Optimal Orders of Spatial Differencing for Various Time Discretizations of $u_t = u_{xx}$ Based on Work to Solve the Linear Systems.

$|E| = 10^{-2}$

Type (space-time)	Order	Diagonals in matrix	Mult. per Mesh Pt.	N	Unaccounted for Canceln.	Work	$\Delta t/h^2$
Explicit-Euler	6	7	7	6.3	yes	2600	.02
Explicit-backward Euler	4	5	11	9.5	no	6500	.04
Explicit-duFort Frankel	12	13	13	3.6	yes	360	.04
Implicit-duFort Frankel, $(LU)^{-1}B$	16	9	19	2.5	yes	240	.03
Explicit-2nd order explicit	6	7	14	5.6	no	410	.15
Spline-2nd order explict, $(LU)^{-1}B$	6	7	30	3.7	yes	560	.07
Implicit-2nd order explicit, $(LU)^{-1}B$	8	5	22	3.6	no	390	.07
Explicit-trapezoidal	4	5	16	8.4	yes	480	.5
Spline-trapezoidal	4	5	16	5.5	no	310	.2
Implicit-trapezoidal	8	5	16	3.6	yes	190	.1
Explicit-Calahan-Zlámal	6	7	42	5.5	yes	480	.4
Spline-Calahan-Zlámal	6	7	42	3.7	no	310	.2
Implicit-Calahan-Zlámal	8	5	28	3.5	yes	200	.2
Explicit-4th order Padé	4	5	68	7.5	no	380	1.7
Spline-4th order Padé	4	5	128	5.0	yes	470	.8
Implicit-4th order Padé	4	3	64	6.0	no	290	1.2

$|E| = 10^{-4}$

Type (space-time)	Order	Diagonals in matrix	Mult. per Mesh Pt.	N	Unaccounted for Canceln.	Work	$\Delta t/h^2$
Explicit-Euler	12	13	13	6.3	yes	440000	.0002
Explicit-backwards Euler	8	9	29	9.4	no	1500000	.0004
Explicit-duFort Frankel	24	25	25	3.8	yes	7800	.004
Implicit-duFort Frankel, $(LU)^{-1}B$	24	13	27	2.8	yes	4200	.003
Explicit-2nd order explicit	12	13	26	5.9	no	6900	.02
Spline-2nd order explicit, $(LU)^{-1}B$	10	11	46	4.0	yes	8200	.01
Implicit-2nd order explicit, $(LU)^{-1}B$	12	7	30	4.2	no	5600	.01
Explicit-trapezoidal	8	9	93	8.7	yes	10700	.06
Spline-trapezoidal	8	9	38	4.9	no	6000	.02
Implicit-trapezoidal	8	5	16	6.2	yes	3200	.03
Explicit-Calahan-Zlámal	8	9	58	8.4	yes	5060	.17
Spline-Calahan-Zlámal	10	11	76	3.9	no	3000	.04
Implicit-Calahan-Zlámal	12	7	42	4.1	yes	1700	.04
Explicit-4th order Padé	6	7	124	11.7	no	3200	1.6
Spline-4th order Padé	6	7	208	6.6	yes	3000	.5
Implicit-4th order Padé	8	5	128	5.9	no	1600	.4

varying function of m, particularly if the optimizing m is large. So a smaller m can often be used without much loss in efficiency. But N may then increse significantly with a corresponding increase in storage requirements. The column labeled "order" is the order of accuracy of the optimizing spatial scheme.

The main thing that stands out in these comparisons is that Euler and backward-Euler time differencing is not competitive. Otherwise, for $|E| = 10^{-2}$ $\{|E| = 10^{-4}\}$, the worst of the time differencing schemes, when optimized over the order of the spatial differencing, takes no more than three {seven} times as much work in solving its linear systems than the best. DuFort-Frankel differencing (3.4) stands out for the low number of intervals per wavelength it requires, for the high order of the spatial differencing that goes with this, and for the small ratio of $\Delta t/h^2$ it uses. 4th-order Padé differencing stands out as the only scheme with $\Delta t/h^2 > \frac{1}{2}$ after being optimized over spatial order of accuracy.

We might point out that at least two generalizations of the diagonal Padé scheme to nonlinear problems are possible [15], and that Watanabe and Flood [20] not only proposed but also computed with one such scheme (5-point explicit spatial differencing combined with collocation in time, using piecewise-cubics, at three Lobatto points). They chose to solve the nonlinear equations with one of Broyden's quasi-Newton methods. They reported striking increases in computational efficiency when compared with their implementation of some second-order explicit methods on certain nonlinear problems.

Many of the optimized schemes in Table 5.3 use less than seven intervals per wavelength with associated errors of 10^{-2} or 10^{-4}. The reader may wonder how to interpolate such accurate yet sparsely spaced function values. In [14, Appendix] there is an application of some of Golomb's work concerning smooth spline interpolation to this end. The reader should be warned that, like much of this paper, it applies directly only to approximations of $\exp(2\pi i\omega x)$ using a uniform mesh.

6. Comparison of some full discretizations for $u_t = u_x$.

Differential-difference schemes and full discretization for the hyperbolic problem

$$(6.1) \quad u_t = u_x, \; u(0,t) = u(1,t), \; u(x,0) = \exp(2\pi i\omega x);$$

$$u(x,t) = \exp[2\pi i\omega(x+t)]$$

were compared [14], [15] in much the same manner as we have just described for the heat equation. But because the case for full discretizations involving implicit spatial differences was skipped there, and because the reader might like to contrast results for these two types of PDE's, we briefly discuss certain results for this problem.

Neither the three centered differential-difference methods for (6.1) nor the stable time discretizations we consider for them change the initial amplitude of the complex exponential initial data - the errors all occur in phase angle. Thus, in this hyperbolic context, E refers to phase error per (time) period. N is, as before, the number of (spatial) intervals per wavelength, but M is the number of (time) intervals per period.

Tables concerning the balancing of N and M for explicit differences+leap-frog, spline+trapezoidal, and spline+4th order Padé discretizations - are found in [14], [15]. However [14] contains optimal pairs (N, M) based on certain nonlinear expressions for $|E|$ like (4.3), while [15] contains pairs based on linearization of the temporal truncation error.

The amplification factors $a(\theta)$ and $b(\theta)$ for Collatz' Mehrstellenverfahren for implicit approximations to d/dx are described in [14], [15]; the $O(h^4)$ tridiagonal and $O(h^8)$ pentadiagonal schemes are explicitly given in Collatz [2, Appendix].

In Table 6.1 we present newly published optimal pairs (N, M) for the trapezoidal and 4th order Padé time discretizations combined with implicit spatial differencing; $|E|$ is treated as in [15]. The spatial schemes are $O(h^{4m})$ accurate for schemes of band width $2m+1$.

Table 6.1

Optimal Pairs (N, M) for Implicit Spatial Differencing and Two Time Discretizations of $u_t = u_x$.

Order of accuracy \ Scheme	$\|E\| = 10^{-2}$ Trapezoidal	$\|E\| = 10^{-2}$ Padé	$\|E\| = 10^{-4}$ Trapezoidal	$\|E\| = 10^{-4}$ Padé
4	(11, 56)	(10.4, 7.2)	(36, 560)	(32, 23)
8	(4.8, 50)	(4.5, 6.7)	(8.2, 510)	(7.7, 21)
12	(3.6, 49)	(3.4, 6.5)	(5.0, 490)	(4.8, 21)
16	(3.1, 48)	(3.0, 6.4)	(3.9, 480)	(3.8, 20)
20	(2.9, 47)	(2.8, 6.3)	(3.4, 470)	(3.3, 20)

Having balanced N and M, we first use some data from [15] to compare two commonly proposed schemes (Table 6.2) with the work based on the solution of linear systems.

Table 6.2

$\|E\|$	Type	Mult. per mesh pt.	N	Work	$\Delta t/h$	Trunc.
10^{-2}	Diamond scheme	2	91	16000	1	$O(h^2 + \Delta t^2)$
	Linear spline-trap. (same as 3 pt. implicit-trap.)	8	11.4	5100	0.2	$O(h^4 + \Delta t^2)$
10^{-4}	Diamond	2	910	1.6×10^6	1	$O(h^2 + \Delta t^2)$
	Linear spline-trap.	8	36	1.6×10^5	0.06	$O(h^4 + \Delta t^2)$

Finally, using Table 6.1 we augment a summary in [15] which, like Table 5.3, is based on picking that order of spatial differencing, for a given temporal discretization, which is optimal relative to the work associated with solving the linear systems (Table 6.3).

Table 6.3

Optimal orders of spatial differencing for various time discretizations of $u_t = u_x$, based on matrix manipulation work.

$|E| = 10^{-2}$

Scheme	Order	Diagonals	Mult. per mesh point	N	Mult. per wavelength per period	$\Delta t/h$
Explicit leap-frog	8	9	8	7.5	4300	0.1
Spline-trap.	6 (quad.)	5	16	5.9	4900	0.1
Implicit-trap.	8	5	16	4.8	3800	0.1
Spline-Padé	6 (quad.)	5	128	5.5	4800	0.8
Implicit-Padé	8	5	128	4.5	3800	0.7
		$\lvert E\rvert = 10^{-4}$				
Explicit-leap-frog	12	13	12	8.3	69,000	0.01
Spline-trap.	8 (cubic)	7	26	7.2	94,000	0.01
Implicit-trap.	12	7	26	5.0	64,000	0.01
Spline-Padé	8 (cubic)	7	208	6.8	30,000	0.3
Implicit-Padé	12	7	208	4.8	20,000	0.2

7. Numerov's method for variable coefficients.

The scheme which we called Numerov-trapezoidal in Table 5.2 was discovered and analyzed by Crandall [4] and Douglas [8]. It has turned out to be within a few percent of being optimal over all space-time discretizations we considered in terms of linear system operations for $|E| = 10^{-2}$. This motivates a pursuit of a similar tridiagonal implicit scheme for the linear heat equation

$$u_t = \alpha u_{xx} + \beta u_x + \gamma u$$

with variable coefficients. Our approach is to concoct a differential-difference equation for

$$(7.1) \qquad u_t - f(u,x,t) = \alpha(x,t)\, u_{xx} + \beta(x,t)\, u_x \equiv Lu .$$

Indeed, suppose that we can find tridiagonal matrices A and B so that

$$(7.2) \qquad A(Lf) - Bf = O(h^p) ,$$

where $f = (f(x_0), f(x_1), \ldots, f(x_J))^T$ and f is sufficiently smooth. Then the differential-difference scheme

$$(7.3) \qquad A[v_t - f(v,x,t)] = Bv$$

will have $O(h^P)$ spatial truncation error.

Now this is precisely the sort of problem Collatz seeks to solve by his Mehrstellenverfahren. His approach is to find a linear relation between, say, r successive values of Lv and of v which vanishes for polynomials of as high a degree as possible. And he applies this not only to ordinary differential operators L but to partial differential operators as well, including such constant coefficient operators as the Laplacian and biharmonic operators [2], the heat equation in one space dimension [2, for 3-level schemes] [3, for 2-level schemes], the heat and wave equations in two and three space variables, and to $Lu = u_t - u_{xxxx}$ [3].

Nevertheless, we shall simply be concerned with implicit approximations (7.2) to our ordinary differential operator (7.1). That is, we seek a linear relation (7.2) between three successive values of Lv and the corresponding values of v which has the same $O(h^4)$ truncation error as Numerov (when $L = d^2/dx^2$). Now, extensions of Numerov's scheme to variable coefficients by various methods are discussed in, e.g., Collatz [2] and in Young and Gregory [21, §10.7]. Birkhoff and Gulati [1] create a Numerov-like scheme for $Lu = (pu')' + qu$ which involves values of the coefficients at half-integral mesh points as well. But we know of no such difference formula involving only mesh values of the coefficients. So we present one.

Our approach to the matter is slightly different than Collatz' - it is via generalized Hermite-Birkhoff interpolation.

To fix ideas, we are given

$$x_- < x_0 < x_+$$

and five numbers: $f(y)$ at the three points and $(Lf)(x_\pm)$. We find the quartic polynomial P which interpolates this data. We then compute

$$(LP)(x_0) = -A_+(Lf)_+ - A_-(Lf)_- + B_+f_+ + B_0f_0 + B_-f_- .$$

Replacing $(LP)(x_0)$ by $(Lf)_0$ gives the desired linear relation. Being exact for polynomials of degree four, it ought to be $O(h^3)$ accurate for 2nd order L. But if $x_\pm = x_0 \pm h$, it is known to be $O(h^4)$ accurate for $L = d/dx$ and for $L = d^2/dx^2$.

The relation found in this fashion may be described as follows: with $x_\pm = x_0 \pm h$, set

$$\text{(7.4a)} \quad w_0 \equiv 4[15\alpha_+\alpha_- - 4h(\alpha_+\beta_- - \beta_+\alpha_-) - h^2\beta_+\beta_-] ,$$

$$\text{(7.4b)} \quad w_+ \equiv 6\alpha_0\alpha_- + h(5\alpha_-\beta_0 - 2\alpha_0\beta_-) - h^2\beta_0\beta_- ,$$

$$\text{(7.4c)} \quad w_- \equiv 6\alpha_0\alpha_+ - h(5\alpha_+\beta_0 - 2\alpha_0\beta_+) - h^2\beta_0\beta_+ .$$

Then with D_+D_- the usual centered 2nd difference quotient and D_0 the corresponding 1st difference quotient, the relation

$$(7.4d)\quad w_+(Lv)_+ + w_0(Lv)_0 + w_-(Lv)_-$$
$$= [w_+\alpha_+ + w_0\alpha_0 + w_-\alpha_- + h(w_+\beta_+ - w_-\beta_-)]D_+D_-v$$
$$+ (w_+\beta_+ + w_0\beta_0 + w_-\beta_-)D_0v$$

has $O(h^4)$ truncation error (7.2) for $v \in C^6[0,1]$ if α and β are bounded. We note it may well cost more to compute these coefficients than to solve the linear system.

This scheme has yielded $O(h^4)$ convergence in actual computation concerning some two point boundary value problems. But we have neither proved its convergence for such problems nor its stability in general for (7.3).

However, we have performed a preliminary analysis for the case $\alpha \equiv \text{const.} > 0$, $\beta \equiv \text{const.}$ Using (7.4) we take

$$\text{row}(A) = (6\alpha^2 - 3h\alpha\beta - h^2\beta^2, 60\alpha^2 - 4h^2\beta^2, 6\alpha^2 + 3h\alpha\beta - h^2\beta^2)$$

$$\text{row}(B) = 72\alpha^3 D_+D_- + (72\alpha^2 - 6h^2\beta^2)\beta D_0$$

for the generic rows of A and B. A Fourier analysis yields

$$\frac{h^2 b(\theta)}{3\alpha a(\theta)} = \frac{24(\cos\theta - 1) + i\, r\,(12 - r^2)\sin\theta}{30 - 2r^2 + (6 - r^2)\cos\theta + i\, 3r\sin\theta},$$

where $\theta = 2\pi\omega h$, and where the crucial parameter is

$$r \equiv h\beta/\alpha.$$

We find A can have a zero eigenvalue:

$$|a(\theta)| = 0 \text{ only when } r^2 = 12(\theta = 0) \text{ or } r^2 = 24(\theta = \pi).$$

However, there exists $\delta_1(\varepsilon) > \delta_2(\varepsilon) > 0$ such that

$$(\cos\theta - 1)\delta_1 \le \mathrm{Re}[h^2 b(\theta)/(\alpha a(\theta))] \le (\cos\theta - 1)\delta_2 \le 0$$

when $r^2 \le 24 - \varepsilon$ and $|r^2 - 12| \ge \varepsilon > 0$.

Hence the differential-difference equation

$$A v_t = B v + \gamma A v$$

is as stable as the constant coefficient differential equation

$$u_t = \alpha u_{xx} + \beta u_x + \gamma u$$

it is supposed to approximate if we take, in particular,

$$(h\beta/\alpha)^2 < 12 - \varepsilon .$$

It should be noted that the relation (7.4) is not the only tridiagonal implicit scheme satisfying (7.2) with $p = 4$ - indeed, there appears to be a one parameter family of them [12].

8. Some other difference approximations.

This Hermite-Birkhoff approach can yield innumerable differential-difference schemes of conjectural usefulness.

Consider implicit approximations for d/dx on a variable mesh. Differentiation of the Hermite interpolant of the data

$$f_{i-m}, f_{i-m+1}, \ldots, f_{i+m};\ f'_{i-m}, \ldots, f'_{i-1};\ f'_{i+1}, \ldots, f'_{i+m};$$

at $x = x_i$ yields an implicit formula with $O(h^{4m})$ truncation error with suitable restrictions on the mesh locally. Or, if we prefer, we can devise differential-difference schemes for, say, $u_t = u_x$, where the matrices are structured precisely like those which arise in using Hermite subspaces (piecewise-polynomials of degree $2m-1$ in $C^{(m-1)}[0,1]$) in Galerkin's method. Thus, let P be the polynomial of degree $5m-1$ interpolating the (possibly unequally spaced) Hermite data

$$f_{i+1}, \ldots, f_{i+3m}; \; f'_{i+1}, \ldots, f'_{i+m}; \; f'_{i+2m+1}, \ldots, f'_{i+3m} .$$

Differentiating P at x_{i+m+1}, at $x_{i+m+2}, \ldots,$ and at x_{i+2m} we obtain m linear relations between $f'_{i+1}, \ldots,$ f'_{i+3m} and $f_{i+1}, \ldots, f_{i+3m}$. We then translate by m mesh points and repeat. The central block of the $m \times m$-block-tridiagonal matrix multiplying $Lf\ (= f')$ is diagonal (although we don't know if that matrix is nonsingular). The truncation error in this generalization of the tridiagonal $O(h^4)$ scheme is $O(h^{5m-1})$ (the Hermite spaces yield $O(h^{2m-1})$). Stability, of course, is completely unanalyzed.

We close with an amusing example of an explicit scheme for the hyperbolic problem

$$u_t = \alpha(x,t)u_x + \beta(x,t)u + \gamma(x,t) \tag{8.1}$$

which we found using these ideas.

First, let us return to the Numerov approximation to d^2/dx^2. It can be derived by interpolating $f_{i-1}, f_i, f_{i+1},$ f''_{i-1}, f''_{i+1} with a quartic which is then differentiated twice at x_i. Such a relation should have only $O(h^{5-2})$ truncation error. However, the error e in interpolating x^5 is a quintic satisfying 0 interpolating conditions. When $x_{i\pm 1} = x_i \pm h$, symmetry makes $e''(x_i) = 0$; hence the $O(h^4)$ accuracy for Numerov.

Now, the usual explicit centered approximations to d/dx arise from interpolating $f_{i-m}, \ldots, f_{i+m}$ and differentiating at x_0. Suppose we do so for three equally spaced $x_i (m=1)$. Differentiating the quadratic at the center yields the usual $O(h^2)$ error. But the error e in interpolating x^3 is a cubic whose derivative vanishes at $x_0 \pm h/\sqrt{3}$. So the derivative of the interpolant at these two points is $O(h^3)$ accurate. How can we use this to obtain an $O(h^3 + \Delta t^2)$ scheme for $u_t = u_x$? For a periodic problem we offset every other time line of equally spaced mesh points by $h/\sqrt{3}$:

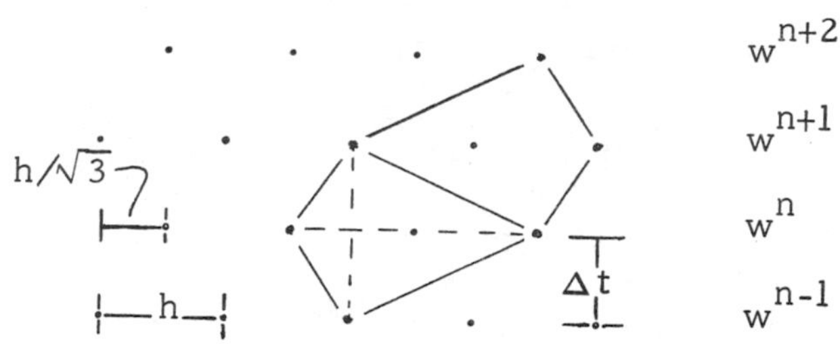

On the next time line we reverse the direction of the kite. To obtain the lower order term in (8.1) we interpolate w^n at the three appropriate points with a quadratic, evaluating it at the point on which the time difference is centered. The resulting truncation error in approximating $u_t = u_x$ is

$$u_t - u_x - Su \approx u_{xxxx} h^3/(36\sqrt{3}) + u_{ttt} \Delta t^2/6 .$$

Keeping $\Delta t^2 \sim h^3$ (see Section 1) we have realized in practice the potential $O(h^3)$ errors in this alternating kite scheme for (8.1) which takes no more work than the diamond scheme it resembles.

Appendix: Optimal Pairs (N, M) for Full Discretizations of the Heat Equation.

We tabulate, for each full space-time discretization of the heat equation, the balanced pair (N, M) which minimizes computational work of the form (4.1) given each of two fixed error levels (3.3). An (*) following a pair indicates that the cancellation of the spatial and temporal truncation errors was avoided through the artifice described in section 4. A pair is underlined if $\Delta t/h^2$ $[= N^2/(4\pi^2 M)] > \frac{1}{2}$. A (†) follows those six pairs in the table which violate the stability condition for that particular space-time difference method.

$|E| = 10^{-2}$

Spatial Scheme	Time Scheme m	Euler	Backward-Euler	duFort-Frankel	2nd order Explicit	Trapezoid	Calahan-Zlámal	4th order Padé
Explicit Differences	1	(31,76)*	(31,74)	(22,61)*	(25,6.2)†	(26,4.1)*	(24,2.5)*	(22,.85)
	2	(9.5,64)*	(9.5,62)	(7.5,19)*	(8.2,5.4)	(8.4,3.6)*	(8.0,2.2)*	(7.5,.75)
	3	(6.3,60)*	(6.3,58)	(5.2,13)*	(5.6,5.1)	(5.7,3.4)*	(5.5,2.1)*	(5.2,.72)
	4	(5.1,58)*	(5.1,56)	(4.3,11)*	(4.6,5.0)	(4.7,3.3)*	(4.6,2.0)*	(4.3,.70)
	5	(4.5,56)*	(4.5,55)	(3.8,8.7)*	(4.1,4.9)	(4.2,3.2)*	(4.1,2.0)*	(3.9,.69)
	6	(4.1,56)*	(4.1,54)	(3.6,7.8)*	(3.8,4.9)	(3.8,3.2)*	(3.7,2.0)*	(3.6,.69)
	7	(3.8,55)*	(3.8,54)	(3.4,7.2)*	(3.6,4.8)	(3.6,3.2)*	(3.5,1.9)*	(3.4,.68)
	8	(3.6,55)*	(3.6,53)	(3.2,6.8)*	(3.4,4.8)	(3.4,3.1)*	(3.4,1.9)*	(3.2,.68)
Spline Galerkin	1	(31,76)†	(31,74)*	(22,107)	(25,6.2)*†	(26,4.1)	(24,2.5)	(22,.82)*
	2	(6.2,62)	(6.2,61)*	(5.0,17)	(5.4,5.3)*	(5.5,3.5)	(5.3,2.1)	(5.0,.73)*
	3	(4.1,57)	(4.0,56)*	(3.5,9.8)	(3.7,5.0)*	(3.8,3.3)	(3.7,2.0)	(3.5,.70)*
	4	(3.4,55)	(3.4,54)*	(3.0,7.6)	(3.2,4.8)*	(3.2,3.2)	(3.1,1.9)	(3.0,.68)*
	5	(3.0,54)	(3.0,53)*	(2.7,6.5)	(2.9,4.7)*	(2.9,3.1)	(2.8,1.9)	(2.8,.67)*
Implicit Centered	1	(7.6,63)*	(7.6,61)	(6.0,16)*	(6.6,5.4)	(6.7,3.6)*	(6.4,2.2)*	(6.0,.75)
	2	(3.9,56)*	(3.9,55)	(3.4,7.7)*	(3.6,4.9)	(3.6,3.2)*	(3.5,2.0)*	(3.4,.69)
	3	(3.1,54)*	(3.1,53)	(2.8,5.8)*	(2.9,4.8)	(3.0,3.1)*	(2.9,1.9)*	(2.8,.67)
	4	(2.8,53)*	(2.8,52)	(2.5,5.0)*	(2.6,4.7)	(2.7,3.1)*	(2.6,1.9)*	(2.6,.66)
	5	(2.6,53)*	(2.6,51)	(2.4,4.5)*	(2.5,4.6)	(2.5,3.0)*	(2.5,1.9)*	(2.4,.66)

$|E| = 10^{-4}$

Spatial Scheme	Time Scheme m	Euler	Backward-Euler	duFort-Frankel	2nd order Explicit	Trapezoid	Calahan-Zlámal	4th order Padé
Explicit Differences	1	(315,7500)*	(315,7500)	(220,6100)*	(260,60)†	(260,41)*	(240,13)*	(220,2.6)
	2	(30,6300)*	(30,6300)	(24,630)*	(27,50)	(27,35)*	(25,11)*	(24,2.3)
	3	(14,5800)*	(14,5800)	(12,290)*	(13,48)	(13,33)*	(12,11)*	(12,2.2)
	4	(9.4,5600)*	(9.4,5600)	(8.1,200)*	(8.7,46)	(8.7,32)*	(8.4,10)*	(8.2,2.1)
	5	(7.4,5500)*	(7.4,5500)	(6.5,160)*	(6.9,45)	(6.9,32)*	(6.7,10)*	(6.5,2.1)
	6	(6.3,5400)*	(6.3,5400)	(5.6,130)*	(5.9,45)	(5.9,31)*	(5.8,10)*	(5.7,2.1)
	8	(5.1,5300)*	(5.1,5300)	(4.6,110)*	(4.9,44)	(4.9,31)*	(4.7,9.9)*	(4.7,2.1)
	10	(4.5,5300)*	(4.5,5300)	(4.1,91)*	(4.3,43)	(4.3,31)*	(4.2,9.8)*	(4.1,2.0)
	12	(4.1,5300)*	(4.1,5300)	(3.8,83)*	(3.9,43)	(3.9,30)*	(3.9,9.8)*	(3.8,2.0)
Spline Galerkin	1	(315,7500)†	(315,7500)*	(220,10600)	(260,58)*†	(260,41)	(235,13)	(220,2.6)*
	2	(18,6200)	(18,6200)*	(15,510)	(16,50)*	(16,35)	(15,11)	(15,2.3)*
	3	(7.7,5800)	(7.7,5800)*	(6.5,200)	(7.1,47)*	(7.1,33)	(6.8,11)	(6.6,2.2)*
	4	(5.2,5500)	(5.2,5500)*	(4.7,130)	(4.9,45)*	(4.9,32)	(4.8,10)	(4.7,2.1)*
	5	(4.2,5400)	(4.2,5400)*	(3.8,100)	(4.0,44)*	(4.0,31)	(3.9,10)	(3.9,2.1)*
	6	(3.7,5300)	(3.7,5300)*	(3.4,85)	(3.5,44)*	(3.6,31)	(3.5,10)	(3.4,2.0)*
Implicit Centered	1	(24,6200)*	(24,6200)	(19,520)*	(21,50)	(21,35)*	(20,11)*	(19,2.3)
	2	(6.7,5600)*	(6.7,5600)	(5.8,150)*	(6.2,46)	(6.2,32)*	(6.0,10)*	(5.9,2.1)
	3	(4.4,5400)*	(4.4,5400)	(4.0,94)*	(4.2,44)	(4.2,31)*	(4.1,10)*	(4.0,2.1)
	4	(3.6,5300)*	(3.6,5300)	(3.3,73)*	(3.4,43)	(3.4,30)*	(3.4,9.8)*	(3.3,2.0)
	5	(3.2,5200)*	(3.2,5200)	(3.0,62)*	(3.1,43)	(3.1,30)*	(3.0,9.7)*	(2.9,2.0)
	6	(2.9,5200)*	(2.9,5200)	(2.8,56)*	(2.8,43)	(2.8,30)*	(2.8,9.6)*	(2.8,2.0)

References.

1. G. Birkhoff and S. Gulati, Optimal few-point discretization of linear source problems, to appear SIAM J. Numer. Anal. 11 (1974).

2. L. Collatz, The Numerical Treatment of Differential Equations, Springer-Verlag, New York, 1966.

3. L. Collatz, Hermitean methods for initial value problems in partial differential equations, in Topics in Numerical Analysis, J. Miller, ed., Academic Press, New York (1974), pp. 41-61.

4. S. Crandall, An optimum implicit recurrence formula for the heat conduction equation, Quart. Appl. Math. 13 (1955), pp. 318-320.

5. W. E. Culham and R. S. Varga, Numerical methods for time-dependent nonlinear boundary value problems, Soc. of Petroleum Engineers Journal 11 (1971), pp. 374-388.

6. J. Dendy, personal communication, 1974.

7. J. Douglas, A survey of methods for parabolic differential equations, in Advances in Computers 2 (1961), Academic Press, New York, pp. 1-54.

8. J. Douglas, The solution of the diffusion equation by a high order correct difference equation, J. Math. Phys. 35 (1956), pp. 145-151.

9. J. Gary, On the optimal time step and computational efficiency of difference schemes for PDE, to appear in J. Comp. Phys.

10. C. W. Gear, Numerical Initial Value Problems in Ordinary Differential Equations, Prentice-Hall, New York, 1971.

11. H. -O. Kreiss and J. Oliger, Comparison of accurate methods for the integration of hyperbolic equations, Tellus 24 (1972), pp. 199-215.

12. H. -O. Kreiss and B. Wendroff, personal communications.

13. I. J. Schoenberg, Contributions to the problem of approximation of equidistant data by analytic functions, Quart. Appl. Math. 4 (1946), pp. 45-99, pp. 112-141.

14. B. Swartz and B. Wendroff, The comparative efficiency of certain finite element and finite difference schemes for a hyperbolic problem, in Conference on the Numerical Solution of Differential Equations, G. A. Watson, ed. Lecture Notes in Mathematics 363, Springer-Verlag, New York, 1974, pp. 153-163.

15. B. Swartz and B. Wendroff, The relative efficiency of finite difference and finite element methods, I. hyperbolic problems and splines, to appear SIAM J. Numer. Anal. 11 (1974).

16. B. Swartz and B. Wendroff, The relation between the Galerkin and collocation methods using smooth splines, to appear SIAM J. Numer. Anal. 11 (1974).

17. V. Thomée, Convergence estimates for semi-discrete Galerkin methods for initial value problems, in Numerische, insbesondere approximationstheoretische, Behandung von Functionalgleichungen, Lecture Notes in Mathematics 333, Springer-Verlag, New York, 1973, pp. 243-262.

18. V. Thomée and B. Wendroff, Convergense estimates for Galerkin methods for variable coefficient-initial-value problems, to appear in SIAM J. Numer. Anal. 11 (1974).

19. V. Thomée, Spline-Galerkin methods for initial-value problems with constant coefficients, in Conference on the Numerical Solution of Differential Equations, G. A. Watson, ed., Lecture Notes in Mathematics 363, Springer-Verlag, New York, (1974), pp. 164-175.

20. D. Watanabe and J. R. Flood, An implicit fourth order difference methods for viscous flows, Math. Comp. 28 (1974), pp. 27-32.

21. D. M. Young and R. T. Gregory, A survey of Numerical Mathematics, vol. II, Addison-Wesley, Reading, Mass., 1973.

22. M. Zlámal, Finite element methods for parabolic equations, in Conference on the Numerical Solution of Differential Equations, G. A. Watson, ed., Lecture Notes in Mathematics 363, Springer-Verlag, New York, 1974, pp. 215-221.

23. M. Zlámal, Finite element methods in parabolic equations, Math. Comp. 28 (1974), pp. 393-404.

Work partly supported by the U. S. Atomic Energy Commission.

Los Alamos Scientific Laboratory
of the University of California
Los Alamos, New Mexico 87544

L_2 Error Estimates for Projection Methods for Parabolic Equations in Approximating Domains

TODD DUPONT

1. Introduction.

Two projective methods for the approximate solution of non-linear parabolic problems with essential (Dirichlet-type) boundary conditions will be defined and analyzed. The methods produce approximate solutions in function spaces which do not satisfy the boundary conditions and the computations are performed on domains which are good approximations, but not necessarily equal, to the original domain. In one of the methods the approximate solution is defined by a sequence of linear algebraic equations even though the differential problem is very nonlinear. For each method an L_2 error estimate of optimal order will be presented.

The motivation for the definitions of these procedures can be found in the work of Dendy [4], Nitsche [8], Douglas-Dupont [5] and Bramble-Dupont-Thomée [2]. The tools used in the analysis are related to those in the work of Bramble-Dupont-Thomée [2], Wheeler [10], and Rachford [9].

Section 2 is devoted to describing the notation, the geometric assumptions on the approximating domains, the properties of the function spaces to be employed, and to a sequence of technical lemmas. In Section 3 the approximate solution of elliptic problems is discussed. The elliptic results are developed for use in treating the parabolic problem, but they are also of independent interest. In Section 4, the parabolic procedures are discussed.

2. Notations and preliminaries.

Let Ω be a bounded domain in $\mathbb{R}^p$ with $p = 2$ or 3 and assume that its boundary $\partial\Omega$ is smooth. Take $\mathcal{h}$ to be a set of positive real numbers with zero as an accumulation point. For each $h \in \mathcal{h}$ let Ω_h be a domain which approximates Ω and let $\mathcal{M}_h$ be a finite dimensional linear space of functions defined on Ω_h. The motivation for the assumptions to be made on Ω_h and $\mathcal{M}_h$ comes from the cases in which $\partial\Omega_h$ is a piecewise polynomial approximation to $\partial\Omega$ and $\mathcal{M}_h$ is a class of continuous piecewise polynomial functions.

For each $h \in \mathcal{h}$ assume that

$$\partial\Omega_h = \bigcup_{j=1}^{J_h} (\partial\Omega_h)^{(j)} , \tag{2.1}$$

where each $(\partial\Omega_h)^{(j)}$ is a closed set. Further assume that for each $j = 1, \ldots, J_h$ there is an open set $\mathcal{O} \supset (\partial\Omega_h)^{(j)}$ and a one-to-one C^∞ function φ , with non-vanishing Jacobian, mapping $\mathcal{O}$ into $\mathbb{R}^p$ in such a fashion that $\varphi((\partial\Omega_h)^{(j)})$ is the union of a polygon and its interior in $\mathbb{R}^{p-1}$. For $x \in \partial\Omega_h$ define $n(x) = n_h(x)$ to be an outward normal; note that $n(x)$ is well-defined except on a set of (p-1)-dimensional measure zero. The outward normal to $\partial\Omega$ will be denoted $\nu(x)$. For each $x \in \partial\Omega_h$ let $\delta(x) = \delta_h(x)$ be such that $x + \delta(x)n(x) \in \partial\Omega$ and $|\delta(x)| = \inf\{|s| : x + sn(x) \in \partial\Omega\}$; it will follow from the assumptions below on $\partial\Omega_h$ that for h sufficiently small $\delta(x)$ exists for each $x \in \partial\Omega_h$ and that δ is a piecewise smooth function.

The sets $\partial\Omega_h$ will be assumed to converge to $\partial\Omega$ in the sense that

$$\sup\{\mathrm{dist}(x, \partial\Omega) : x \in \partial\Omega_h\} = O(h^2) \tag{2.2}$$

as $h \to 0$. Further assume that there is a constant C , an open cover $\{\mathcal{O}_j\}_{j=1}^{N_{\mathcal{O}}}$ of $\partial\Omega$, and a collection of C^∞, one-to-one, onto functions $\varphi_j : \mathcal{O}_j \to (-1, 1)^p$ with the following properties:

$$\text{i)} \quad \varphi_j(\partial\Omega \cap \mathcal{O}_j) = \{(x',0) : x' \in (-1,1)^{p-1}\},$$

$$(2.3) \quad \text{ii)} \quad \varphi_j(\Omega \cap \mathcal{O}_j) = \{(x',y) : x' \in (-1,1)^{p-1}, y > 0\},$$

$$\text{iii)} \quad \varphi_j(\Omega_h \cap \mathcal{O}_j) = \{(x',y) : x' \in (-1,1)^{p-1}, \ y > g_{h,j}(x')\},$$

where $g_{h,j}$ is a continuous, piecewise smooth function with values in $(-1,1)$ such that $|\nabla g_{h,j}| \leq Ch$, and the second derivatives of $g_{h,j}$ are bounded by C, where $g_{h,j}$ is smooth.

Note that these properties and (2.2) imply that there is a C' so that as $h \to 0$

$$(2.4) \qquad \bar{\delta} = \sup\{|\delta(x)| : x \in \partial\Omega_h\} \leq C'h^2$$

and

$$(2.5) \qquad \sup\{|n(x) - \nu(x + \delta(x)n(x))| : x \in \partial\Omega_h\} \leq C'h.$$

Henceforth assume that all $h \in \mathcal{h}$ are sufficiently small that δ_h is defined and $\delta\Omega_h$ is covered by the $\mathcal{O}_j$'s.

For each $h \in \mathcal{h}$ let

$$(2.6) \qquad \Omega^{(j)} = \{x + sn(x) : s \in I_h(x), \ x \in (\partial\Omega_h)^{(j)}\},$$

where $I_h(x)$ is the closed interval with end points 0 and $\delta(x)$. Assume that there is an M_1 independent of h such that

$$(2.7) \qquad \Big\| \sum_{j=1}^{J_h} \chi_{\Omega^{(j)}} \Big\|_{L_\infty(\mathbb{R}^p)} \leq M_1,$$

where χ_A denotes the characteristic function of the set A.

For a measurable set $A \subset \mathbb{R}^p$ and a nonnegative integer s let

$$(2.8a) \qquad \|\varphi\|^2_{H^s(A)} = \sum_{|\alpha| \leq s} \|D^\alpha \varphi\|^2_{L_2(A)},$$

where the α's in the sum are p-tuples of nonnegative integers $\alpha = (\alpha_1, \ldots, \alpha_p)$, $D^\alpha\varphi = (\frac{\partial}{\partial x_1})^{\alpha_1} \ldots (\frac{\partial}{\partial x_p})^{\alpha_p}\varphi$, and $|\alpha| = \alpha_1 + \ldots + \alpha_p$. The space $H^s(A)$ is the closure of $C^\infty(\mathbb{R}^p)$ functions in the above norm. For nonnegative inteters s and $\varphi \in L_2(A)$,

$$\|\varphi\|_{H^{-s}(A)} = \sup\{\int_A \varphi\psi dx : \|\psi\|_{H^s(A)} = 1\}. \tag{2.8b}$$

Also adopt the notations

$$\begin{aligned}
&\|\varphi\| = \|\varphi\|_{L_2(\Omega)}, && \|\varphi\|_h = \|\varphi\|_{L_2(\Omega_h)},\\
&\|\varphi\|_s = \|\varphi\|_{H^s(\Omega)}, && \|\varphi\|_{h,s} = \|\varphi\|_{H^s(\Omega_h)},\\
&|\varphi| = \|\varphi\|_{L_2(\partial\Omega)}, && |\varphi|_h = \|\varphi\|_{L_2(\partial\Omega_h)},\\
&(f,g) = \int_\Omega fg dx, && (f,g)_h = \int_{\Omega_h} fg\, dx,\\
&\langle f,g\rangle = \int_{\partial\Omega} fg\, d\sigma, && \langle f,g\rangle_h = \int_{\partial\Omega_h} fg\, d\sigma.
\end{aligned} \tag{2.9}$$

For each $h \in \mathcal{H}$ let $\mathcal{N}_h$ be the space of functions $v \in H^1(\Omega_h)$ such that for almost every point $x \in \partial\Omega_h$ there is an open ball B about x such that v is in $H^2(\Omega_h \cap B)$ and such that $\frac{\partial v}{\partial n}$ is in $L_2(\partial\Omega_h)$. Note in particular that $H^2(\Omega_h) \subset \mathcal{N}_h \subset H^1(\Omega_h)$. Define a norm on $\mathcal{N}_h$ by

$$|||\varphi|||^2 = \|\varphi\|_{h,1}^2 + h\left|\frac{\partial\varphi}{\partial n}\right|_h^2 + h^{-1}|\varphi|_h^2. \tag{2.10}$$

If X is a normed space with norm $\|\;\|_X$ and $\varphi:[0,T] \to X$, then

$$(2.11)\qquad \begin{aligned} \|\varphi\|_{L_s(X)} &= \Big[\int_0^T \|\varphi(t)\|_X^s dt\Big]^{1/s}, \qquad 1 \le s < \infty, \\ \|\varphi\|_{L_\infty(X)} &= \sup_{0 \le t \le T} \|\varphi(t)\|_X ; \end{aligned}$$

we say $\varphi \in L_s(X)$ if $\|\varphi\|_{L_s(X)} < \infty$.

It is convenient to be able to extend functions in $H^s(\Omega)$ or $H^s(\Omega_h)$ to be functions in $H^s(\mathbb{R}^p)$ in such a fashion that the extension process is a continuous linear operator. There are two types of these extensions that will be useful. The first we shall call a Lions extension. Given an integer $k \ge 0$ there is a map E_k, defined for any $\varphi:\Omega \to \mathbb{R}$, such that $E_k\varphi:\mathbb{R}^p \to \mathbb{R}$, $E_k\varphi = \varphi$ on Ω, and there is a C_{E_k} such that for $-1 \le s \le k$

$$(2.12)\qquad \|E_k\varphi\|_{H^s(\mathbb{R}^p)} \le C_{E_k} \|\varphi\|_{H^s(\Omega)} ;$$

E_k is the Lions extension of order k. The construction of this map E_k is reasonably elementary and can be found in [7].

At one point in the construction of the Lions extension it is necessary to use a C^k change of variables to flatten a piece of $\partial\Omega$; this makes this extension operator unsuitable for extending functions on $H^s(\Omega_h)$. For such functions we shall employ an extension process due to Calderón. Notice that it follows from the assumptions (2.3) that there is an open cover $\{\tilde{\mathcal{O}}_j\}_{j=1}^{N_{\mathcal{O}}}$ of $\partial\Omega \cup \bigcup_{h\in\mathcal{h}} \partial\Omega_h$, where these sets $\tilde{\mathcal{O}}_j$ may be slightly smaller than the sets $\mathcal{O}_j$, and a corresponding set of cones $\{\tilde{C}_j\}_{j=1}^{N_{\mathcal{O}}}$, each with the origin as vertex, such that

$$(2.13)\qquad \begin{aligned} &\text{i) } x \in \tilde{\mathcal{O}}_j \cap \Omega \text{ implies } x + \tilde{C}_j \subset \Omega, \\ &\text{ii) } x \in \tilde{\mathcal{O}}_j \cap \Omega_h \text{ implies } x + \tilde{C}_j \subset \Omega_h . \end{aligned}$$

It then follows from the Calderón extension theorem [1, 3] that for each positive integer s there exist continuous linear extension operators $\mathcal{E}_s$ and $\mathcal{E}_{h,s}$ such that

$$\text{(2.14)} \qquad \begin{aligned} &\mathcal{E}_s : H^s(\Omega) \to H^s(\mathbb{R}^p) , \\ &\mathcal{E}_{h,s} : H^s(\Omega_h) \to H^s(\mathbb{R}^p) . \end{aligned}$$

The proof of the Calderón extension theorem [1] and (2.13) show that the norms of the operators $\mathcal{E}_{h,s}$ can be bounded independently of h, for each fixed s. At one point it will be convenient to extend a function defined on $\Omega \cup \Omega_h$ to $\mathbb{R}^p$ via a Calderón extension. Note that (2.13) also implies that the norm of such an extension operator can be bounded independently of h.

It is not the case that the Calderón extension of $H^s(\Omega)$ to $H^s(\mathbb{R}^p)$ is continuous as a map into $H^k(\mathbb{R}^p)$ from $H^s(\Omega) \cap H^k(\Omega)$ equipped with the $H^k(\Omega)$ norm for $0 \leq k < s$; it is this difference between the Lions and Calderón extensions that prompts us to use the Lions extension for certain functions in $H^s(\Omega)$ instead of the Calderón extension.

The "uniform restricted cone property" exhibited in (2.13) also has the consequence that there is a constant C, independent of h, such that if $v \in H^2(\Omega_h)$, then

$$\text{(2.15)} \qquad \|v\|_{L_\infty(\Omega_h)} \leq C \|v\|_{h,2} .$$

This can be seen easily by extending v to $H^2(\mathbb{R}^p)$ and using the imbedding result on $\mathbb{R}^p$.

Several lemmas will now be presented that will be used frequently in the computations of the following sections.

Lemma 2.1.

For each $k = 1, 2, \ldots$ there is a constant C such that if $\varphi \in H^k(\mathbb{R}^p)$ and $\varphi \equiv 0$ in Ω then

$$\text{(2.16)} \qquad \|\varphi\|_h \leq C \, \bar{\delta}^k \|\varphi\|_{H^k(\mathbb{R}^p)} .$$

Proof.

It is clearly sufficient to assume that φ is a smooth function with its support (compactly) contained in one of the sets $\mathcal{O}_j$ used to cover $\partial\Omega$ in (2.3). By change of variables we see that there is no loss of generality in assuming that $\partial\Omega$ coincides near the support of φ with the hyperplane $\{x = (x_1, \ldots, x_p) : x_p = 0\}$. With $x = (x', x_p)$, Taylor's formula implies that

$$\varphi(x) = \frac{1}{(k-1)!} \int_0^{x_p} (x_p - t)^{k-1} \left(\frac{\partial}{\partial x_p}\right)^k \varphi(x', t)\,dt .$$

Hence

$$\varphi^2(x', x_p) \leq ((k-1)!)^{-2} (2k-1)^{-1} |x_p|^{2k-1} \int_{\mathbb{R}} \left(\left(\frac{\partial}{\partial x_p}\right)^k \varphi\right)^2 (x', t)\,dt.$$

If this inequality is integrated for x_p in the interval $\tilde{I}(x')$ with end points 0 and $\bar{x}_p$, where $(x', \bar{x}_p) \in \partial\Omega_h$, it follows that

$$\int_{\tilde{I}(x')} \varphi^2(x', x_p)\,dx_p \leq C |\bar{x}_p|^{2k} \int_{\mathbb{R}} \left(\left(\frac{\partial}{\partial x_p}\right)^k \varphi\right)^2 (x', t)\,dt$$

$$\leq C\, \bar{\delta}^{2k} \int_{\mathbb{R}} \left(\left(\frac{\partial}{\partial x_p}\right)^k \varphi\right)^2 (x', t)\,dt .$$

Integrating this inequality with respect to x' gives the conclusion.

Lemma 2.2.

There is a constant C such that, for h sufficiently small and $\varphi \in L_2(\mathbb{R}^p)$,

$$\int_{\partial\Omega_h} \int_{I_h(x)} \varphi^2(x + n(x)t)\,dt\, d\sigma(x) \leq C \|\varphi\|^2_{L_2(\Omega\Delta\Omega_h)}, \tag{2.17}$$

where $I_h(x)$ is defined immediately after (2.6) and $\Omega\Delta\Omega_h$ denotes the symmetric difference of Ω and Ω_h.

Proof.

This lemma follows from integrating over each $(\partial\Omega_h)^{(j)}$, using (2.3 iii), summing for $j = 1, \ldots, J_h$, and using (2.7). In fact, $C \leq M_1(1 + O(h))$.

Assume henceforth that all $h \in \mathcal{h}$ are so small that (2.17) holds.

Lemma 2.3.

There is a constant C such that if $\varphi \in H^1(\mathbb{R}^p)$, then

$$\|\varphi\|_{L_2(\Omega \Delta \Omega_h)} \leq C \bar{\delta}^{\frac{1}{2}} \|\varphi\|_{H^1(\mathbb{R}^p)} . \tag{2.18}$$

Proof.

Just as in the proof of Lemma 2.1 we may assume that φ has its support in a particular $\mathcal{O}_j$, and that $\partial\Omega$ coincides with the hyperplane $x_p = 0$. Note that for any $x = (x', x_p)$

$$\varphi^2(x) \leq \int_{\mathbb{R}} [\varphi^2 + (\frac{\partial}{\partial x_p} \varphi)^2](x', t)dt .$$

Thus for each x', with $\tilde{I}(x')$ as in the proof of Lemma 2.1,

$$\int_{\tilde{I}(x')} \varphi^2(x', x_p)dx_p \leq C \bar{\delta} \int_{\mathbb{R}} [\varphi^2 + (\frac{\partial}{\partial x_p})^2](x', t)dt .$$

If we integrate both sides with respect to x' we get the conclusion.

Lemma 2.4.

There is a constant C independent of h such that for $\varphi \in H^1(\Omega_h)$ and $\psi \in H^1(\Omega)$

(2.19)
$$|\varphi|_h \le C[\|\varphi\|_h \|\varphi\|_{h,1}]^{\frac{1}{2}} \le C\|\varphi\|_{h,1} ,$$
$$|\psi| \le C[\|\psi\| \|\psi\|_1]^{\frac{1}{2}} \le C\|\psi\|_1 .$$

Proof.

The inequality for ψ is well known [7]. In order to see that C is independent of h we reproduce the usual proof. We can assume that φ has its support in a single $\mathcal{O}_j$ and that $\partial\Omega$ is the $x_p = 0$ hyperplane in $\mathcal{O}_j$. We also assume without loss of generality that if $x = (x', x_p)$ and $x = (x', x_p) \in \partial\Omega_h \cap \mathcal{O}_j$ then there is a point $\bar{x} = (x', \bar{x}_p)$ with $\bar{x}_p > x_p$ such that the line from x to $\bar{x}$ is in Ω_h and $\varphi(\bar{x}) = 0$. Note that

$$\varphi^2(x', x_p) \le \int_{x_p}^{\bar{x}_p} 2(\varphi \frac{\partial\varphi}{\partial x_p})(x', t)dt .$$

Integrating with respect to x' and using the boundedness of the functions $|\nabla g_{h,j}|$ of (2.3 iii) gives the first inequality for φ.

Lemma 2.5.

There exists a constant C such that for $\varphi \in H_0^1(\Omega) \cap H^2(\mathbb{R}^p)$

(2.20)
$$\left|\varphi + \delta \frac{\partial\varphi}{\partial n}\right|_h \le C\,\bar{\delta}^{3/2}\|\varphi\|_{H^2(\Omega\Delta\Omega_h)} .$$

Proof.

As before we can assume that φ is a smooth function with compact support and that, near the support of φ, $\partial\Omega$ coincides with the hyperplane $x_p = 0$. Since for $x \in \partial\Omega_h$, $\varphi(x + \delta(x)n(x)) = 0$, Taylor's theorem implies that for $x \in \partial\Omega_h$

$$|(\varphi + \delta \frac{\partial\varphi}{\partial n})(x)|^2 = |\int_0^{\delta(x)} (\delta(x) - t)\varphi''(x+tn(x))dt|^2$$

$$\leq \frac{1}{3} |\delta(x)|^3 \int_{I_h(x)} (\varphi'')^2(x + tn(x))dt ,$$

where φ'' denotes the second derivative with respect to t of the function $\varphi(x + tn(x))$. If $|\varphi''|$ is estimated using the second partial derivatives of φ and the resultant inequality is integrated over $\partial\Omega_h$, we see that

$$\left|\varphi + \delta \frac{\partial\varphi}{\partial n}\right|_h^2 \leq C\bar{\delta}^3 \int_{\partial\Omega_h} \int_{I_h(x)} \sum_{|\alpha|=2} |D^\alpha \varphi|^2(x+tn(x))dt\, d\sigma(x).$$

Lemma 2.2 then implies the conclusion.

Lemma 2.6.

<u>There is a constant</u> C <u>such that for</u> $\varphi \in H_0^1(\Omega) \cap H^3(\mathbb{R}^p)$

$$(2.21) \qquad \left|\varphi + \delta \frac{\partial\varphi}{\partial n}\right|_h \leq C\,\bar{\delta}^2 \|\varphi\|_{H^3(\mathbb{R}^p)} .$$

Proof.

Apply Lemma 2.5 and then estimate $\|\varphi\|_{H^2(\Omega\Delta\Omega)}$ using Lemma 2.3.

Lemma 2.7.

<u>There is a constant</u> C_0, <u>independent of</u> $h \in \mathcal{h}$, <u>such that for</u> $\varphi \in H^1(\Omega)$ <u>and</u> $\psi \in H^1(\Omega_h)$

$$(2.22) \qquad \begin{aligned} \|\varphi\|_1^2 &\leq C_0[\|\nabla\varphi\|^2 + |\varphi|^2], \\ \|\psi\|_{h,1}^2 &\leq C_0[\|\nabla\psi\|_h^2 + |\psi|_h^2]. \end{aligned}$$

Proof.

Let D be a bounded open set in $\mathbb{R}^p$ with a piecewise smooth boundary ∂D; i.e., ∂D can be described as in (2.1) and the sentence following it. Let $2d = \sup\{|x_p - y_p| : (x', x_p) \in D, (y', y_p) \in D\}$. The lemma will be proved if it is shown that for $v \in H^1(D)$

$$\|v\|^2_{L_2(D)} \le 4d^2 \left\|\frac{\partial v}{\partial x_p}\right\|^2_{L_2(D)} + 2d|v|^2_{L_2(\partial D)}. \tag{2.23}$$

There is clearly no loss in generality in assuming that $D \subset \{(x', x_p) : |x_p| < d,\ x' \in \mathbb{R}^{p-1}\}$. Now note that

$$\|v\|^2_{L_2(D)} = \int_D v^2(x) \frac{\partial}{\partial x_p} x_p \, dx$$

$$= -\int_D x_p \frac{\partial}{\partial x_p}(v^2)(x)dx + \int_{\partial D} v^2(x) x_p n_p(x) d\sigma(x),$$

where for $x \in \partial D$, $n_p(x)$ is the p-th component of the outward normal to ∂D. From this it follows that

$$\|v\|^2_{L_2(D)} \le 2d\|v\|_{L_2(D)} \left\|\frac{\partial v}{\partial x_p}\right\|_{L_2(D)} + d\|v\|^2_{L_2(\partial D)}$$

$$\le \tfrac{1}{2}\|v\|^2_{L_2(D)} + 2d^2\left\|\frac{\partial v}{\partial x_p}\right\|^2_{L_2(D)} + d\|v\|^2_{L_2(\partial D)}.$$

This implies (2.23) and thus completes the proof.

Lemma 2.8.

There is a constant C_1, independent of $h \in \mathcal{h}$ such that for $\varphi \in \mathcal{K}_h$, $\psi \in H^1(\Omega)$

$$\|\varphi\|_{L_2(\Omega_h \backslash \Omega)} \le C_1[\bar{\delta}^{\frac{1}{2}}|\varphi|_h + \bar{\delta}\|\varphi\|_{H^1(\Omega_h \backslash \Omega)}], \tag{2.24a}$$

$$\|\varphi\|_{L_2(\Omega_h \backslash \Omega)} \le C_1(\bar{\delta} h)^{\frac{1}{2}} |||\varphi||| , \tag{2.24b}$$

(2.24c) $\qquad \|\psi\|_{L_2(\Omega\setminus\Omega_h)} \le C_1[\bar{\delta}^{\frac{1}{2}}|\psi| + \bar{\delta}\,\|\psi\|_{H^1(\Omega\setminus\Omega_h)}]$.

Proof.

The proof of (2.24c) is a simplification of the proof of (2.24a) and will be omitted. There is no loss in generality in assuming that, near the support of φ, $\partial\Omega$ is the hyperplane $x_p = 0$ and that points in $\partial\Omega_h$ are of the form $(x', g(x'))$, where g is as in (2.3 iii). For each x' for which $g(x') > 0$ we write (for $0 \le x_p \le g(x')$)

$$\varphi^2(x', x_p) \le \varphi^2(x', g(x')) + \int_0^{g(x')} [\tfrac{1}{\varepsilon}\varphi^2 + \varepsilon(\tfrac{\partial\varphi}{\partial x_p})^2](x', s)ds.$$

Thus,

$$\int_0^{g(x')} \varphi^2(x', x_p)dx_p$$

$$\le \bar{\delta}[\varphi^2(x', g(x')) + \int_0^{g(x')} [\tfrac{1}{\varepsilon}\varphi^2 + \varepsilon(\tfrac{\partial}{\partial x_p}\varphi)^2](x', s)ds].$$

If we integrate with respect to x' and use the fact that ∇g is bounded near the support of φ, we see that

$$\|\varphi\|^2_{L_2(\Omega_h\setminus\Omega)} \le C\bar{\delta}\,[\,|\varphi|^2_h + \tfrac{1}{\varepsilon}\,\|\varphi\|^2_{L_2(\Omega_h\setminus\Omega)} + \varepsilon\|\varphi\|^2_{H^1(\Omega_h\setminus\Omega)}].$$

Choose $\frac{1}{\varepsilon} C\bar{\delta} = \frac{1}{2}$; i.e., $\varepsilon = 2C\bar{\delta}$. Then

$$\|\varphi\|^2_{L_2(\Omega_h\setminus\Omega)} \le C[\bar{\delta}\,|\varphi|^2_h + \bar{\delta}^2\,\|\varphi\|^2_{H^1(\Omega_h\setminus\Omega)}].$$

This implies (2.24a) which implies (2.24b).

For each $h \in \mathscr{h}$ let $\mathscr{M}_h$ be a finite dimensional subspace of $\mathscr{V}_h$. Assume that functions in $\mathscr{M}_h$ have bounded gradients and that there is a constant C_2, independent of h, such that for $V \in \mathscr{M}_h$

(2.25a) $$\|V\|_{L_\infty(\Omega_h)} + h\|\nabla V\|_{L_\infty(\Omega_h)} \le C_2 h^{-p/2}\|V\|_h ,$$

(2.25b) $$h^{\frac{1}{2}}\left|\frac{\partial V}{\partial n}\right|_h \le C_2\|V\|_{h,1} ,$$

(2.25c) $$h|||V||| \le C_2\|V\|_h ;$$

these are called "inverse assumptions" on the family of spaces $\{\mathcal{M}_h\}$. Suppose in addition that, for some integer $r \ge 2$, C_2 and $\{\mathcal{M}_h\}$ are such that for $2 \le k \le r+1$ and $\varphi \in H^k(\Omega_h)$

(2.26) $$\inf\{\|\varphi - V\|_h + h|||\varphi - V||| : V \in \mathcal{M}_h\} \le C_2 h^k \|\varphi\|_{h,k} .$$

In light of the remarks about the Calderón extensions $\mathcal{E}_{h,k}$ the approximation assumption is equivalent to the condition that for all $\varphi \in H^k(\mathbb{R}^p)$ with $2 \le k \le r+1$

(2.27) $$\inf\{\|\varphi - V\|_h + h|||\varphi - V||| : V \in \mathcal{M}_h\} \le Ch^k \|\varphi\|_{H^k(\mathbb{R}^p)} .$$

The spaces will also be required to approximate well in the max norm in the sense that, for $3 \le k \le r+1$, if $\varphi \in H^k(\Omega_h)$ then

(2.28) $$\inf\{\|\varphi - V\|_h + h|||\varphi - V||| + h^{p/2}(\|\varphi - V\|_{L_\infty(\Omega_h)} + h\|\nabla(\Omega - V)\|_{L_\infty(\Omega_h)}) : V \in \mathcal{M}_h\} \le C_2 h^k \|\varphi\|_{h,k} .$$

For later convenience the constant C_2 will be taken so large that in addition to the above relations we have that C_2 bounds the norms of the operators $\mathcal{E}_s$, $\mathcal{E}_{h,s}$ of (2.14) for $0 \le s \le r+1$ and that $C_{E_k} \le C_2$ for $0 \le k \le r+1$.

3. Elliptic theory.

This section contains elliptic error estimates related to those of Bramble-Dupont-Thomée [2]; these estimates are more general in some ways and less general in others than those of [2]. These elliptic results will be used to prove error bounds for the parabolic procedures of the next section.

Let $a(x)$ be a smooth function on $\bar{\Omega}$. Assume that $a(x)$ is uniformly bounded above and below by positive numbers. Let $\tilde{a}(x)$ be an extension of $a(x)$ to $\bar{\Omega} \cup \bigcup_{h \in \mathscr{h}} \bar{\Omega}_h$ such that $\tilde{a}$ is bounded above and below by the positive constants $\bar{a}$ and $\underline{a}$ respectively.

For each $h \in \mathscr{h}$ let $\rho = \rho_h$ be an approximation to $\delta = \delta_h$ defined on $\partial\Omega_h$. This approximation ρ will be used in the numerical methods to correct for the perturbation of the domain. It seems reasonable to choose ρ in one of two fundamentally different ways. The choice $\rho \equiv 0$ corresponds to not correcting for the perturbation of the domain. To take full advantage of the correction process the choice of ρ should be such that $\rho - \delta$ is of the same order as the fundamental error in the approximation process. To simplify some of the computations we assume throughout that

$$\overline{\delta - \rho} = \sup_{x \in \partial\Omega_h} |\delta(x) - \rho(x)| \leq C_3 h^3 \tag{3.1}$$

and that

$$\bar{\delta} \leq C_3 h^2, \quad \bar{\rho} = \sup_{x \in \partial\Omega_h} |\rho(x)| \leq C_3 h^2 . \tag{3.2}$$

Let γ be a positive number and let φ and ψ be elements of $\mathscr{N}_h$. Set

$$N(\varphi, \psi) = (\tilde{a} \nabla\varphi, \nabla\psi)_h - \langle \tilde{a} \frac{\partial\varphi}{\partial n}, \psi \rangle_h - \langle \varphi + \rho \frac{\partial\varphi}{\partial n}, \tilde{a} \frac{\partial\psi}{\partial n} - \gamma h^{-1}(\psi + \rho \frac{\partial\psi}{\partial n}) \rangle_h . \tag{3.3}$$

Lemma 3.1.

There is a $\gamma_0 = \gamma_0(\underline{a}, \bar{a}, c_0, C_1, C_2, C_3)$ such that for $\gamma \geq \gamma_0$ there are positive constants h_γ and C_4 such that if $h \leq h_\gamma$, then

$$\|\!| V |\!\|^2 \leq C_4 N(V,V), \qquad V \in \mathcal{M}_h \ . \tag{3.4}$$

Proof.

It follows from (3.2) and (3.3) that

$$\begin{aligned} N(V,V) &\geq \underline{a} \| \nabla V \|_h^2 + \gamma h^{-1} \left| V + \rho \frac{\partial V}{\partial n} \right|_h^2 \\ &\qquad - \bar{a} \left| \frac{\partial V}{\partial n} \right|_h \left(2|V|_h + C_3 h^2 \left| \frac{\partial V}{\partial n} \right|_h \right) . \end{aligned} \tag{3.5}$$

Thus for any $\varepsilon > 0$

$$\begin{aligned} N(V,V) &\geq \underline{a} \| \nabla V \|_h^2 + \gamma h^{-1} \left| V + \rho \frac{\partial V}{\partial n} \right|_h^2 \\ &\qquad - \left[\frac{\bar{a}^2}{\varepsilon} h^{-1} |V|_h^2 + (\varepsilon + \bar{a} C_3 h) h \left| \frac{\partial V}{\partial n} \right|_h^2 \right] \\ &\geq \underline{a} \| \nabla V \|_h^2 + \left(\tfrac{1}{2} \gamma - \frac{\bar{a}^2}{\varepsilon} \right) h^{-1} |V|_h^2 \\ &\qquad - (\varepsilon + \tilde{a} C_3 h + \gamma C_3^2 h^2) h \left| \frac{\partial V}{\partial n} \right|_h^2 . \end{aligned} \tag{3.6}$$

For $V \in \mathcal{M}_h$, (2.25b) and Lemma 2.7 imply that

$$h \left| \frac{\partial V}{\partial n} \right|_h^2 \leq C_0 C_2^2 [\| \nabla V \|_h^2 + |V|_h^2] .$$

Hence (3.6) shows that for $V \in \mathcal{M}_h$

$$\begin{aligned} N(V,V) &\geq (\underline{a} - C_0 C_2^2 [\varepsilon + \bar{a} C_3 h + \gamma C_3^2 h^2]) \| \nabla V \|_h^2 \\ &\qquad + \left(\tfrac{1}{2} \gamma - \left(\frac{\bar{a}^2}{\varepsilon} + C_0 C_2^2 [\varepsilon + \bar{a} C_3 h + \gamma C_3^2 h^2] \right) \right) |V|_h^2 . \end{aligned}$$

Take $\varepsilon = \underline{a}/(2C_0C_2^2)$ and $\gamma \geq \gamma_0 = \underline{a}(1 + 8(\bar{a}/\underline{a})^2 C_0 C_2^2)$. Then for h sufficiently small and $V \in \mathcal{M}_h$

(3.7) $$N(V, V) \geq \tfrac{1}{4}\underline{a}(\|\nabla V\|_h^2 + |V|_h^2).$$

The conclusion now follows from (3.7), Lemma 2.7 and (2.25b).

Henceforth assume that $\gamma \geq \gamma_0$ is fixed and that $h \leq h_\gamma$. The following lemma is trivial.

Lemma 3.2.

There is a constant $C_5 = C_5(\bar{a}, C_3)$ such that for $\varphi, \psi \in \mathcal{N}_h$

(3.8) $$N(\varphi, \psi) \leq C_5 |||\varphi||| \, |||\psi||| .$$

It is convenient to present here three lemmas that will be applied in several different situations to derive results needed to handle the parabolic problem.

Lemma 3.3.

Suppose that $\zeta \in \mathcal{N}_h$ and that B is a continuous linear functional on $\mathcal{M}_h$ with norm $|||B|||$ with respect to the norm $|||\cdot|||$ on $\mathcal{M}_h$. If

(3.9) $$N(\zeta, V) = B(V), \quad V \in \mathcal{M}_h,$$

then

(3.10) $$|||\zeta||| \leq (C_4C_5 + 1)\inf\{|||\zeta - \chi||| : \chi \in \mathcal{M}_h\} + C_4 |||B||| .$$

Proof.

For any $\chi \in \mathcal{M}_h$, Lemmas 3.1 and 3.2 imply that

$$\begin{aligned}|||\chi|||^2 &\le C_4[N(\chi - \zeta, \chi) + N(\zeta, \chi)] \\ &= C_4[N(\chi - \zeta, \chi) + B(\chi)] \\ &\le C_4[C_5 |||\chi - \zeta||| + |||B|||]\,|||\chi||| .\end{aligned}$$

From this we see that

$$\begin{aligned}|||\zeta||| &\le |||\zeta - \chi||| + |||\chi||| \\ &\le (1 + C_4 C_5)|||\zeta - \zeta||| + C_4 |||B||| .\end{aligned}$$

Taking the infimum over all $\chi \in \mathcal{M}_h$ gives the conclusion.

Lemma 3.4.

Suppose that $\zeta \in \mathcal{V}_h$ and that B is a continuous linear functional on $\mathcal{V}_h$ with norm $|||B|||$ with respect to the $|||\cdot|||$ norm. Suppose also that

$$N(\zeta, V) = B(V) , \qquad V \in \mathcal{M}_h . \tag{3.11}$$

Let

$$|B|_\ell = \sup\{ |B(\varphi)| : \varphi \in H^{2+\ell}(\mathbb{R}^p), \varphi \equiv 0 \text{ on } \partial\Omega, \quad \|\varphi\|_{H^{2+\ell}(\mathbb{R}^p)} = 1\}. \tag{3.12}$$

Then there is a constant C_6, depending on $\bar{a}, \underline{a}, C_0, \ldots, C_5, \Omega$, and $\|\tilde{a}\|_{H^2(\Omega \cup \Omega_h)}$, such that

$$\|\zeta\|_{h,-\ell} \le C_6\{h^{\ell+1}(|||\zeta||| + |||B|||) + |B|_\ell\}, \quad \ell = 0 \text{ or } 1. \tag{3.13}$$

Proof.

Take $\psi \in H^\ell(\Omega_h)$ and let $\tilde{\psi} = \mathcal{E}_{h,\ell}\psi$, where $\mathcal{E}_{h,\ell}$ is the Calderón extension operator of (2.14). Take

$\varphi \in H_0^1(\Omega) \cap H^{\ell+2}(\Omega)$ such that

$$-\nabla \cdot a \nabla \varphi = \tilde{\psi} \quad \text{in } \Omega . \tag{3.14}$$

Let $\tilde{\varphi} = \mathcal{E}_{\ell+2}\varphi \in H^{\ell+2}(\mathbb{R}^p)$. Then there is a constant C, depending only on the permitted quantities, such that

$$\|\tilde{\varphi}\|_{H^{\ell+2}(\mathbb{R}^p)} \leq C\|\psi\|_{h,\ell}, \qquad \ell = 0 \text{ or } 1. \tag{3.15}$$

Note that

$$\begin{aligned}(\zeta,\psi)_h &= -(\zeta, \nabla\cdot \tilde{a}\nabla\tilde{\varphi})_h + (\zeta, \psi + \nabla\cdot\tilde{a}\nabla\tilde{\varphi})_h \\ &= N(\zeta,\tilde{\varphi}) + B_1(\tilde{\varphi}) + (\zeta,\psi+\nabla\cdot\tilde{a}\nabla\tilde{\varphi})_h ,\end{aligned} \tag{3.16}$$

where

$$B_1(\tilde{\varphi}) = \left\langle \tilde{a}\frac{\partial\zeta}{\partial n} - \gamma h^{-1}\left(\zeta + \rho\frac{\partial\zeta}{\partial n}\right), \tilde{\varphi} + \rho\frac{\partial\tilde{\varphi}}{\partial n}\right\rangle_h . \tag{3.17}$$

Let $a^* = \mathcal{E}_{h,2}\tilde{a}$. (It is of no concern to us that a^* might not agree with a on $\Omega\backslash\Omega_h$.) Then estimate the last term in (3.16) using Lemmas 2.1 and 2.8 as follows:

$$\begin{aligned}(\zeta,\psi+\nabla\cdot\tilde{a}\nabla\tilde{\varphi})_h &\leq \|\zeta\|_{L_2(\Omega_h\backslash\Omega)}\|\psi+\nabla\cdot\tilde{a}\nabla\tilde{\varphi}\|_h \\ &\leq C_1(\bar{\delta}h)^{\frac{1}{2}}|||\zeta|||\,C\bar{\delta}^{\ell}\|\tilde{\psi}+\nabla\cdot a^*\nabla\tilde{\varphi}\|_{H^{\ell}(\mathbb{R}^p)} \\ &\leq Ch^{\ell+1}|||\zeta|||\;\|\psi\|_{h,\ell} .\end{aligned} \tag{3.18}$$

The $B_1(\tilde{\varphi})$ term is bounded, using Lemmas 2.5 and 2.4, as follows:

$$\begin{aligned}B_1(\tilde{\varphi}) &\leq Ch^{-\frac{1}{2}}|||\zeta|||\left(\left|\tilde{\varphi}+\delta\frac{\partial\tilde{\varphi}}{\partial n}\right|_h + \left|(\rho-\delta)\frac{\partial\tilde{\varphi}}{\partial n}\right|_h\right) \\ &\leq Ch^{-\frac{1}{2}}|||\zeta|||(\bar{\delta}^{3/2}+C_3h^3)\|\tilde{\varphi}\|_{H^2(\mathbb{R}^p)} \\ &\leq Ch^{5/2}|||\zeta|||\;\|\psi\|_h \leq Ch^{\ell+1}|||\zeta|||\;\|\psi\|_{h,\ell} .\end{aligned} \tag{3.19}$$

Using (3.11), Lemma 3.2, (2.26), (3.12), and (3.15) it can be seen that for appropriate $V \in \mathcal{M}_h$

$$N(\zeta,\tilde{\varphi}) = N(\zeta,\tilde{\varphi}-V) + B(V-\tilde{\varphi}) + B(\tilde{\varphi})$$

$$\text{(3.20)} \qquad \leq C_5 |||\zeta|||\ |||\tilde{\varphi}-V||| + |||B|||\ |||\tilde{\varphi}-V||| + |B|_\ell \|\tilde{\varphi}\|_{H^{\ell+2}(\mathbb{R}^p)}$$

$$\leq C[(|||\zeta||| + |||B|||)h^{\ell+1} + |B|_\ell]\|\psi\|_{h,\ell} .$$

Combining (3.16), (3.18), (3.19), (3.20), and (2.8b) gives the conclusion.

Lemmas 3.3 and 3.4 give the immediate corollary.

Lemma 3.5.

If the hypotheses of Lemma 3.4 hold, the constant C_6 (dependent on only the same quantities) can be taken so large that

$$\text{(3.21)} \qquad \|\zeta\|_{h,-\ell} \leq C_6\{h^{\ell+1}[\,|||B||| + \inf\{|||\zeta-\chi||| : \chi\in\mathcal{M}_h\}] + |B|_\ell\},$$

$$\ell = 0 \ \underline{\text{or}}\ 1 .$$

Lemmas 3.3 and 3.5 allow us to read off easily error estimates for approximate solutions of a Dirichlet problem. Suppose that $u \in H^k(\Omega)$ for some k such that $3 \leq k \leq r+1$ and that u satisfies

$$\text{(3.22)} \qquad \begin{aligned} -\nabla\cdot(a\nabla u) &= f \quad \text{on } \Omega , \\ u &= g \quad \text{on } \partial\Omega . \end{aligned}$$

Assume that $f \in H^{k-2}(\Omega)$ and that $\tilde{f} = \tilde{f}_h$ is an extension of f to $\mathbb{R}^p$. For $x \in \partial\Omega_h$ let $\tilde{g}(x)$ be such that

$$\text{(3.23)} \qquad \tilde{g}(x) = g(x + \delta(x)n(x)) .$$

Define $W \in \mathcal{M}_h$ by

$$N(W, V) = (\tilde{f}, V)_h - \langle \tilde{g}, \tilde{a}\frac{\partial V}{\partial n} - \gamma h^{-1}(V + \rho\frac{\partial V}{\partial n})\rangle_h, \quad V \in \mathcal{M}_h \,. \tag{3.24}$$

The existence and uniqueness of W follow from Lemma 3.1.

Letting $\tilde{u}$ be a Lions extension of u of order at least k, we see that for $V \in \mathcal{M}_h$

$$N(\tilde{u}, V) = (-\nabla\cdot \tilde{a}\nabla\tilde{u}, V)_h - \langle \tilde{u} + \rho\frac{\partial \tilde{u}}{\partial n}, \tilde{a}\frac{\partial V}{\partial n} - \gamma h^{-1}(V + \rho\frac{\partial V}{\partial n})\rangle_h. \tag{3.25}$$

Thus for $\zeta = W - \tilde{u}$ we see that

$$N(\zeta, V) = B(V), \qquad V \in \mathcal{M}_h, \tag{3.26}$$

where

$$B(\psi) = (f + \nabla\cdot(\tilde{a}\nabla\tilde{u}), \psi)_h + \langle \tilde{u} + \rho\frac{\partial\tilde{u}}{\partial n} - \tilde{g}, \tilde{a}\frac{\partial\psi}{\partial n} - \gamma h^{-1}(\psi + \rho\frac{\partial\psi}{\partial n})\rangle_h. \tag{3.27}$$

Lemmas 2.1 and 2.8 imply that

$$|(\tilde{f} + \nabla\cdot(\tilde{a}\nabla\tilde{u}), \psi)_h| \le Ch^{\frac{1}{2}}\bar{\delta}^{\ell+\frac{1}{2}}\|\tilde{f} + \nabla\cdot(\tilde{a}\nabla\tilde{u})\|_{h,\ell}|||\psi|||.$$

Lemmas 2.6 and 2.4 together with (2.12) imply that

$$|\tilde{u} + \rho\frac{\partial\tilde{u}}{\partial n} - \tilde{g}|_h \le C(\bar{\delta}^2 + \overline{\delta - \rho})\|u\|_3 \,. \tag{3.28}$$

Thus we see that, for B given by (3.27),

$$|||B||| \le C\{h^{\frac{1}{2}}\bar{\delta}^{\ell+\frac{1}{2}}\|\tilde{f} + \nabla\cdot(\tilde{a}\nabla\tilde{u})\|_{h,\ell} + (\bar{\delta}^2 + \overline{\delta - \rho})\|u\|_3 h^{-\frac{1}{2}}\}. \tag{3.29}$$

If $\varphi \in H^2(\mathbb{R}^p)$ and $\varphi \equiv 0$ on $\partial\Omega$, then

$$|(\tilde{f} + \nabla\cdot(\tilde{a}\nabla\tilde{u}), \varphi)_h| \le \bar{\delta}^{5/2}\|\tilde{f} + \nabla\cdot\tilde{a}\nabla\tilde{u}\|_{h,1}\|\varphi\|_{H^2(\mathbb{R}^p)} \tag{3.30}$$

by Lemmas 2.1, 2.8, and 2.3. In addition Lemmas 2.4 and 2.5 imply that

$$|\tilde{a}\frac{\partial\varphi}{\partial n} - \gamma h^{-1}(\varphi + \rho\frac{\partial\varphi}{\partial n})|_h$$

$$\leq C\|\varphi\|_{H^2(\mathbb{R}^p)} + Ch^{-1}(\bar{\delta}^{3/2} + \overline{\delta - \rho})\|\varphi\|_{H^2(\mathbb{R}^p)} \tag{3.31}$$

$$\leq C\|\varphi\|_{H^2(\mathbb{R}^p)} .$$

Thus (3.28), (3.30) and (3.31) give

$$|B|_1 \leq |B|_0 \leq C[\bar{\delta}^{5/2}\|\tilde{f}+\nabla\cdot\tilde{a}\nabla\tilde{u}\|_{h,1} + (\bar{\delta}^2 + \overline{\delta - \rho})\|u\|_3]. \tag{3.32}$$

These estimates can be combined with Lemmas 3.3 and 3.5 and (2.26) to give the following Theorem.

Theorem 3.1.

Suppose that $u \in H^k(\Omega)$ satisfies (3.22) and that $3 \leq k \leq r+1$. Let $\tilde{u}$ be the Lions extension of order $k+1$ of u and suppose that $\tilde{f} + \nabla\cdot\tilde{a}\nabla\tilde{u}$ is in $H^\ell(\Omega_h)$ where $\ell = 1$ if $k \leq 4$ and $\ell = 2$ if $k > 4$. Then there is a constant C_7, dependent on Ω, $\bar{a}, \underline{a}, C_0, \ldots, C_5, \|\tilde{a}\|_{H^2(\Omega \cup \Omega_h)}$, $\|\tilde{f} + \nabla\cdot\tilde{a}\nabla\tilde{u}\|_{h,\ell}$, and $\|u\|_k$, such that for $\zeta = W - \tilde{u}$,

$$|||\zeta||| \leq C_7(h^{k-1} + h^{-\frac{1}{2}}(\bar{\delta}^2 + \overline{\delta - \rho})),$$

$$\|\zeta\|_h \leq C_7(h^k + \bar{\delta}^2 + \overline{\delta - \rho}), \tag{3.33}$$

$$\|\zeta\|_{h,-1} \leq C_7(h^{k+1} + \bar{\delta}^2 + \overline{\delta - \rho}).$$

There are two cases of interest in which $\tilde{f} + \nabla\cdot\tilde{a}\nabla\tilde{u}$ vanishes identically. In the first case $\Omega_h \subset \Omega$. In the second case we do not actually compute the elliptic approximation but use it only as an analytical tool; this allows us to set $\tilde{f} = -\nabla\cdot\tilde{a}\nabla\tilde{u}$ after choosing $\tilde{a}$ and $\tilde{u}$. It is this second case we shall study next. Notice that in both these cases we can choose C_7 to depend on Ω, $\bar{a}, \underline{a}, C_0, \ldots, C_5$, and $\|\tilde{a}\|_{H^2(\Omega \cup \Omega_h)}$ such that

$$(3.33') \quad \begin{aligned} |||\zeta||| &\le C_7\{h^{k-1} + h^{-\frac{1}{2}}(\bar{\delta}^2 + \overline{\delta-\rho})\}\|u\|_k \\ \|\zeta\|_{h,-\ell} &\le C_7\{h^{k+\ell} + \bar{\delta}^2 + \overline{\delta-\rho}\}\|u\|_k, \quad \ell = 0,1. \end{aligned}$$

For the remainder of this section let a be a function of $(x,t) \in \Omega \times [0,T]$. Assume that $\tilde{a}$ is an extension of a to $\tilde{\Omega} \times [0,T] = (\bar{\Omega} \cup (\bigcup_{h \in \mathcal{h}} \bar{\Omega}_h)) \times [0,T]$ such that $\tilde{a}$ is bounded above and below by positive constants $\bar{a}$ and $\underline{a}$, respectively. The function that sends $t \in [0,T]$ into $\tilde{a}(\cdot,t)$ will be assumed to be a boundedly differentiable map of $[0,T]$ into $H^2(\tilde{\Omega})$. Take $u(x,t)$ to be defined on $\Omega \times [0,T]$ and to be such that, for some k_0 with $3 \le k_0 \le r+1$, $u(\cdot,t)$ belongs to $H^{k_0}(\Omega)$ for each $t \in [0,T]$. Also assume that, for some k_1 satisfying $3 \le k_1 \le k_0$, $u(\cdot,t)$ is a differentiable function of $[0,T]$ into $H^{k_1}(\Omega)$. Let

$$(3.34) \quad \tilde{g}(x,t) = u(x + \delta(x)n(x), t), \; (x,t) \in \partial\Omega_h \times [0,T].$$

Take $\tilde{u}$ to be a Lions extension of order $\ge k_0$ of u into $H^{k_0}(\mathbb{R}^p)$; note that $\frac{\partial}{\partial t}(\tilde{u}) = \tilde{u}_t$ defines an extension of u_t $H^{k_1}(\mathbb{R}^p)$.

For each $t \in [0,T]$ define $W(t) \in \mathcal{M}_h$ by

$$(3.35) \quad N(W,V) = (-\nabla\cdot(\tilde{a}\nabla\tilde{u}), V)_h - \langle \tilde{g}, \tilde{a}\frac{\partial V}{\partial n} - \gamma h^{-1}(V + \rho\frac{\partial V}{\partial n})\rangle_h, \quad V \in \mathcal{M}_h,$$

where $\tilde{a}$ is evaluated at t. Thus, with $\eta(t) = W(t) - \tilde{u}(\cdot,t)$, (3.33') implies that there exists C_7, dependent on Ω, $\bar{a}$, $\underline{a}$, $C_0,\ldots,C_5$, and $\|\tilde{a}\|_{L_\infty(H^2(\tilde{\Omega}))}$, such that for $0 \le t \le T$

$$(3.36) \quad \begin{aligned} &h|||\eta(t)||| + \|\eta\|_h + h^{-1}\|\eta\|_{h,-1} \\ &\le C_7[h^{k_0} + \bar{\delta}^2 + \overline{\delta-\rho}]\|u(\cdot,t)\|_{k_0}. \end{aligned}$$

Differentiating (3. 35) with respect to t gives

$$N(\eta_t, V) = B_1(V) \ , \quad V \in \mathcal{M}_h \ , \tag{3.37}$$

where

$$\begin{aligned} B_1(V) &= \widetilde{B}(V) + B_2(V) + B_3(V) \ , \\ \widetilde{B}(V) &= \langle \widetilde{u}_t + \rho \frac{\partial}{\partial n} \widetilde{u}_t - \widetilde{g}_t, \widetilde{a} \frac{\partial V}{\partial n} - \gamma h^{-1}(V + \rho \frac{\partial V}{\partial n}) \rangle_h \ , \\ B_2(V) &= -(a_t \nabla \eta, \nabla V)_h + \langle \widetilde{a}_t \frac{\partial \eta}{\partial n}, V \rangle_h + \langle \eta + \rho \frac{\partial \eta}{\partial n}, \widetilde{a}_t \frac{\partial V}{\partial n} \rangle_h , \\ B_3(V) &= \langle \widetilde{u} + \rho \frac{\partial \widetilde{u}}{\partial n} - \widetilde{g}, \widetilde{a}_t \frac{\partial V}{\partial n} \rangle_h \ . \end{aligned} \tag{3.38}$$

The form $\widetilde{B}$ is exactly the form B of (3. 27) with $\widetilde{u}$ replaced by $\widetilde{u}_t$, and of course $\widetilde{f} + \nabla \cdot \widetilde{a} \nabla \widetilde{u} \equiv 0$. Thus (3. 29) and (3. 32) show that

$$\begin{aligned} |||\widetilde{B}||| &\le C h^{-\frac{1}{2}}(\bar{\delta}^2 + \overline{\delta - \rho}) \, \|u_t(\cdot, t)\|_3 \ , \\ |\widetilde{B}|_1 &\le |\widetilde{B}|_0 \le C(\bar{\delta}^2 + \overline{\delta - \rho}) \, \|u_t(\cdot, t)\|_3 \ . \end{aligned} \tag{3.39}$$

It is easily seen that, with C proportional to $\|\widetilde{a}_t(\cdot,t)\|_{H^2(\Omega)}$,

$$\begin{aligned} |||B_2||| &\le C \, |||\eta||| \\ |||B_3||| &\le C h^{-\frac{1}{2}}[\bar{\delta}^2 + \overline{\delta - \rho}] \, \|u(\cdot, t)\|_3 \ . \end{aligned} \tag{3.40}$$

Rewriting B_2 as

$$B_2(V) = (\eta, \nabla \cdot \widetilde{a}_t \nabla V)_h + \langle \frac{\partial \eta}{\partial n}, \widetilde{a}_t(V + \rho \frac{\partial V}{\partial n}) \rangle_h \tag{3.41}$$

shows that

$$\begin{aligned} |B_2|_0 &\le C[\|\eta\|_h + h^{5/2} |||\eta|||] \ , \\ |B_2|_1 &\le C[\|\eta\|_{h,1} + h^{5/2} |||\eta|||] \ , \end{aligned} \tag{3.42}$$

where C is proportional to $\|\tilde{a}_t(\cdot,t)\|_{H^2(\tilde{\Omega})}$. It is straightforward that

(3.43) $$|B_3|_1 \le |B_3|_0 \le C[\bar{\delta}^2 + \overline{\delta-\rho}]\,\|u(\cdot,t)\|_3 .$$

Combining (3.38), (3.39), (3.40), (3.42), and (3.43) gives

$$|||B_1||| \le C[\,|||\eta(t)||| + h^{-\frac{1}{2}}(\bar{\delta} + \overline{\delta-\rho})(\,\|u(\cdot,t)\|_3 + \|u_t(\cdot,t)\|_3)] ,$$

$$|B_1|_\ell \le C[\|\eta(t)\|_{h,-\ell} + h^{5/2}\,|||\eta(t)||| + (\bar{\delta}^2 + \overline{\delta-\rho})(\,\|u(\cdot,t)\|_3 + \|u_t(\cdot,t)\|_3)] .$$

Thus, (3.36) and Lemmas 3.3 and 3.5 imply that there exists $C_8 \ge C_7$, dependent on Ω, $\bar{a}, \underline{a}, C_0, \ldots, C_5$, $\|\tilde{a}\|_{L_\infty(H^2(\tilde{\Omega}))}$, and $\|\tilde{\mathbf{a}}_t\|_{L_\infty(H^2(\tilde{\Omega}))}$, such that

(3.44) $$h\,|||\eta_t||| + \|\eta_t\|_h + h^{-1}\|\eta_t\|_{h,-1} \le C_8[h^{k_1} + \bar{\delta}^2 + \overline{\delta-\rho}][\,\|u(\cdot,t)\|_{k_1} + \|u_t(\cdot,t)\|_{k_1}].$$

Theorem 3.2.

Let W be defined by (3.35) with $\tilde{g}$ given by (3.34). Assume that $\tilde{a}$ and $\tilde{a}_t$ are boundedly in $H^2(\tilde{\Omega})$ and that for some k satisfying $3 \le k \le r$,

$$\|u\|_{L_\infty(H^{k+1}(\Omega))} + \|u_t\|_{L_2(H^k(\Omega))} < \infty .$$

Then, with C_8 as in (3.44) $\eta = W - \tilde{u}$, and $\tilde{u}$ a Lions extension of order $k+1$,

$$h\|\eta\|_{L_\infty(\mathcal{N})} + \|\eta\|_{L_\infty(L_2(\Omega_h))} + h^{-1}\|\eta\|_{L_\infty(H^1(\Omega_h))}$$

$$(3.45)\quad + h^2\|\eta_t\|_{L_2(\mathcal{N}_h)} + h\|\eta_t\|_{L_2(L_2(\Omega_h))} + \|\eta_t\|_{L_2(H^{-1}(\Omega_h))}$$

$$\leq C_8[h^{k+1} + \bar{\delta}^2 + \overline{\delta - \rho}][\|u\|_{L_\infty(H^{k+1}(\Omega))} + \|u_t\|_{L_2(H^k(\Omega))}].$$

A minor reworking of the above arguments, using Lemma 2.5 instead of Lemma 2.6 to avoid $\|u\|_3$ terms, gives the estimates

$$(3.46)\quad \begin{aligned} &\|\eta(\cdot,t)\|_h \leq C[h^2 + \bar{\delta}^{3/2} + \overline{\delta - \rho}]\|u(\cdot,t)\|_2, \\ &\|\eta_t(\cdot,t)\|_h \leq C[h^2 + \bar{\delta}^{3/2} + \overline{\delta - \rho}][\|u(\cdot,t)\|_2 + \|u_t(\cdot,t)\|_2]. \end{aligned}$$

Theorem 3.2, (3.46), (2.28), (2.25) and (3.2) imply the following corollary.

Corollary 3.1.

If, in addition to the hypotheses of Theorem 3.2, $\|u_t\|_{L_\infty(H^2(\Omega))}$ is finite, then there is a C_9 such that

$$(3.47)\quad \begin{aligned} &\|W\|_{L_\infty(L_\infty(\Omega_h))} + \|\nabla W\|_{L_\infty(L_\infty(\Omega_h))} + \|W_t\|_{L_\infty(L_\infty(\Omega_h))} \\ &\quad + \|\nabla W_t\|_{L_\infty(L_2(\Omega_h))} + \left\|\frac{\partial}{\partial n} W_t\right\|_{L_\infty(L_2(\partial\Omega_h))} < C_9. \end{aligned}$$

4. Approximate solution of parabolic equations.

In this section two methods for approximate solution of nonlinear parabolic equations will be presented and analyzed. The first method is a continuous-time approximation in which the approximate solution is defined by a system of nonlinear ordinary differential equations, and the second is a discrete-time approximation in which the approximate solution is defined by a sequence of linear algebraic equations.

Let $u(x,t)$ be the solution of

$$c(x,u)\frac{\partial u}{\partial t} - \nabla\cdot(a(x,u)\nabla u) = f(x,t,u) + \nabla\cdot F(x,t,u),\ (x,t)\in \Omega\times(0,T],$$

$$(4.1)\qquad u(x,t) = g(x,t)\ ,\quad (x,t)\in \partial\Omega\times(0,T]\ ,$$

$$u(x,0) = u_0(x)\ ,\quad x\in\Omega\ .$$

Assume that c and a are bounded above and below by positive constants which are independent of their arguments. Also assume that a and F are three times continuously differentiable, that c and f are twice continuously differentiable, and that all these derivatives are bounded uniformly on $\Omega\times\mathbb{R}$ or $\Omega\times\mathbb{R}\times[0,T]$ as is appropriate. (Actually, these assumptions need only hold in a neighborhood of the solution u, since we can derive uniform convergence from the error estimates below.)

Let c, a, f and F be extended to $\mathbb{R}^p\times\mathbb{R}$ or $\mathbb{R}^p\times\mathbb{R}\times[0,T]$ in such a fashion that these extended functions retain the smoothness assumed above. The extended functions c and a will be assumed to be bounded below by positive constants $\underline{c}$ and $\underline{a}$, respectively.

For $\chi\in H^1(\Omega_h)$ and $\varphi,\psi\in \mathcal{N}_h$ define $N(\chi,\varphi,\psi)$ by

$$(4.2)\qquad \begin{aligned} N(\chi,\varphi,\psi) &= (a(\chi)\nabla\varphi,\nabla\psi)_h - \langle a(\chi)\frac{\partial\varphi}{\partial n},\psi\rangle_h \\ &\quad - \langle \varphi+\rho\frac{\partial\varphi}{\partial n},\ a(\chi)\frac{\partial\psi}{\partial n} - \gamma h^{-1}(\psi+\rho\frac{\partial\psi}{\partial n})\rangle_h\ , \end{aligned}$$

where ρ and γ are as in Section 3. Note in particular that the choice of γ in Section 3 which gives the coerciveness of N does not depend on χ. Throughout this section it is assumed that γ is fixed and taken so large that (3.4) holds and that all $h\in\mathcal{h}$ are dominated by h_γ.

Take $\tilde{g}$ to be as in (3.34) and define a continuous-time approximation to be a differentiable map $U:[0,T]\to\mathcal{M}_h$ satisfying

$$(c(U)U_t, V)_h + N(U, U, V) = (f(U) + \nabla \cdot F(U), V)_h$$

(4.3a)

$$- \langle g, a(U)\frac{\partial V}{\partial n} - \gamma h^{-1}(V + \rho\frac{\partial V}{\partial n})\rangle_h \;,$$

$$V \in \mathcal{M}_h, \quad 0 \leq t \leq T.$$

The initial condition $U(0)$ will be chosen so that $U(0) - u_0$ is small. For this approximation the following theorem will be proved later in this section.

Theorem 4.1.

Suppose that u and U satisfy (4.1) and (4.3a), respectively, and that, for some k satisfying $3 \leq k \leq r+1$,

$$\|u\|_{L_\infty(H^{k+1}(\Omega))} + \|u_t\|_{L_2(H^k(\Omega))} + \|u_t\|_{L_\infty(H^2(\Omega))} < \infty.$$

Let u be extended if necessary by the Lions extension operator of order $k+1$ and assume that $U(0) \in \mathcal{M}_h$ is chosen so that

$$\|U(0) - u(\cdot, 0)\|_h \leq C[h^{k+1} + \bar{\delta}^2 + \overline{\delta - \rho}\,]. \tag{4.3b}$$

Then there is a constant C such that for h sufficiently small

$$\|U - u\|_{L_\infty(L_2(\Omega_h))} \leq C[h^{k+1} + \bar{\delta}^2 + \overline{\delta - \rho}]. \tag{4.4}$$

In the discrete-time case we adopt the following notation. Let M be a positive integer and let $\Delta t = T/M$. For functions Z defined at the discrete times $t_m = m\Delta t$, $0 \leq m \leq M$, let

$$Z_m = Z(t_m)\,, \quad Z_{m+\frac{1}{2}} = \tfrac{1}{2}(Z_{m+1} + Z_m)\,,$$

(4.5) $$\partial_t Z_m = (Z_{m+1} - Z_m)/\Delta t,$$

$$E^m Z = \begin{cases} Z_{m+\frac{1}{2}}\,, & m = 0,1, \\ 2\,Z_{m-\frac{1}{2}} - Z_{m-\frac{3}{2}}\,, & 2 \le m \le M\,. \end{cases}$$

The discrete-time approximate solution is a sequence $\{U_m\}_{m=0}^{M}$ in $\mathcal{M}_h$ satisfying

(4.6) $$\begin{aligned} &(c(E^m U)\,\partial_t U_m, V)_h + N(E^m U, U_{m+\frac{1}{2}}, V) \\ &= (f(E^m U) + \nabla\cdot F(E^m U), V)_h \\ &\quad - \langle \tilde{g}_{m+\frac{1}{2}}, a(E^m U)\frac{\partial V}{\partial n} - \gamma h^{-1}(V + \rho\frac{\partial V}{\partial n})\rangle_h\,, \\ &\qquad V \in \mathcal{M}_h,\quad 0 \le m < M\,, \end{aligned}$$

where f and F have $t_{m+\frac{1}{2}}$ as the value of their t-argument. Note that, for $m \ge 2$, (4.6) gives a linear set of equations for U^{m+1} in terms of previous U^k's. The two initial steps involve the solution of nonlinear algebraic equations; in practice these would probably be replaced by predictor-corrector equations, but we shall not treat that additional complication here. The choice of uniform time steps is for convenience; the arguments below are almost unchanged provided each time step is no larger than a fixed multiple of the previous one.

For this particular discrete-time approximation the following theorem will be proved later in this section.

Theorem 4.2.

Suppose that u satisfies (4.1) and that, for some k satisfying $3 \le k \le r+1$,

$$\|u\|_{L_\infty(H^{k+1}(\Omega))} + \|u_t\|_{L_2(H^k(\Omega))} + \|u_t\|_{L_\infty(H^{k-1}(\Omega))}$$

(4.7)

$$+ \|u_{tt}\|_{L_2(H^1(\Omega))} + \|u_{tt}\|_{L_\infty(L_2(\Omega))} + \|u_{ttt}\|_{L_2(H^{-1}(\Omega))} < \infty .$$

Let u *be extended, if necessary, by the Lions extension of order* $k+1$. *Then for any* $U_0 \in \mathcal{M}_h$ *there exists* $\{U_m\}_{m=0}^M$ *(possibly non-unique) satisfying* (4.6). *For any* $\tau_1 > 0$ *there exists constants* C *and* $\tau_2 > 0$ *such that any solution* $\{U_m\}_{m=0}^M$ *of* (4.6), *with* $0 < h < \tau_2$, *satisfies*

$$\max_{0 \le m \le M} \|U_m - u(\cdot, t_m)\|_h \le C[h^{k+1} + \bar{\delta}^2 + \overline{\delta - \rho} + (\Delta t)^2], \tag{4.8}$$

provided

(4.9)

$$h^{-2}\Delta t + h^{-4}\overline{\delta - \rho} \le \tau_1 ,$$

$$\|U_0 - u(\cdot, 0)\|_h \le C[h^{k+1} + \bar{\delta}^2 + \overline{\delta - \rho}] .$$

The norms on u in Theorems 4.1 and 4.2 are naturally related and can even be argued to be optimal in certain cases. In particular, the analysis of [6] can be modified in a minor way to show that if f and F vanish, c and a are independent of u and sufficiently smooth, and if u_0 satisfies certain compatibility conditions on the boundary, then the norms of u appearing in these theorems are bounded by $C\|u_0\|_{k+1}$.

For the remainder of this section fix k as in the above theorems and take u to be extended by the Lions extension of order $k+1$. In the proof of both of these results we use the function $W:[0,T] \to \mathcal{M}_h$ which is defined by (3.35) with $a(x,t)$ given by $a(x, u(x,t))$. Note that because of the assumed smoothness on a and u we see that $a(\cdot, u(\cdot, t))$ and $\frac{\partial}{\partial t}(a(\cdot, u(\cdot, t)))$ are boundedly in $H^2(\mathbb{R}^p)$. Hence the conclusions of Theorem 3.2 and Corollary 3.1 can be used to estimate $\eta = W - u$.

Proof of Theorem 4.1.

The proof is carried out by using an energy estimate on $\vartheta = W - U$ to show that U is close to W. Note that for $V \in \mathcal{M}_h$, ϑ satisfies

$$
\begin{aligned}
&(c(U)\vartheta_t, V)_h + N(U, \vartheta, V) \\
&\quad = \{((c(U) - C(u))W_t, V)_h + (c(u)\eta_t, V)_h\} \\
&\qquad + [N(U, W, V) - N(u, W, V) - \langle \tilde{g}, (a(u) - a(U))\frac{\partial V}{\partial n}\rangle_h] \qquad (4.10) \\
&\qquad + (Z, V)_h + (f(u) - f(U) + \nabla\cdot(F(u) - F(U)), V)_h ,
\end{aligned}
$$

where

$$
Z(x, t) = c(u)u_t - \nabla\cdot(a(u)\nabla u) - f(u) - \nabla\cdot F(u); \qquad (4.11)
$$

since u has been extended to $\Omega_h \backslash \Omega$, it is not necessarily the case that $Z \equiv 0$. Since $W_t, c(u)$, and $\nabla(c(u))$ are bounded on $\Omega_h \times [0, T]$, the term in braces in (4.10) can be bounded as

$$
\{\cdots\} \leq C[\|\eta\|_h + \|\vartheta\|_h + \|\eta_t\|_{h,-1}]\,|||V||| . \qquad (4.12)
$$

With $b = a(U) - a(u)$, the term in brackets is bounded as follows:

$$
\begin{aligned}
[\cdots] &= (b\nabla W, \nabla V)_h - \langle b\frac{\partial W}{\partial n}, V\rangle_h - \langle W + \rho\frac{\partial W}{\partial n} - \tilde{g}, b\frac{\partial V}{\partial n}\rangle_h \\
&\leq C[\|b\|_h\|\nabla V\|_h + (|b|_h h^{\frac{1}{2}})(h^{-\frac{1}{2}}|V|_h) \qquad (4.13) \\
&\quad + (h^{-1}\|W + \rho\frac{\partial W}{\partial n} - \tilde{g}\|_{L_\infty(\partial\Omega_h)})(|b|_h h^{\frac{1}{2}})(h^{\frac{1}{2}}|\frac{\partial V}{\partial n}|_h)] \\
&\leq C[\|\eta\|_h + \|\vartheta\|_h + h|||\eta||| + h|||\vartheta|||]\,|||V||| .
\end{aligned}
$$

In deriving (4.13) we used that

$$|b(x,t)| \leq C(|\eta(x,t)| + |\vartheta(x,t)|)$$

and that

$$\begin{aligned} \left|W + \rho\frac{\partial W}{\partial n} - \tilde{g}\right|_{L_\infty(\partial\Omega_h)} &\leq \|W-u\|_{L_\infty(\Omega_h)} + \bar{\rho}\,\|\nabla W\|_{L_\infty(\Omega_h)} \\ &\quad + \|u - \tilde{g}\|_{L_\infty(\partial\Omega_h)} \\ &\leq Ch^2 \leq Ch\,. \end{aligned} \tag{4.14}$$

Since $Z \equiv 0$ in Ω, Lemma 2.1 and term by term examination of Z show that

$$\|Z\|_h^2 \leq C(1 + \|u\|_4^2)(1 + \|u_t\|_2^2)\,\bar{\delta}^4\,.$$

Hence, since $\|u\|_{L_\infty(H^4(\Omega))}$ and $\|u_t\|_{L_\infty(H^2(\Omega))}$ are finite,

$$(Z,V)_h \leq C\bar{\delta}^2\,\|V\|_{L_2(\Omega_h\backslash\Omega)} \leq C\bar{\delta}^2\,|||V|||\,. \tag{4.15}$$

The last inner product in (4.10) is bounded as follows:

$$\begin{aligned} &(f(u)-f(U),V)_h - (F(U)-F(U),\nabla V)_h + \langle (F(u)-F(U))\cdot n, V\rangle_h \\ &\quad \leq C[\|\eta\|_h + \|\vartheta\|_h + h\,|||\eta||| + |||\vartheta|||]\,|||V|||\,. \end{aligned} \tag{4.16}$$

Thus (4.10)-(4.16) and Lemma 3.1 imply that for h sufficiently small

$$\begin{aligned} (c(U)\vartheta_t, \vartheta) + \frac{1}{2C_4}\,|||\vartheta|||^2 &\leq C[\|\eta\|_h^2 + h^2\,|||\eta|||^2 \\ &\quad + \|\eta_t\|_{h,-1}^2 + \bar{\delta}^4 + \|\vartheta\|_h^2]\,. \end{aligned} \tag{4.17}$$

In order to integrate (4.17) we introduce

$$R(x, b, \ell) = \int_0^{\ell - b} \tau c(x, \tau - b) d\tau \;;$$

see [10, Sec. 3.4]. Note that

$$\frac{\partial}{\partial t} R(x, W, U) = c(x, U)\vartheta_t \vartheta - W_t \int_0^{\vartheta} c_u(x, \tau - w)\tau \, d\tau.$$

Hence, since W_t is bounded,

$$\text{(4.18)} \qquad \frac{d}{dt}(R(\cdot, W, U), 1)_h \leq (c(U)\vartheta_t, \vartheta)_h + C\|\vartheta\|_h^2 .$$

Also note that

$$\text{(4.19)} \qquad (R(\cdot, W, U), 1)_h \geq \tfrac{1}{2} \underline{c} \|\vartheta\|_h^2 .$$

Thus, (4.17), (4.18), (4.19), the fact that $\vartheta(0)$ is bounded as in (4.3b), Gronwall's lemma, and Theorem 3.2 imply that

$$\text{(4.20)} \qquad \|\vartheta\|_{L_\infty(L_2(\Omega_h))} + \|\vartheta\|_{L_2(\mathscr{V}_h)} \leq C[h^{k+1} + \bar{\delta}^2 + \overline{\delta - \rho}].$$

This estimate is used with the triangle inequality and Theorem 3.2 to complete the proof.

Proof of Theorem 4.2.

It is clear from Lemma 3.1 that for any U_m and $Y \in \mathcal{M}_h$ there exists U_{m+1} satisfying

$$\text{(4.21)} \qquad \begin{aligned} &(c(Y)\partial_t U_m, V)_h + N(Y, U_{m+\frac{1}{2}}, V) \\ &= (f(Y) + \nabla\cdot F(Y), V)_h - \langle \tilde{g}, a(Y)\frac{\partial V}{\partial n} - \gamma h^{-1}(V + \rho \frac{\partial V}{\partial n})\rangle_h . \end{aligned}$$

With $V = U_{m+\frac{1}{2}}$ we see that

$$\text{(4.22)} \qquad \frac{1}{2\Delta t}(\underline{c}\|U_{m+1}\|_h^2 - \bar{c}\|U_m\|_h^2) + \frac{1}{2C_4}|||U_{m+\frac{1}{2}}|||^2 \leq Ch^{-1} .$$

Thus the map $Y \mapsto U_{m+1}$ sends $\mathcal{M}_h$ into a ball. Thus, since the map is clearly continuous, the Brouwer fixed point theorem implies the existence of U_1 and U_2 satisfying (4.6). The existence of $U_3, \ldots, U_m$ follows directly from the existence of solutions of (4.21) once U_0, U_1, and U_2 are given.

Adopt the following notation

$$e_s = h^{k+1} + \bar{\delta}^2 + \overline{\delta - \rho},$$

$$t^*_{m-2} = \begin{cases} t_{m-2}, & m \geq 2, \\ 0, & m = 0, 1, \end{cases}$$

$$\mu_m^2 = \frac{1}{\Delta t} \int_{t^*_{m-2}}^{t_{m+1}} (1 + \|u_t\|^2_{h,k} + \|u_{tt}\|^2_{h,1} + \|u_{ttt}\|^2_{h,-1}) dt,$$

$$\vartheta_m = W(t_m) - U_m .$$

The error estimate (4.8) will be proved by showing that there exist constants C and $\tau_2 > 0$ such that if (4.9) holds, $h \leq \tau_2$ and

$$\|\vartheta_k\|_h \leq h^{3.5}, \qquad 0 \leq k \leq m, \tag{4.24}$$

then

$$\max_{0 \leq k \leq m+1} \|\vartheta_k\| \leq C[e_s + (\Delta t)^2]. \tag{4.25}$$

Note that this proves the theorem since by decreasing τ_2 we can obtain

$$C[e_s + (\Delta t)^2] \leq Ch^4 < h^{3.5} \tag{4.26}$$

and thus (4.25) holds with $m = M-1$.

In order to establish this result we need to derive a difference inequality satisfied by ϑ. Let Z be given by (4.11) and note that

$$Z_{m+\frac{1}{2}} = c(E^m U)\partial_t W + \rho_{m,1} - (\nabla \cdot a(u)\nabla u)_{m+\frac{1}{2}} \tag{4.27}$$
$$- (f(E^m U) + \rho_{m,2}) - \nabla \cdot (F(E^m U) + \rho_{m,3}) ,$$

where

$$\rho_{m,1} = (c(u)u_t)_{m+\frac{1}{2}} - c(E^m U)\partial_t W ,$$
$$\rho_{m,2} = f(u)_{m+\frac{1}{2}} - f(t_{m+\frac{1}{2}}, E^m U) , \tag{4.28}$$
$$\rho_{m,3} = f(u)_{m+\frac{1}{2}} - F(t_{m+\frac{1}{2}}, E^m U) ,$$

and f and F in (4.27) are evaluated at $(x, t_{m+\frac{1}{2}}, E^m U(x))$. Each of the ρ's needs to be estimated. Using the fact that

$$(ab)_{m+\frac{1}{2}} = a_{m+\frac{1}{2}} b_{m+\frac{1}{2}} + (\frac{\Delta t}{2})^2 (\partial_t a_m)(\partial_t b_m) , \tag{4.29}$$

we see that

$$\rho_{m,1} = \{c(u)_{m+\frac{1}{2}}((u_t)_{m+\frac{1}{2}} - \partial_t u_m)\} + \{(\frac{\Delta t}{2})^2 \partial_t c(u)_m \partial_t u_m\}$$
$$+ \{[c(u)_{m+\frac{1}{2}} - c(E^m u)]\partial_t W_m\} + \{[c(E^m u) - c(E^m U)]\partial_t W_m\} \tag{4.30}$$
$$+ \{c(u)_{m+\frac{1}{2}} \partial_t (u-W)_m \}$$
$$= \rho_{m,1,1} + \rho_{m,1,2} + \rho_{m,1,3} + \rho_{m,1,4} + \rho_{m,1,5} ,$$

where the braces indicate the decomposition of $\rho_{m,1}$. Since $\nabla(c(u))$ is bounded,

$$\| \rho_{m,1,1} \|_{h,-1} \le C(\Delta t)^2 \mu_m \le C(\Delta t)^{3/2} \tag{4.30a}$$
$$\| \rho_{m,1,5} \|_{h,-1} \le Ce_s \mu_m, \quad \| \rho_{m,1,5} \|_h \le Ch^{-2} e_s .$$

Since $\|u_t\|_{L_\infty(L_\infty(\Omega))}$ and $\|u_{tt}\|_{L_\infty(L_2(\Omega_h))}$ are finite,

$$(4.30b)\qquad \begin{aligned} &\|\rho_{m,1,2}\|_h \le c(\Delta t)^2 , \\ &\|\rho_{m,1,3}\|_h \le c(\Delta t)^2 . \end{aligned}$$

Also

$$(4.30c)\qquad \begin{aligned} \|\rho_{m,1,4}\|_h &\le C\|E^m u - E^m U\|_h \le C[\|E^m\vartheta\|_h + \|E^m\eta\|_h] \\ &\le C[e_s + \|E^m\vartheta\|_h] . \end{aligned}$$

The term $\rho_{m,2}$ can be decomposed as

$$(4.31)\qquad \begin{aligned} \rho_{m,2} &= \rho_{m,2,1} + \rho_{m,2,2} \\ &= \{f(u)_{m+\frac{1}{2}} - f(u)(t_{m+\frac{1}{2}}) + f(u)(t_{m+\frac{1}{2}}) - f(t_{m+\frac{1}{2}}, E^m u)\} \\ &\qquad + \{f(t_{m+\frac{1}{2}}, E^m u) - f(t_{m+\frac{1}{2}}, E^m U)\}. \end{aligned}$$

These are estimated easily to see that

$$(4.31a)\qquad \begin{aligned} &\|\rho_{m,2,1}\|_h \le C(\nabla t)^2 , \\ &\|\rho_{m,2,2}\|_h \le C[e_s + \|E^m\vartheta\|_h] . \end{aligned}$$

Similarly

$$\begin{aligned} \rho_{m,3} &= \rho_{m,3,1} + \rho_{m,3,2} \\ &= \{F(u)_{m+\frac{1}{2}} - F(u)(t_{m+\frac{1}{2}}) + F(u)(t_{m+\frac{1}{2}}) - F(t_{m+\frac{1}{2}}, E^m u)\} \\ &\qquad + \{F(t_{m+\frac{1}{2}}, E^m u) - F(t_{m+\frac{1}{2}}, E^m U)\} , \end{aligned}$$

where

$$\|\rho_{m,3,1}\|_{h,1} \leq C(\Delta t)^2 \mu_m \leq C(\Delta t)^{3/2},$$

$$(4.32a)\quad \begin{aligned} \|\rho_{m,3,2}\|_h &\leq C[e_s + \|E^m \vartheta\|_h], \\ |\rho_{m,3,2}|_h &\leq C[|E^m \eta|_h + |E^m \vartheta|_h] \\ &\leq Ch^{-\frac{1}{2}}[e_s + \|E^m \vartheta\|_h]. \end{aligned}$$

Next use (3.38) (the definition of W) to see that, for $V \in \mathcal{M}_h$,

$$(4.33)\quad \begin{aligned} -((\nabla \cdot a(u)\nabla u)_{m+\frac{1}{2}}, V)_h &= N(E^m U, W_{m+\frac{1}{2}}, V) \\ &+ \langle \tilde{g}_{m+\frac{1}{2}}, a(E^m U)\frac{\partial V}{\partial n} - \gamma h^{-1}(V + \rho\frac{\partial V}{\partial n})\rangle_h \\ &+ (\rho_{m,4}, \nabla V)_h - \langle \rho_{m,5}, V\rangle_h + \langle \rho_{m,6}, \frac{\partial V}{\partial n}\rangle_h, \end{aligned}$$

where

$$(4.34)\quad \begin{aligned} \rho_{m,4} &= [a(u)_{m+\frac{1}{2}} - a(E^m u) + a(E^m u) - a(E^m U)]\nabla W_{m+\frac{1}{2}} \\ &\quad + (\frac{\Delta t}{2})^2(\partial_t a(u)_m)(\partial_t \nabla W_m), \\ \rho_{m,5} &= [a(u)_{m+\frac{1}{2}} - a(E^m u) + a(E^m u) - a(E^m U)]\frac{\partial}{\partial n} W_{m+\frac{1}{2}} \\ &\quad + (\frac{\Delta t}{2})^2(\partial_t a(u)_m)(\partial_t(\frac{\partial W}{\partial n})_m), \\ \rho_{m,6} &= (W + \rho\frac{\partial W}{\partial n} - \tilde{g})_{m+\frac{1}{2}}[a(u)_{m+\frac{1}{2}} - a(E^m u) + a(E^m u) - a(E^m U)] \\ &\quad + (\frac{\Delta t}{2})^2(\partial_t(W + \rho\frac{\partial W}{\partial n} - \tilde{g})_m(\partial_t a(u)_m). \end{aligned}$$

These can be bounded as

$$
(4.35)\quad
\begin{aligned}
\|\rho_{m,4}\|_h &\le C((\Delta t)^2 + e_s + \|E^m\vartheta\|_h)\\
|\rho_{m,5}|_h &\le C((\Delta t)^2\mu_m + h^{-\frac{1}{2}}e_s + |E^m\vartheta|_h)\\
&\le C((\Delta t)^2\mu_m + h^{-\frac{1}{2}}(e_s + \|E^m\vartheta\|_h))\ ,\\
|\rho_{m,6}|_h &\le Ch^2((\Delta t)^2\mu_m + h^{-\frac{1}{2}}(e_s + \|E^m\vartheta\|_h))\ ,
\end{aligned}
$$

where we used (4.14) and

$$
\begin{aligned}
|\partial_t(W+\rho\frac{\partial W}{\partial n}-\tilde{g})|_h &\le |\partial_t\eta_m|_h + \bar{\rho}|\partial_t(\frac{\partial\eta}{\partial n})_m|_h\\
&\quad + |\partial_t(u+\rho\frac{\partial u}{\partial n}-\tilde{g})_m|_h\\
&\le Ch^{2.5}\mu_m\ .
\end{aligned}
$$

Combining (4.27) and (4.33) gives

$$
(4.36)\quad
\begin{aligned}
&(c(E^mU)\partial_t\vartheta_m, V)_h + N(E^mU, \vartheta_{m+\frac{1}{2}}, V)\\
&= (Z_{m+\frac{1}{2}} - \rho_{m,1} + \rho_{m,2} + \nabla\cdot\rho_{m,3}, V)_h\\
&\quad - (\rho_{m,4}, \nabla V)_h + \langle\rho_{m,5}, V\rangle_h - \langle\rho_{m,6}, \frac{\partial V}{\partial n}\rangle_h .
\end{aligned}
$$

We first use (4.36) and (4.24) to derive a bound for $\|\partial_t\vartheta_k\|_{L_\infty(\Omega_h)}$ (and therefore for $\|\partial_t U_k\|_{L_\infty(\Omega_h)}$) for $0 \le k \le m$. It suffices to treat the case $k = m$. Using $V = \partial_t\vartheta_m$ in (4.36) we see that

$$
(4.37)\quad
\begin{aligned}
&\underline{c}\|\partial_t\vartheta_m\|_h^2 + \frac{\Delta t}{2}N(E^mU, \partial_t\vartheta_m, \partial_t\vartheta_m)\\
&\le -N(E^mU, \tfrac{1}{2}\vartheta_m, \partial_t\vartheta_m) + (Z_{m+\frac{1}{2}} - \rho_{m,1} + \rho_{m,2} + \nabla\cdot\rho_{m,3}, \partial_t\vartheta_m)_h\\
&\quad -(\rho_{m,4}, \nabla\partial_t\vartheta_m)_h + \langle\rho_{m,5}, \partial_t\vartheta_m\rangle - \langle\rho_{m,6}, \partial_t\frac{\partial}{\partial n}\vartheta_m\rangle_h\\
&\le C\|\partial_t\vartheta_m\|_h[h^{3.5-2} + h^{-2}e_s + h^{-1}((\Delta t)^{3/2} + \|E^m\vartheta\|_h)],
\end{aligned}
$$

where we used the bound for Z in (4.15). Thus

$$\|\partial_t \vartheta_m\|_h \leq C[h^{3/2} + h^{-1}\|E^m \vartheta\|] . \tag{4.38}$$

For $m \geq 0$ we have by the induction hypothesis (4.24)

$$\|E^m \vartheta\|_h \leq C[h^{3.5} + \Delta t \|\partial_t \vartheta_m\|_h] . \tag{4.39}$$

Hence, (4.39) and (4.38) imply that

$$\|\partial_t \vartheta_m\|_{L_\infty(\Omega)} \leq Ch^{-3/2}\|\partial_t \vartheta_m\|_h \leq C .$$

Hence

$$\|\partial_t U_m\|_{L_\infty(\Omega)} \leq C . \tag{4.40}$$

Note that with $E^{-1}U = E^0(U)$ we see that for $0 \leq m < M$

$$\begin{aligned}(c(E^m U)\partial_t \vartheta_m, \vartheta_{m+\frac{1}{2}})_h = {} & \frac{1}{2\Delta t}[(c(E^m U)\vartheta_{m+1}, \vartheta_{m+1})_h \\ & - (c(E^{m-1}U)\vartheta_m, \vartheta_m)_h] - \tfrac{1}{2}((\partial_t(EU))_{m-1}\vartheta_m, \vartheta_m)_h .\end{aligned} \tag{4.41}$$

Hence $V = \vartheta_{m+\frac{1}{2}}$ in (4.36) gives

$$\begin{aligned}\frac{1}{2\Delta t}[(c(E^m U)\vartheta_{m+1}, \vartheta_{m+1})_h - (c(E^{m-1}U)\vartheta_m, \vartheta_m)_h] + \frac{1}{C_4}|||\vartheta_{m+\frac{1}{2}}|||^2 \\ \leq \frac{1}{2C_4}|||\vartheta_{m+\frac{1}{2}}|||^2 + C\|\partial_t(EU)_{m-1}\|_{L_\infty(\Omega)}\|\vartheta_m\|_h^2 + C\Big[\|Z_{m+\frac{1}{2}}\|_h^2 \\ + \|\rho_{m,1}\|^2_{h,-1} + \|\rho_{m,2}\|_h^2 + \|\rho_{m,3}\|_h^2 + h|\rho_{m,3}|_h^2 \\ + \|\rho_{m,4}\|_h^2 + h|\rho_{m,5}|_h^2 + h^{-1}|\rho_{m,6}|_h^2\Big] \\ \leq \frac{1}{2C_4}|||\vartheta_{m+\frac{1}{2}}|||^2 + C\Big[\mu_m^2((\Delta t)^4 + e_s^2) + \|\vartheta_m\|_h^2 + \|E^m \vartheta\|_h^2\Big] .\end{aligned}$$

Thus the discrete analog of Gronwall's lemma gives the existence of C and τ_2 such that (4.25). This completes the proof.

References

1. S. Agmon, Lectures on Elliptic Boundary Value Problems, Van Nostrand, Princeton, N. J., 1965.

2. J. H. Bramble, T. Dupont, and V. Thomée, Projection methods for Dirichlet's problem in approximating polygonal domains with boundary value corrections, Math. Comp. 26 (1972), 869-879.

3. A. P. Calderón, Lebesgue spaces of differentiable functions and distributions, Partial Differential Equations, Proc. Symp. Pure Math. Vol. 4, AMS, 1961.

4. J. E. Dendy, Thesis, Rice University, 1971; also to appear in SIAM Numer. Anal.

5. J. Douglas, Jr., and T. Dupont, Galerkin methods for parabolic equations, SIAM Numer. Anal. 4 (1970), 575-626.

6. T. Dupont, Some L^2 error estimates for parabolic Galerkin methods, The Mathematical Foundations of the Finite Element Method with Applications to Partial Differential Equations, A. K. Aziz (ed.), Academic Press, New York 1972.

7. J. L. Lions and E. Magenes, Problèmes aux limites non homogènes et applications, Vol. 1, Dunod, Paris, 1968.

8. J. Nitsche, Über ein Variationsprinzip zur Lösung von Dirichlet-Problemen bei Verwendung von Teilräumen, die keinen Randbedingungen unterworfen sind, Abh. Math. Sem. Univ. Hamburg, 36 (1971), 9-15.

9. H. H. Rachford, Jr., Two-level discrete time Galerkin approximations for second order parabolic differential equations, SIAM Numer. Anal. 10 (1973), 1010-1026.

10. M. F. Wheeler, A priori L_2 error estimates for Galerkin approximations to parabolic partial differential equations, SIAM Numer. Anal. 10 (1973), 723-759.

Department of Mathematics
University of Chicago
Chicago, Illinois 60637

An H^{-1}-Galerkin Procedure for the Two-Point Boundary Value Problem

H. H. RACHFORD, JR. AND MARY F. WHEELER

1. Introduction.

In this paper we shall define a Galerkin procedure for the two-point boundary value problem which admits the use of nonconforming, i.e., discontinuous, subspaces for the approximate solution. Use of such subspaces reduces the coupling between elements and leads to error estimates that are local in the sense that they are of optimum order with respect to local smoothness of the solution plus a perturbation term of higher order depending on the global smoothness. Also we present computed results using nonconforming subspaces for a problem whose solution has derivatives that are large only in a small interval. The results qualitatively confirm the conclusions from analysis.

We shall examine the problem

$$\begin{aligned} Lu &\equiv (a(x)u')' - b(x)u' - c(x)u = f(x), \quad x \in I, \\ u(0) &= u(1) = 0, \end{aligned} \tag{1.1}$$

where $I = (0,1)$. Assume that a, a', b, b', and c are in $L_\infty(I)$, and that there exist constants a_0 and a_1 such that

$$0 < a_0 \le a(x) \le a_1, \quad x \in I.$$

Also assume (1.1) has a unique solution for $f \in C(I)$.

The H^{-1} Galerkin procedure is to be defined for approximate solutions in the subspace $\mathcal{M}(k,s,\Delta)$ consisting of all piecewise polynomial functions in $C^{(k)}$ of degree s with knots at the points of the partition $\Delta: 0 = x_0 < x_1 < \ldots < x_N = 1$ of $[0,1]$. Set $I_j = [x_{j-1}, x_j]$ and $h_j = x_j - x_{j-1}$. Let

$$h = \max_{1 \le j \le N} h_j,\ \mathcal{M}(-1,s,\Delta) = \{v \mid v \in P_s(I_j),\ j = 1,2,\ldots,N\},$$

and

$$\mathcal{M}(k,s,\Delta) = \{v \mid v \in C^k(I),\ v \in P_s(I_j),\ \ j = 1,\ldots,N\} \text{ for } k \ge 0.$$

A function defined on I belongs to $P_s(E)$ for $E \subset I$ if its restriction to E agrees with a polynomial of degree not greater than s. For $k \ge 0$ we assume that the partition Δ is quasiuniform, i.e., there exists a constant δ_0 such that

$$\max_{i,j} h_i h_j^{-1} \le \delta_0 .$$

In [2], Douglas, Dupont, and Wahlbin assume a quasiuniformity condition. We shall make use of their results. Briefly, these are as follows. Let y be a function on I such that $y^{(q)} \in L_p(I)$, $0 \le q \le s+1$, and let $\beta(x) \in L_\infty(I)$ such that $0 < \beta_0 \le \beta(x) \le \beta_1$. If Y is the weighted L_2 projection of y into $\mathcal{M}(k,s,\Delta)$ with the weight function β, then there exists a constant $C > 0$ such that

$$\|y-Y\|_{L_p(I)} \le Ch^q \sum_{\ell \le q} \|y^{(\ell)}\|_{L_p(I)} .$$

For brevity, we shall use the notation

$$\mathcal{M}_k^s = \mathcal{M}(k,s,\Delta) ,$$

and shall later need the space we denote by

$$\mathcal{N}_k^s = \mathcal{M}(k+2,s+2,\Delta) \cap \{v \mid v(0) = v(1) = 0\} .$$

Also let

$$(\varphi, \psi) = \int_0^1 \varphi(x)\psi(x)dx ,$$

$$(\varphi, \psi)_E = \int_E \varphi(x)\psi(x)dx, \qquad E \subset I .$$

The H^{-1} Galerkin procedure is defined as follows: Let $U \in \mathcal{M}_k^s$ satisfy

$$(1.2) \qquad (U, L^* \varphi) = (f, \varphi) , \quad \varphi \in \mathcal{N}_k^s ,$$

where L^* is the adjoint of the operator L.

With considerable effort it can be shown that (1.2) is a special case of a general variational procedure of Babuška and Aziz [1], for whose solution were established existence, uniqueness and L_2 error estimates with h small. However, such results are easier to obtain directly than by showing that (1.2) fits the general setting of Ref. [1]. Hence, we note that such properties follow immediately from the more general results of Theorem 1 and its corollary below.

The paper is in six sections. In §2 error estimates for $u - U$ will be established in L_p and in W_p^{-m} for $1 \leq p \leq \infty$ and $0 \leq m \leq s+1$. The most interesting choice, $k = -1$, is treated in detail in §3 and §4. In §3 the error is shown to be almost local in the sense that

$$\|U-u\|_{L_p(I_j)} \leq C \inf_{\chi \in \mathcal{M}_{-1}^s} \|u-\chi\|_{L_p(I_j)} + O(h^{2s+2}) ,$$

for sufficiently smooth u. In the next section, several superconvergence results are discussed. It is shown that the error at certain Gauss points is $O(h^{s+2})$ and that a trivial a posteriori calculation can be made to provide approximations to u and u' at the knots that are accurate to within $O(h^{2s+2})$ for smooth u; it is also the case that gains of h and h^{s+1}, respectively, in the error estimates hold for nonsmooth solutions. In §5, examples of numerical studies are presented to illustrate the local character of the error. The last section indicates briefly some extensions.

2. Estimates in L_p and W_p^{-m}.

For an open interval E, let $H^m(E)$ and $W_p^m(E)$ denote the closure of $C^\infty(\bar{E})$ in the norms

$$\|v\|_{H^m(E)} = \left(\sum_{i=0}^{m} \|v^{(i)}\|^2_{L_2(E)}\right)^{\frac{1}{2}}$$

and

$$\|v\|_{W_p^m(E)} = \sum_{i=0}^{m} \|v^{(i)}\|_{L_p(E)} ,$$

respectively. When $E = I$, we drop the dependency on the interval. For convenience, we set

$$\|v\| = \|v\|_{L_2(I)} .$$

Since $\mathcal{M}_k^s$ and $\mathcal{N}_k^s$ are of the same dimension, existence of U satisfying (1.2) follows from uniqueness. The following lemma results immediately from Theorem 1 below.

Lemma 1.

Suppose $W \in \mathcal{M}_k^s$ satisfies

$$(2.1) \qquad (W, L^* \varphi) = 0 , \quad \varphi \in \mathcal{N}_k^s .$$

Then, for h sufficiently small, $W \equiv 0$.

We now derive an error estimate for $U-u$ which is optimal in the sense that the exponent on h cannot be increased and the norm on the solution cannot be weakened.

Theorem 1.

Let u be the solution to (1.1). For h sufficiently small, there exists $U \in \mathcal{M}_k^s$ satisfying (1.2) and a constant $C_r > 0$ such that

$$(2.7) \qquad \|U - \hat{U}\|_{L_r(I)} \leq C_r \|u - \hat{U}\|_{L_r(I)} , \quad 1 \leq r \leq \infty ,$$

where $\hat{U}$ *is the* L_2 *projection of* u *into* $\mathcal{M}_k^s$.

Corollary.

If $u \in W_r^q(I)$, $\max\{0,k\} \leq q \leq s+1$, *then*

$$\|u-U\|_{L_r(I)} \leq C\|u\|_{W_r^q} h^q .$$

Proof:

The corollary follows from Theorem 1 and Lemma 4.2 of [2] for $k = 0,1,\ldots,s-1$, or the observation that the projection is local for $k = -1$, [4].

Proof of Theorem 1:

The argument will be given for $r = \infty$; interchanging one and infinity gives the proof for $r = 1$, and interpolation completes the proof.

Let $p \in \mathcal{M}_k^s$ and determine φ as the solution of the adjoint problem

$$\text{(2.7)} \qquad \begin{aligned} L^*\varphi &= p, \quad x \in I, \\ \varphi(0) &= \varphi(1) = 0 . \end{aligned}$$

Define $\varphi^* \in \mathcal{N}_k^s$ by

$$\text{(2.8)} \qquad (a(\varphi-\varphi^*)'', v'') = 0, \quad v \in \mathcal{N}_k^s .$$

Since

$$(U-u, L^*\varphi^*) = 0$$

and

$$(U, a(\varphi-\varphi^*)'') = (\hat{U}, a(\varphi-\varphi^*)'') = 0 ,$$

we have

$$
\begin{aligned}
(U-u, p) &= (U-u, L^* \varphi) = (U-u, L^*(\varphi-\varphi^*)) \\
&= (\hat{U}-u, a(\varphi-\varphi^*)'') + (U-u, a'(\varphi-\varphi^*)') \qquad (2.9) \\
&\quad + (U-u, (b(\varphi-\varphi^*))' - c(\varphi-\varphi^*)) .
\end{aligned}
$$

Note that $(\varphi^*)''$ is the L_2-projection with the weight function a of φ'' into $\mathcal{M}_k^s$. Letting $w = \varphi'$ and $w^* = (\varphi^*)'$, one can easily verify that

$$(a(w-w^*)', \zeta') = 0 , \quad \zeta \in \mathcal{M}(k+1, s+1, \Delta).$$

Using the L_∞ and L_1 estimates obtained in [2] and [4], we have that

$$\|\varphi-\varphi^*\|_{W_1^2} \leq C\|\varphi\|_{W_1^2} \leq C\|p\|_{L_1} ,$$

$$\|\varphi-\varphi^*\|_{W_1^1} \leq C h \|p\|_{L_1} ,$$

and

$$\|\varphi-\varphi^*\|_{L_1} \leq C h^2 \|p\|_{L_1} .$$

Hence we see for $p \in \mathcal{M}_k^s$ that

$$|(U-\hat{U}, p)| = |(U-u, p)| \leq C\{\|\hat{U}-u\|_{L_\infty} + h\|U-u\|_{L_\infty}\}\|p\|_{L_1} .$$

Thus,

$$
\begin{aligned}
\|U-\hat{U}\|_{L_\infty} &\leq C\{\|\hat{U}-u\|_{L_\infty} + h\|U-u\|_{L_\infty}\} \\
&\leq C\{\|\hat{U}-u\|_{L_\infty} + h(\|U-\hat{U}\|_{L_\infty} + \|\hat{U}-u\|_{L_\infty})\} .
\end{aligned}
$$

So, for h sufficiently small,

$$\|U-\hat{U}\|_{L_\infty} \le C\|\hat{U}-u\|_{L_\infty},$$

as desired.

Optimal estimates in W_r^{-m} also easily follow from the foregoing analysis. If we define $W_r^{-m} = (W_{r^*}^m)'$ using the norm

$$\|\varphi\|_{W_r^{-m}} = \sup_{0\neq\psi\in W_{r^*}^m} \frac{(\varphi,\psi)}{\|\psi\|_{W_{r^*}^m}}, \quad \frac{1}{r}+\frac{1}{r^*} = 1,$$

we may obtain the following.

Theorem 2.

Let u be the solution to (1.1). If $u \in W_r^q$, where $1 \le r \le \infty$ and $0 \le q \le s+1$, then

$$\|U-u\|_{W_r^{-\ell}} \le C\|u\|_{W_r^q} h^{q+\ell}, \quad 0 \le \ell \le s+1,$$

for h sufficiently small.

Proof:

The argument will be given for $r = 2$. A similar proof holds for the general case.

From (2.9), for $p \in \mathcal{M}_k^s$,

$$|(U-u,p)| \le (\|\hat{U}-u\| + h\|U-u\|)\|p\|.$$

Trivially,

$$\|\hat{U}-u\| \le C\|u\|_{H^q} h^q, \quad 0 \le q \le s+1.$$

It follows from Theorem 1 that

$$\|U-\hat{U}\| \le C\|u-\hat{U}\|$$

for h sufficiently small; thus,

$$\|u-U\| \leq C\|u\|_{H^q} h^q, \quad 0 \leq q \leq s+1 .$$

Let $\psi \in H^\ell$, $0 \leq \ell \leq s+1$. We consider the boundary value problem

$$L^* \varphi = \psi , \quad x \in I ,$$

$$\varphi(0) = \varphi(1) = 0 .$$

We have

$$(U-u, \psi) = (U-u, L^* \varphi) = (U-u, L^*(\varphi-\varphi^*)), \quad \varphi^* \in \mathcal{N}_k^s .$$

Thus, for $0 \leq \ell \leq s+1$ and $0 \leq q \leq s+1$,

$$\begin{aligned} |(U-u,\psi)| &\leq C\|U-u\| \|\varphi\|_{H^{\ell+2}} h^\ell \\ &\leq C\|U-u\| \|\psi\|_{H^\ell} h^\ell \\ &\leq C\|u\|_{H^q} \|\psi\|_{H^\ell} h^{q+\ell} . \end{aligned}$$

3. <u>Local behavior for</u> $k = -1$.

In the following three sections of this paper we shall assume $k = -1$. For convenience let $\mathcal{M} = \mathcal{M}_{-1}^s$ and $\mathcal{N} = \mathcal{N}_{-1}^s$. We wish to modify Theorem 1 to show the local nature of the error $U-u$, and to that end we require now that

$$a, b \in W_1^{s+2}(I_j) , \quad c \in W_1^{s+1}(I_j), \quad j = 1, 2, \ldots, N .$$

<u>Theorem 3.</u>

<u>Let</u> u <u>be the solution to</u> (1.1) <u>and assume</u> $u \in W_\ell^r(I_j) \cap W_\ell^q$, $1 \leq \ell \leq \infty$, $0 \leq q \leq r \leq s+1$.

<u>For</u> h <u>sufficiently small, there exists</u> $U \in \mathcal{M}$ <u>satisfying</u> (1.2) <u>such that</u>

$$(3.1) \quad \|U-\hat{U}\|_{L_\ell(I_j)} \le C(\|u\|_{W_\ell^r(I_j)} h^{r+1} + \|u\|_{W_\ell^q} h^{s+q+1}),$$

<u>where</u> $\hat{U}$ <u>is the</u> L_2 <u>projection of</u> u <u>into</u> $\mathcal{M}$.

<u>Corollary.</u>

<u>If</u> $u \in W_\ell^{s+1}(I_j) \cap W_\ell^q$, $1 \le \ell \le \infty$, $0 \le q \le s+1$, <u>then</u>

$$\|U-\hat{U}\|_{L_\ell(I_j)} \le C(\|u\|_{W_\ell^{s+1}(I_j)} + \|u\|_{W_\ell^q} h^{q-1})h^{s+2}$$

<u>and</u>

$$\|U-u\|_{L_\ell(I_j)} \le C(\|u\|_{W_\ell^{s+1}(I_j)} + \|u\|_{W_\ell^q} h^q)h^{s+1}.$$

<u>Proof:</u>

The corollary follows from Theorem 3 and the Peano kernel theorem [3].

<u>Proof of Theorem 3:</u>

We obtain estimate (3.1) for $\ell = \infty$. Inequality (3.1) for $\ell = 1$ can be derived in the same manner with ∞ and 1 interchanged, and the result for general ℓ then follows by interpolation.

Let $p \in P_s(I_j)$, and define φ by

$$(L^*\varphi)(x) = \begin{cases} p(x), & x \in I_j, \\ 0, & x \in I \setminus I_j, \end{cases}$$

$$\varphi(0) = \varphi(1) = 0.$$

Observe that

$$(U-u, p)_{I_j} = (U-u, L^*\varphi) = (U-u, L^*(\varphi-\chi)), \ \chi \in \mathcal{N}.$$

Since $\hat{U}$ is the local L_2 projection of u into $\mathcal{M}$,

$$|(U-\hat{U},p)_{I_j}| \leq \|U-u\|_{L_\infty(I_j)} \|L^*(\varphi-\chi)\|_{L_1(I_j)}$$

$$+ \|U-u\|_{L_\infty(I)} \|L^*(\varphi-\chi)\|_{L_1(I\setminus I_j)}, \ \chi \in \mathcal{N}.$$

Now,

$$\|\varphi\|_{W_1^{s+3}(I\setminus I_j)} \leq C\|p\|_{L_1(I_j)} .$$

Since $\varphi \in W_1^2$, $\varphi \in C^1(I)$. Thus, we can choose $\varphi^* \in \mathcal{N}$ on $I\setminus I_j$ and on I_j independently with φ^* and $\varphi^{*\prime}$ interpolating φ and φ' at $x = x_{j-1}$ and x_j. On $I\setminus I_j$, we can construct φ^* such that

$$\|L^*(\varphi-\varphi^*)\|_{L_1(I\setminus I_j)} \leq C\|p\|_{L_1(I_j)} h^{s+1} .$$

Next, consider φ on I_j. Since $p \in P_s(I_j)$,

$$\varphi \in W_1^{s+3}(I_j)$$

and

$$\|\varphi\|_{W_1^{s+3}(I_j)} \leq C\|p\|_{W_1^s(I_j)} .$$

Consequently, we can find φ^* on I_j interpolating φ and φ' at x_{j-1} and x_j such that

$$\|L^*(\varphi-\varphi^*)\|_{L_1(I_j)} \leq C\|p\|_{W_1^s(I_j)} h_j^{s+1}$$

$$\leq C\|p\|_{L_1(I_j)} h_j .$$

In the last inequality we used homogeneity in the interval length for the polynomial p on I_j. It follows that

$$(1-Ch_j)\,\|U-\hat{U}\|_{L_\infty(I_j)} \leq C(\|u-\hat{U}\|_{L_\infty(I_j)} h_j + \|u-U\|_{L_\infty(I)} h^{s+1}).$$

Thus, for h sufficiently small,

$$\begin{aligned}\|U-\hat{U}\|_{L_\infty(I_j)} &\leq C(\|u-\hat{U}\|_{L_\infty(I_j)} h_j + \|u\|_{W^q_\infty} h^{s+q+1}) \\ &\leq C(\|u\|_{W^r_\infty(I_j)} h^{r+1} + \|u\|_{W^q_\infty} h^{s+q+1}),\end{aligned}$$

and the proof is finished.

4. Some superconvergence results for $k = -1$.

In this section, we establish superconvergence at certain Gauss points. We also define local quadratures which give $O(h^{2s+2})$ estimates for both the function and the flux at the knots.

Let $\xi_1, \xi_2, \ldots \xi_{s+1}$ be $s+1$ Gauss points on I such that for a set of positive weights $\{A_j\}$

$$\int_I p\,dx = \sum_{j=1}^{s+1} A_j p(\xi_j), \qquad p \in P_{2s+1}.$$

One can easily verify that

$$(4.1)\quad |(\hat{U}-u)(\xi_{ij})| \leq C\|u\|_{W_1^{s+2}(I_i)} h^{s+2}, \quad \xi_{ij} = x_{i-1} + h_i\xi_j.$$

Then, Theorem 3 and (4.1) imply the following theorem.

Theorem 4.

Let $u \in W_1^{s+2}(I_i)$ and $u \in W_1^q(I)$, $0 \leq q \leq s+1$. Then

$$(4.2)\quad |(U-u)(\xi_{ij})| \leq C[\|u\|_{W_1^{s+2}(I_i)} + \|u\|_{W_1^q(I)} h^{q-1}]h^{s+2},$$

where ξ_{ij} is the jth Gauss point on I_i.

We now define simple local quadratures which allow computation of very accurate approximations to the function and the flux at the knots. Let ν_j, $j = 1, 2, \ldots, N-1$, be the approximation to $u(x_j)$ given by

$$(4.3)\qquad \nu_j = \begin{cases} \frac{1}{a(x_j)} [-(U, L^*\varphi_j)_{I_{j+1}} + (f, \varphi_j)_{I_{j+1}}], & x_j \geq \frac{1}{2}, \\ \frac{1}{a(x_j)} [(U, L^*\theta_j)_{I_j} - (f, \theta_j)_{I_j}], & x_j < \frac{1}{2}, \end{cases}$$

where

$$(4.4)\qquad \begin{aligned} \varphi_j &= h_{j+1}^{-2} (x-x_j)(x-x_{j+1})^2, & x \in I_{j+1}, \\ \theta_j &= h_j^{-2} (x-x_{j-1})^2 (x-x_j), & x \in I_j. \end{aligned}$$

Equation (4.3) is motivated by considering $(Lu, \varphi_j)_{I_{j+1}}$ or $(Lu, \theta_j)_{I_j}$ and integrating by parts. An approximation γ_j to $u'(x_j)$, $j = 0, 1, \ldots, N$, is defined by

$$(4.5a)\qquad \gamma_0 = \frac{1}{a(0)} [(U, L^*\psi_0)_{I_1} - (f, \psi_0)_{I_1}],$$

$$(4.5b)\qquad \gamma_j = \frac{1}{a(x_j)} [(f, \psi_j)_{I_j} - (U, L^*\psi_j)_{I_j} + b(x_j)\nu_j], \quad j = 1, 2, \ldots, N,$$

where

$$(4.6a)\qquad \psi_0(x) = h_1^{-3} (x-x_1)^2 (2x+h_1), \quad x \in I_1,$$

and

$$(4.6b)\qquad \psi_j(x) = h_j^{-3} (x-x_{j-1})^2 [h_j + 2(x_j - x)], \quad x \in I_j.$$

Equations (4.5a) and (4.5b) are motivated by considering $(Lu, \psi_0)_{I_1}$ and $(Lu, \psi_j)_{I_j}$, respectively.

Theorem 5.

Let $u \in W_1^q$, $0 \leq q \leq s+1$. *Then, for* h *sufficiently small,*

$$(4.7a) \quad |\nu_j - u(x_j)| \leq C\|u\|_{W_1^q} h^{q+s+1}, \quad j = 1, 2, \ldots, N-1,$$

and

$$(4.7b) \quad |\gamma_j - u'(x_j)| \leq C\|u\|_{W_1^q} h^{q+s+1}, \quad j = 0, 1, \ldots, N.$$

Proof:

Let $x_j \geq \frac{1}{2}$ and let Φ_j be defined by

$$\Phi_j(x) = \begin{cases} \dfrac{x(x-x_j)}{x_j}, & 0 \leq x < x_j, \\ \varphi_j(x), & x_j \leq x \leq x_{j+1}, \\ 0, & x_{j+1} < x \leq 1. \end{cases}$$

We observe that $\Phi_j \in \mathcal{N}$; hence,

$$(U-u, L^* \Phi_j) = 0.$$

It is easy to deduce that

$$(4.8) \quad \begin{aligned} |u(x_j) - \nu_j| &= \frac{1}{a(x_j)} |(U-u, L^* \Phi_j)_{I_{j+1}}| \\ &= \frac{1}{a(x_j)} |(u-U, L^* \Phi_j)_{(0,x_j)}|. \end{aligned}$$

Let z be defined by

$$L^* z = \begin{cases} L^* \Phi_j, & x \in (0, x_j] , \\ 0 \quad , & x \in (x_j, 1) , \end{cases}$$

$$z(0) = z(1) = 0 .$$

Now,

$$\begin{aligned} (u-U, L^* \Phi_j)_{(0,x_j)} &= (u-U, L^* z) \\ &= (u-U, L^*(z-Z)), \quad Z \in \mathcal{N}. \end{aligned}$$

Thus, for Z suitably chosen,

$$\begin{aligned} &|(u-U, L^* \Phi_j)_{(0,x_j)}| \\ (4.9) \quad &\leq \|u-U\|_{L_1(0,x_j)} \|L^*(z-Z)\|_{L_\infty(0,x_j)} \\ &\quad + \|u-U\|_{L_1(x_j,1)} \|L^*(z-Z)\|_{L_\infty(x_j,1)} \\ &\leq C \|u\|_{W_1^q} \sum_{\ell=0}^{s+1} \|(L^* \Phi_j)^{(\ell)}\|_{L_\infty(0,x_j)} h^{q+s+1} \\ &\quad + C \|u\|_{W_1^q} \|L^* \Phi_j\|_{L_\infty(0,x_j)} h^{q+s+1} \\ &\leq C \|u\|_{W_1^q} h^{q+s+1} . \end{aligned}$$

Estimate (4.7a) follows from (4.8) and (4.9). A similar argument can be given if $x_j < \frac{1}{2}$.

Let Ψ_j be defined by

$$(4.10) \qquad \Psi_j = \begin{cases} 0 , & 0 \leq x < x_{j-1} , \\ \psi_j , & x_{j-1} \leq x \leq x_j , \\ 1 , & x_j < x \leq 1 . \end{cases}$$

We see that

$$(4.11) \qquad (U-u, L^*(\Psi_j - x)) = 0 .$$

Note that

$$|\gamma_j - u'(x_j)| = \frac{1}{a(x_j)} |[(u-U, L^* \Psi_j)_{I_j} + b(x_j)(\nu_j - u(x_j)]| ,$$

for $j > 0$. Thus,

$$(4.12) \quad |\gamma_j - u'(x_j)| \leq \frac{1}{a(x_j)} [|(u-U, L^* \Psi_j)_{I_j}| + |b(x_j)| |\nu_j - u(x_j)|].$$

From (4.10) and (4.11) we have

$$(u-U, L^* \Psi_j)_{I_j} = (u-U, L^* x)_{(0, x_j)} - (u-U, L^*(1-x))_{(x_j, 1)} .$$

Applying an argument similar to that for (4.9), we obtain the inequality

$$(4.13) \qquad |(u-U, L^* \Psi_j)_{I_j}| \leq C \|u\|_{W_1^{s+1}} h^{q+s+1} .$$

Using (4.7a), (4.12), and (4.13) we see that inequality (4.7b) for $j = 1, 2, \ldots, N$ has been demonstrated.

A similar proof shows that γ_0 is an $O(h^{q+s+1})$ approximation to $u'(0)$.

5. Discussion and examples.

The primary motivation for the H^{-1} procedure introduced for the two-point boundary value problem in section 1 is to permit relaxing all continuity constraints on the approximating subspace. It was hoped thereby to achieve good local approximation of solutions whose nature precludes good global approximation in a subspace of reasonable dimension.

The analysis in section 3 suggests that we expect to meet this goal asymptotically. The prime purpose of this section is to test this property experimentally. The tests are twofold in character. First, to treat a "smooth" problem with a simple formulation of the procedure and to observe experimentally the asymptotic decay of the error as predicted by the analyses of sections 3 and 4. The second set of tests will deal with a "rough" problem, i.e., one whose solution, though analytic, changes locally so rapidly that it cannot be approximated globally at all well in the subspaces used. The purpose of these tests is to observe, if possible, whether good convergence takes place locally in intervals removed from the point of ill behavior.

The two-point boundary value problem studied is to find $u(x)$ such that

$$(5.1)\qquad \begin{aligned} (au_x)_x &= -2(1+\alpha(x-\bar{x})(\arctan \alpha(x-\bar{x}) + \arctan \alpha\bar{x})), \quad x \in I, \\ u(0) &= u(1) = 0, \end{aligned}$$

where $a(x) = \alpha^{-1} + \alpha(x-\bar{x})^2$, $\alpha > 0$. The solution is

$$u(x) = (1-x)(\arctan \alpha(x-\bar{x}) + \arctan \alpha\bar{x}),$$

which for large values of α has a sharp knee close to $\bar{x}$, a constant which will be taken to lie in $(0,1)$.

The smooth problem will be that for which $\alpha = 5$ and $\bar{x} = 0.2$; a plot of its solution is shown in Figure 1. The rough problem studied will be that for $\alpha = 100$, and $\bar{x} = 0.36388$. This value of $\bar{x}$ is such that $\lim_{\alpha \to \infty} u(\bar{x}^+) \cong 2$. A plot of the solution for the rough problem is shown in Figure 5.

In all cases, we computed $U \in \mathcal{M}^1_{-1}$ satisfying (1.2) for Δ a uniform partition of spacing h. The set of h studied corresponds to reciprocals of integers, N, from 4 through 20. Figures 2 through 4 give the dependence upon $\log(\frac{1}{N})$ of several measures of error for the approximation of the smooth problem. Figure 2 shows on a logarithmic scale the L_2 error (as open circles) and the maximum absolute error at any superconvergence point (which is any Gauss point of the intervals, i.e., a point $x_j + h(\frac{1}{2} \pm \frac{1}{\sqrt{12}})$, $j = 0, 1, \ldots, N-1$). Also plotted for reference are lines with slope 2 and 3. Although not plotted, the L_∞ error was computed and found to be about 5 times the L_2 error for all smooth cases calculated. These results very clearly agree with the predictions of the corollary to Theorem 1, since we would expect quadratic convergence for $s=1$ in both of the L_∞ and L_2 norms. Also we certainly interpret the convergence rate at the superconvergence (Gauss) points to be cubic as predicted by Theorem 4 in Section 4.

To illustrate the results of Theorem 5, Figure 3 shows two quantities plotted versus $\frac{1}{N}$ on a log-log scale. One is the maximum error at the points x_j in the local quadrature approximating $u(x_j)$, i.e., ν_j of Equation (4.3), $j=1, \ldots, N-1$. The other quantity is the maximum error in the local quadrature approximating $u_x(x_j)$, i.e., γ_j of Equations (4.5a) and (4.5b) for $j = 0, 1, \ldots, N$. We note in both cases that Theorem 5 predicts convergence at the rate $O(h^{2s+2})$, or $O(h^4)$ for this case $(s=1)$. The computational results agree well with this estimate as is observed by comparing the plotted results with a reference line of slope 4 on Figure 3.

Comparison of the magnitudes of the errors in the local quadrature approximation with those of the piecewise linear solution, U, is also helpful in revealing the value of performing the extra computation. We note that even for an N as small as 5 the maximum error in approximating the function at the knots by ν_j is about 0.002, whereas the L_∞-error in U is about 0.08, or some 40 times greater for this case.

In considering the use of the results of Theorem 5, it should again be emphasized that once U has been computed the computation for each ν_j or γ_j is local, i.e., is independent of the computation at any other value of the index. Thus, if the estimate for u or u_x is needed accurately or in detail in a particular interval, only the values of ν and γ associated with that interval need be computed. For example if in our computation with s=1 we compute $\nu_{j-1}, \gamma_{j-1}, \nu_j$ and γ_j for $O(h^4)$ estimates for $u(x_{j-1})$, $u_x(x_{j-1})$, $u(x_j)$, and $u_x(x_j)$, respectively, and then use these to define a Hermite cubic interpolate for I_j, say $\tilde{U}$, then $\tilde{U}$, which is trivial to compute, should be an $O(h^4)$ approximation for u on I_j if u is sufficiently smooth on I_j.

It is more difficult to test the local convergence rate predicted by the Corollary to Theorem 3. If we choose a specific interval, say I_j for j fixed, and compute

$$\|U-u\|_{L_\infty(I_j)}$$

for h decreasing, then I_j will correspond to intervals of progressively lower values of x, thereby changing the x-location for which we are examining the local error. Therefore, instead we choose to examine the maximum norm of the error near a fixed value of x, i.e., in the interval which contains the fixed value. In Figure 4 we have plotted $\|U-u\|_{L_\infty(I_{(.8)})}$ versus $\frac{1}{N}$, (again on a log-log scale) where $I_{(.8)}$ means the interval with highest index such that it contains the point x = 0.8. As a matter of interest, the maximum error at the superconvergence points in $I_{(.8)}$ is also shown in Figure 4.

For $\mathcal{M}^1_{-1}$, we have s=1 in the corollary to Theorem 3, so we expect

$$\|U-u\|_{L_\infty(I_{(.8)})} \leq C\{\|u\|_{W^2_\infty(I_{(.8)})} + \|u\|_{W^2_\infty} h^2\}h^2,$$

i.e., local quadratic convergence with a fourth order

perturbation term. Thus we could expect an asymptotic rate which is only quadratic, but might expect an apparent rate up to fourth order for intermediate values of h. This is not inconsistent with the observations in Figure 3, although it is clear that for the smooth problem the results are inconclusive. It is of interest however, that

$$\|U-u\|_{L_\infty} / \|U-u\|_{L_\infty(I_{(.8)})}$$

is over 12 for $N > 6$, suggesting that there is a significant dependence of the local error upon the local properties of the solution. This conclusion is strengthened by examining the computed results once more for the consistency of the values of x for which

$$\|U-u\|_{L_\infty} / \|U-u\|_{L_\infty(I_{(x)})} = 1 .$$

For all N studied $(4,5,\ldots,20)$ this occurs for x satisfying $.3 \leq x \leq .333\ldots$.

The results of Figure 4 for the convergence at the Gauss points in $I_{(.8)}$ also appear consistent with the results of Theorem 4, it is of interest that the errors at the Gauss points are about 1% of the maximum error in $I_{(.8)}$.

We turn our attention to results from the rough problem. Figure 6 shows $\log\|U-u\|_{L_2}$ versus $\log\frac{1}{N}$, and strongly suggests that $N = 20$ is insufficient to provide good global approximation, since $\|U-u\|_{L_2} = .047$ for this case. However, we still may hope for reasonable local approximation even though we are far from representing the solution well globally. Figure 7 is a plot of $\log\|U-u\|_{L_\infty(I_{(.8)})}$ versus $\log\frac{1}{N}$. It suggests that even though the global approximation is poor, the local approximation is reasonable. The largest value of $\|U-u\|_{L_\infty(I_{(.8)})}$ for any N is just over $.01$, and it occurs for $N = 6$. There appears to be a strong trend toward convergence with $\frac{1}{N}$ at a rate which appears larger than quadratic for $N > 5$. It appears typical of the method that for some preferred values of N the piecewise linear functions can fit the solution more readily. This no doubt depends upon where the region with large derivatives in u lies relative to the boundaries between the piecewise linear functions. Actually, therefore, very little more can be concluded from Figure 7 than that the results are not inconsistent with the predictions of the corollary to Theorem 3.

Before concluding, it appears worthwhile to examine the approximate solutions for several values of N for some qualitative conclusions. Figures 8-13 show plots of U compared with u for $N = 4, 5, 7, 8, 17$, and 18, respectively. It will be noted that when 0.36388 lies close to the boundary between the intervals associated with $\mathcal{M}^1_{-1}$, a significant jump discontinuity appears, e.g., for $N = 5$ with an interval boundary at 0.4, $N = 8$ with a boundary at 0.375, $N = 14$ with a boundary at $.357$, $N = 17$ with a boundary at $.352$, or $N = 19$ with a boundary at $.368$. If the point 0.36388 lies near the middle third of an interval, the solution in $\mathcal{M}^1_{-1}$ appears reasonably close to continuous. It is curious to note that the errors in $I_{(.8)}$ on Figure 6 tend to be abnormally low when large jumps occur (except for $N=17$), and the magnitude by which the error is low seems related in a positive way to how closely the point 0.36388 approximates an interval boundary. This observation might lead one to speculate that there may be useful a posteriori information to be derived from the values of the jumps in the H^{-1} solution. It is also curious that $\|U-u\|_{L_\infty}$ does not appear to correlate with the jump in U in the same way as

$$\|U-u\|_{L_\infty(I_{(.8)})}$$

does.

The last comparison we shall make is with results from the standard L_2-Galerkin procedure in $\mathring{\mathcal{M}}^2_0 = \mathcal{M}^2_0 \cap H^1_0(I)$. This comparison is interesting in that the algebraic problem requires almost an identical amount of computer work as the H^{-1} procedure for a given value of N. We know that asymptotically the L_2-procedure in $\mathring{\mathcal{M}}^2_0$ must exhibit cubic

L_2-convergence, while the best rate we expect in $\mathcal{M}^1_{-1}$ from the H^{-1} procedure is quadratic. However, here we are not interested in the asymptotic behavior, but rather in values for N for which global approximation is poor.

Figure 14 shows the $\mathring{\mathcal{M}}^2_0$ L_2-Galerkin approximation for $N = 8$ along with the H^{-1} approximation in $\mathcal{M}^1_{-1}$ for the same value of N. It is clear that the H^{-1} approximation is superior for this case. Similar plots for $N = 5$ and 7 also lead to this conclusion.

6. Extensions.

The results obtained in the previous sections also hold when the flux or mixed boundary conditions are specified instead of Dirichlet conditions.

References.

1. I. Babuška and A. K. Aziz, Survey lectures on the mathematical foundations of the finite element method, The Mathematical Foundations of the Finite Element Method with Applications to Partial Differential Equations, A. K. Aziz ed., Academic Press, 1972.

2. J. Douglas, Jr., T. Dupont, and L. Wahlbin, Optimal L_∞ error estimates for Galerkin approximations to solutions of two point boundary value problems, to appear in Math. of Comp.

3. B. Wendroff, First Principles of Numerical Analysis, Addison-Wesley, Reading, Mass., 1969.

4. M. F. Wheeler, An optimal L_∞ error estimate for Galerkin approximations to solutions of two point boundary value problems, SIAM J. Numer. Anal. 10 (1973), 914-917.

Department of Mathematical Science
Rice University
Houston, Texas 77001

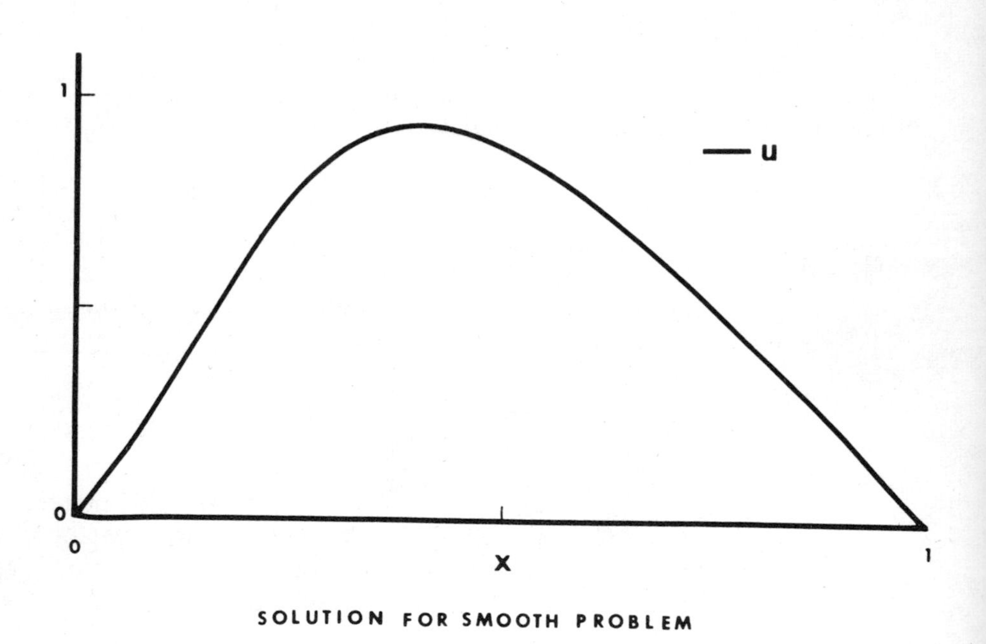

Figure 1.

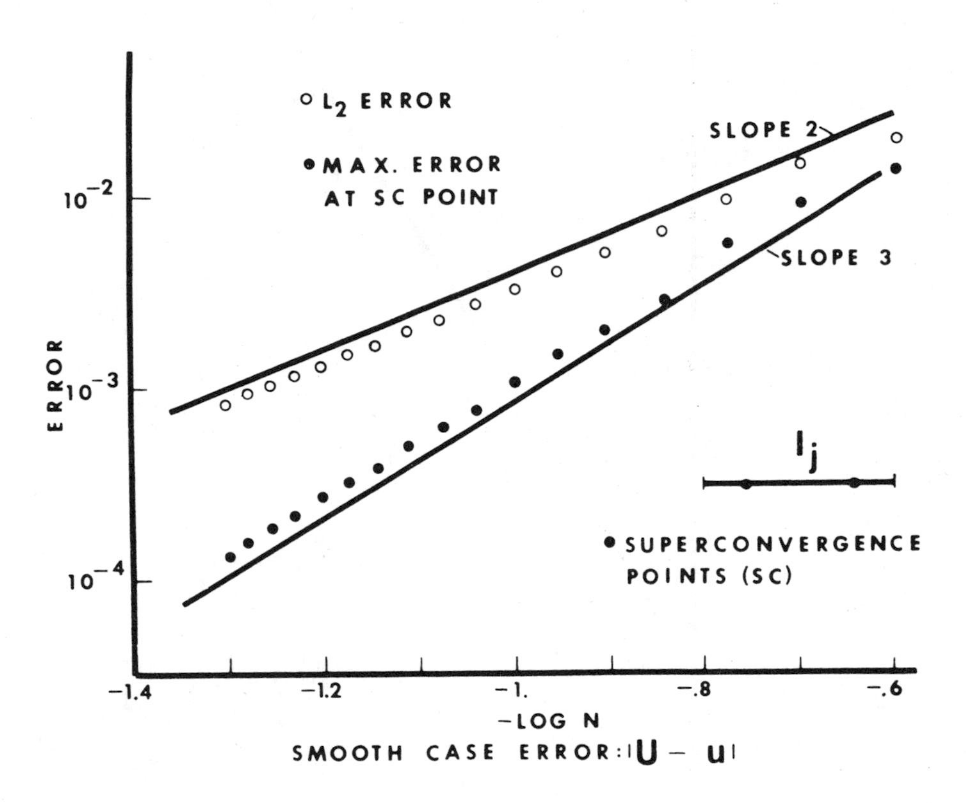
○ L2 ERROR
● MAX. ERROR AT SC POINT
SLOPE 2
SLOPE 3
ERROR
10−2
10−3
10−4
Ij
● SUPERCONVERGENCE POINTS (SC)
−1.4
−1.2
−1.
−.8
−.6
−LOG N
SMOOTH CASE ERROR: |U − u|

Figure 2

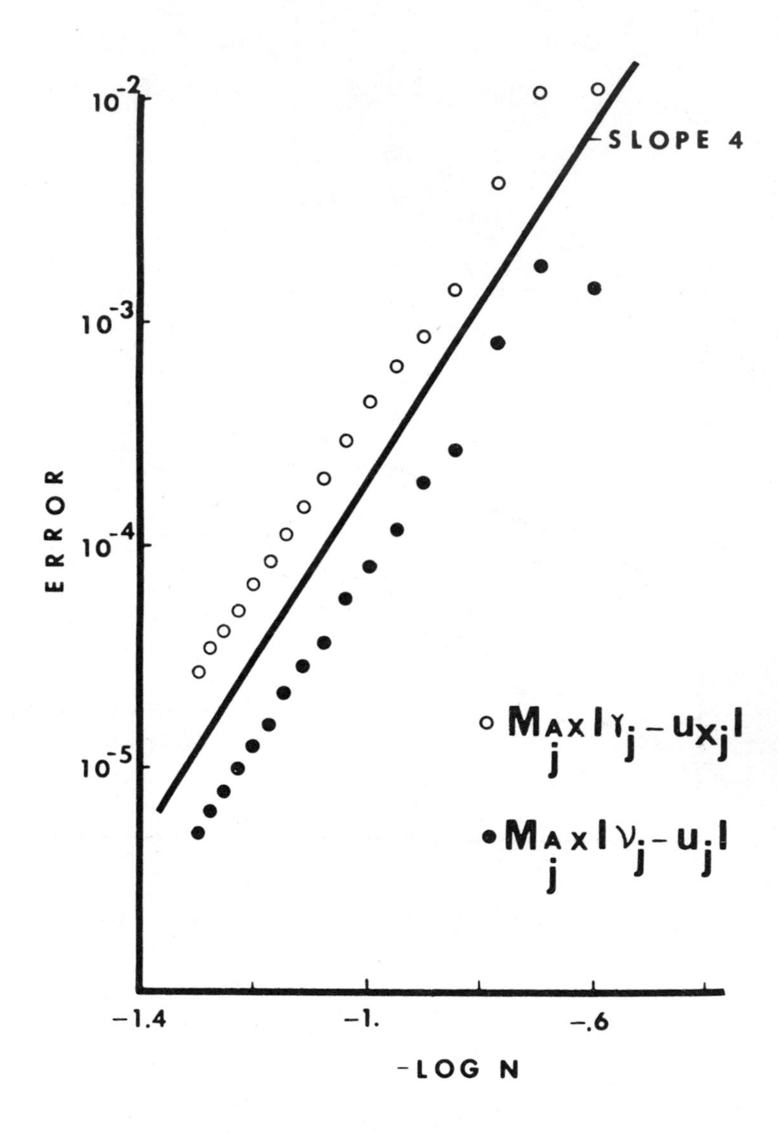

SUPERCONVERGENCE AT KNOTS

Figure 3

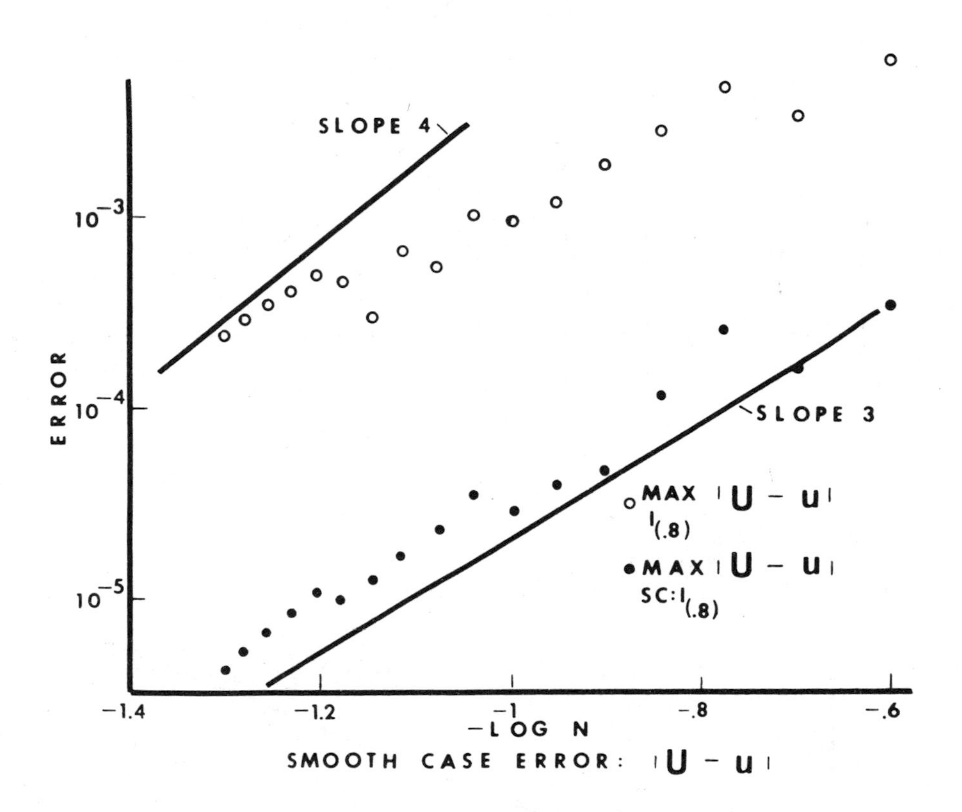
SLOPE 4
SLOPE 3
10^{-3}
10^{-4}
10^{-5}
ERROR
MAX |U − u|
I(.8)
MAX |U − u|
SC:I(.8)
−1.4
−1.2
−1
−.8
−.6
−LOG N
SMOOTH CASE ERROR: |U − u|

Figure 4

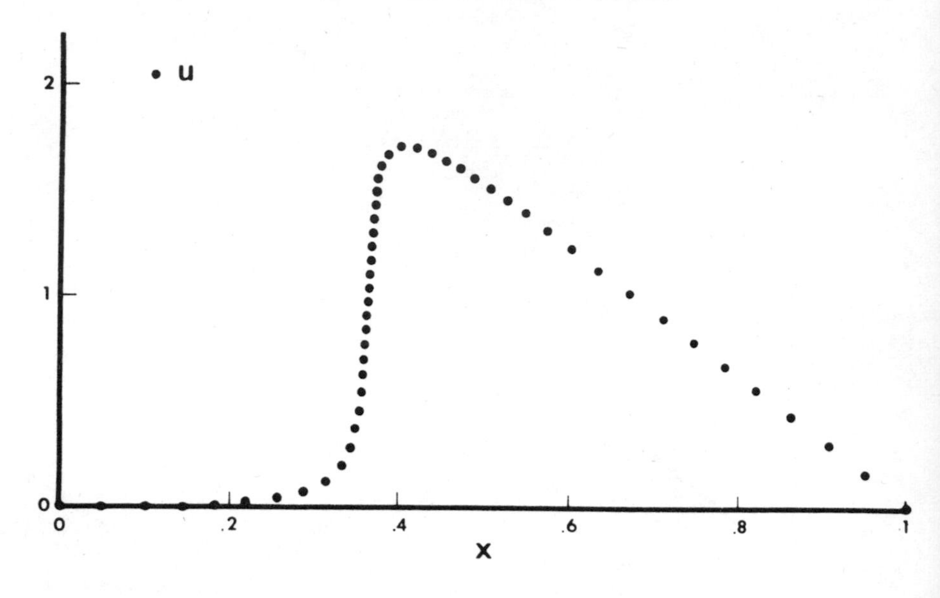

Figure 5

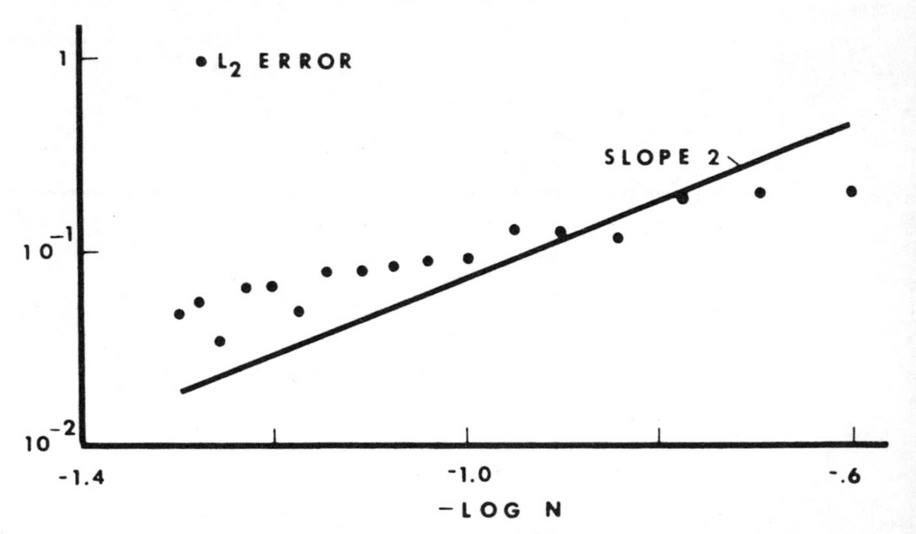

Figure 6

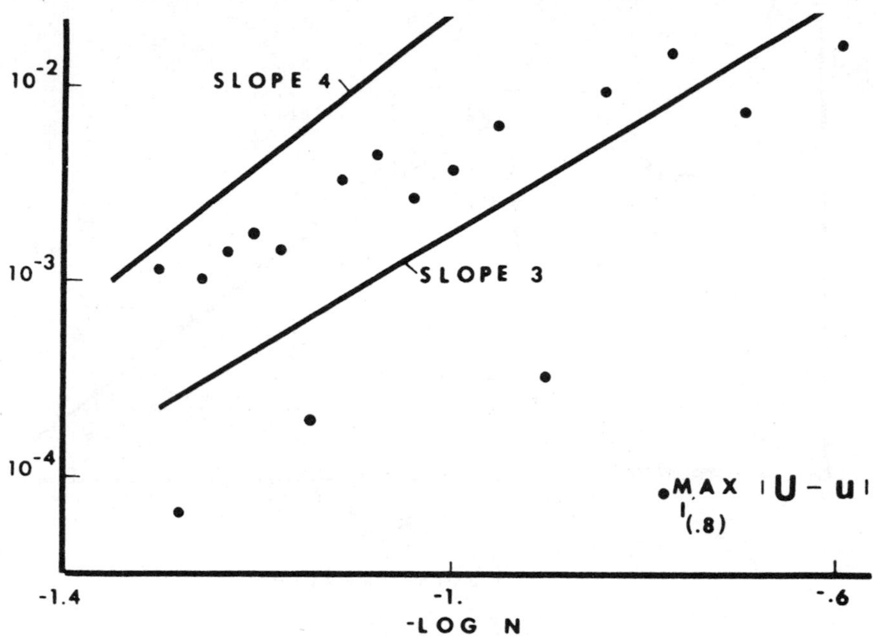

Figure 7

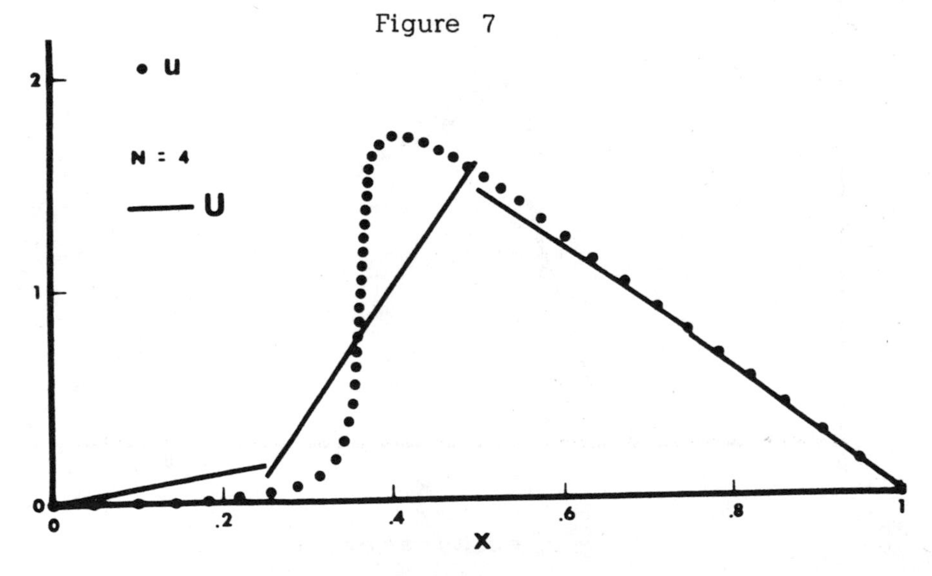

Figure 8

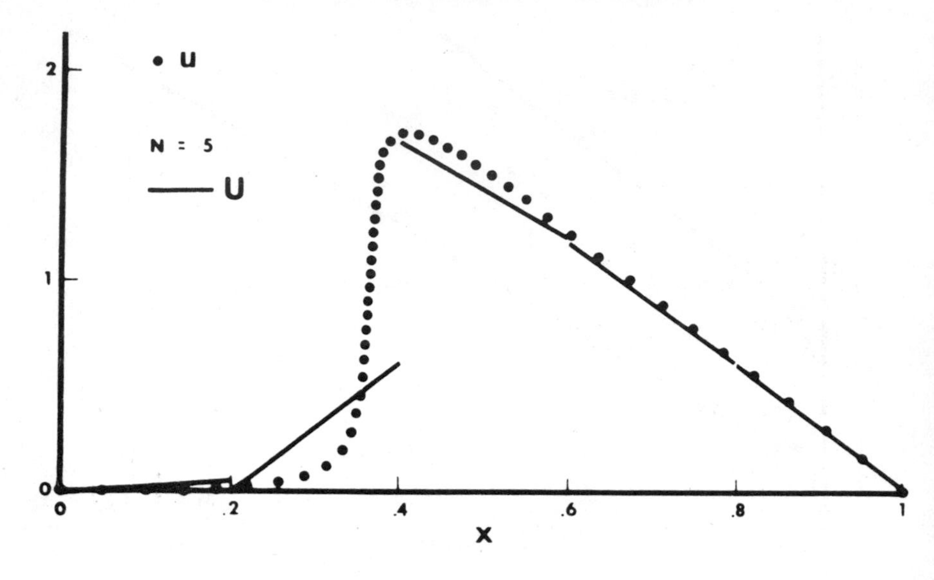

Figure 9

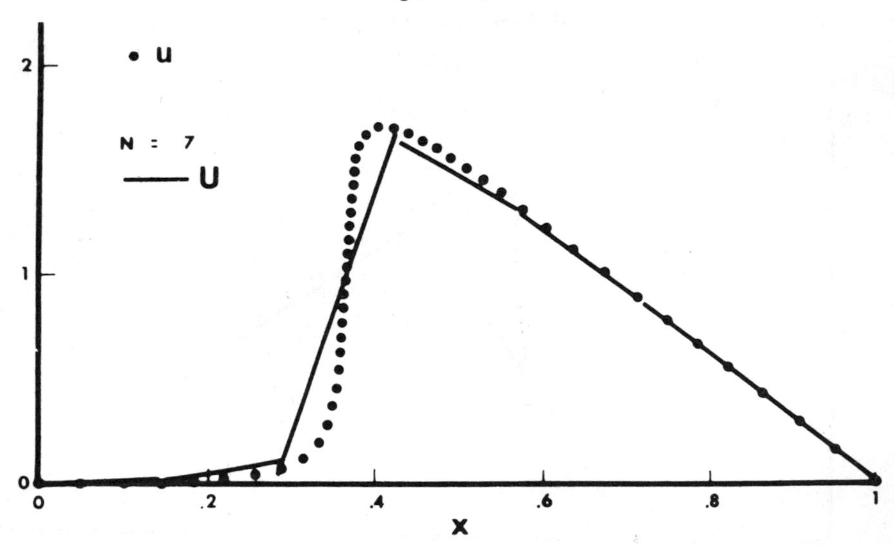

Figure 10

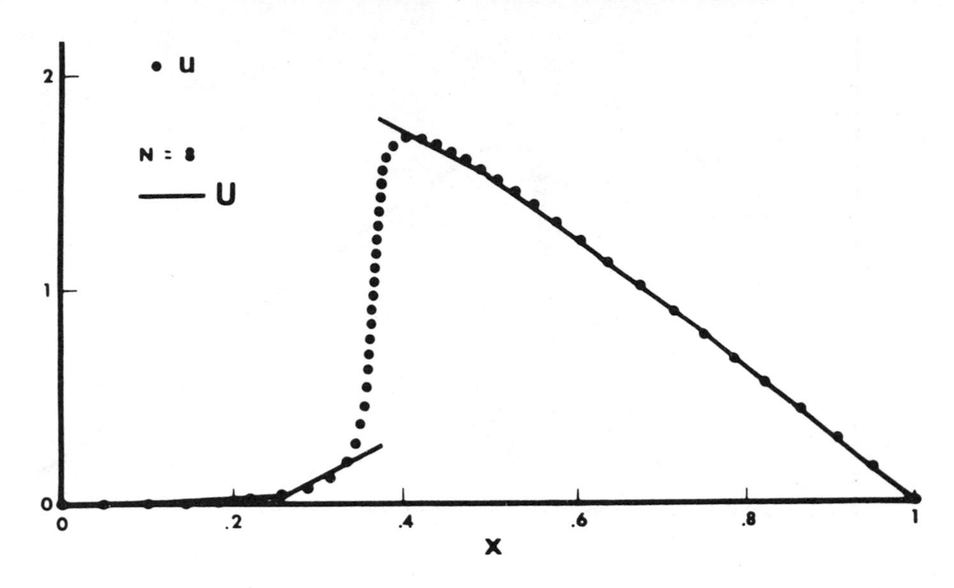

Figure 11

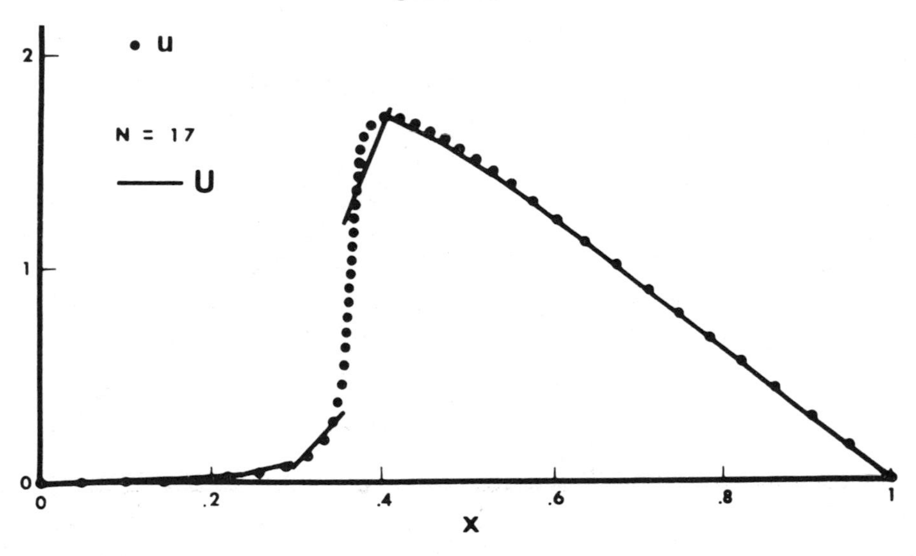

Figure 12

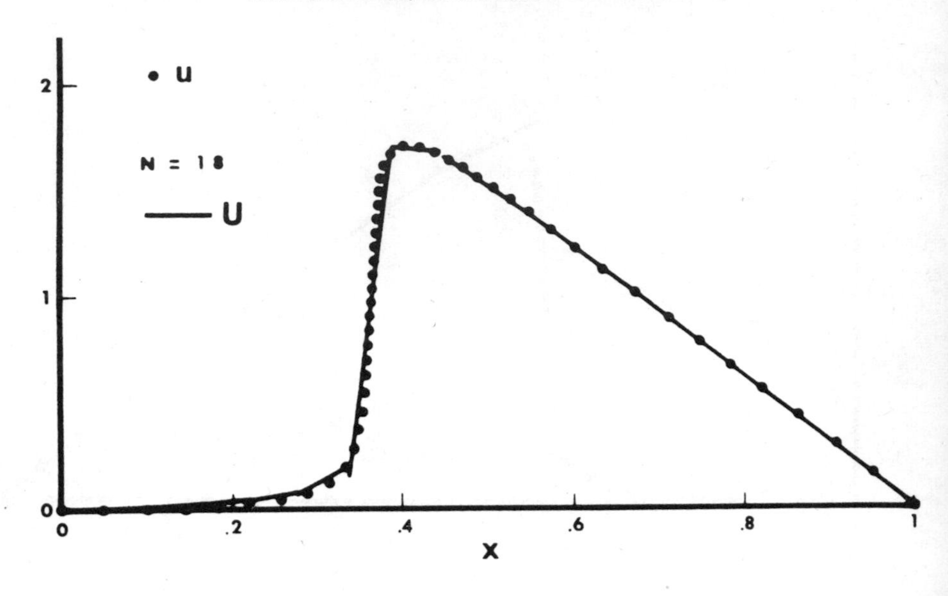

Figure 13

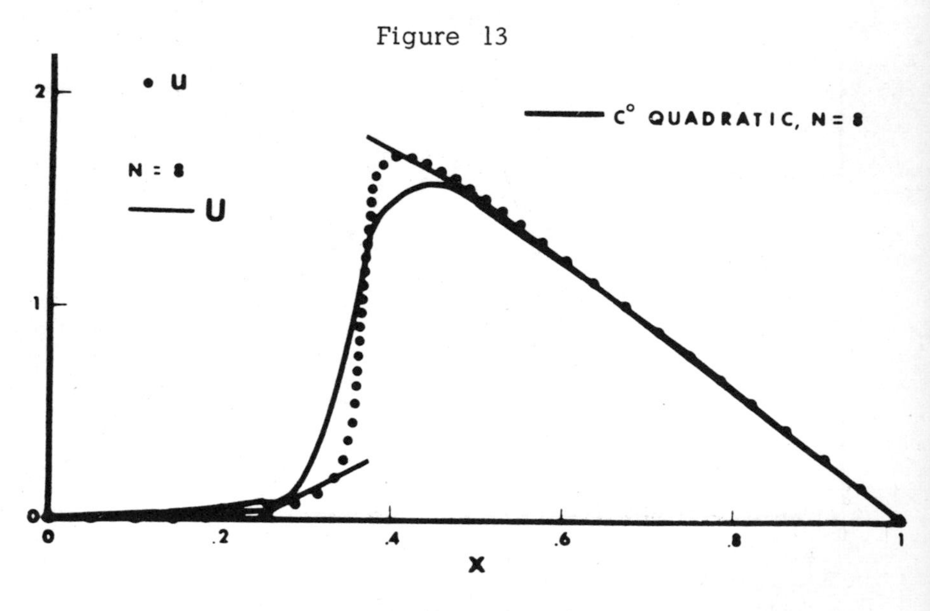

Figure 14

H^1-Galerkin Methods for the Laplace and Heat Equations

JIM DOUGLAS, JR., TODD DUPONT, AND MARY F. WHEELER

1. Introduction.

Let Ω be a bounded domain in $\mathbb{R}^n$ with a smooth boundary $\partial\Omega$, and assume that $\partial\Omega$ is the finite union of non-intersecting compact manifolds without boundary. We shall consider the numerical solution of the two boundary value problems

$$\text{(1.1)} \qquad \begin{aligned} \Delta w &= f(x), \quad x \in \Omega, \\ w &= g(x), \quad x \in \partial\Omega, \end{aligned}$$

and

$$\text{(1.2)} \qquad \begin{aligned} \frac{\partial u}{\partial t} - \Delta u &= -f(x,t), \quad (x,t) \in \Omega \times J, \qquad J = [0, T_0], \\ u &= g(x,t), \quad (x,t) \in \partial\Omega \times J, \\ u &= u_0(x), \quad x \in \Omega, \; t = 0, \end{aligned}$$

by Galerkin methods based on the use of the inner product in $H^1(\Omega)$ rather than that in $L_2(\Omega)$. Let $\mathcal{M}_h, 0 < h < 1$, be a finite-dimensional subspace of $\mathcal{N} = H^2(\Omega) \cap H^2(\partial\Omega)$ [15] drawn from a family of such subspaces having the approximation property that

$$\begin{aligned}(1.3)\qquad &\inf_{x \in \mathcal{M}_h} \sum_{j=0}^{2} \{h^j \|v-x\|_j + h^{j+\frac{1}{2}} |v-x|_j\} \\ &\qquad \le C \|v\|_p h^p , \quad \frac{5}{2} \le p \le r+1 ,\end{aligned}$$

for any function $v \in H^p(\Omega)$, where

$$\begin{aligned}(1.4)\qquad &\|v\| = \|v\|_0, \quad \|v\|_p = \|v\|_{H^p(\Omega)} = (\sum_{|\alpha| \le p} \|D^\alpha v\|^2)^{\frac{1}{2}}, \\ &|v| = |v|_0 , \quad |v|_p = \|v\|_{H^p(\partial\Omega)} .\end{aligned}$$

We shall denote the (real) inner products on $L_2(\Omega)$ and $L_2(\partial\Omega)$ by (u, v) and $\langle u, v\rangle$, respectively.

If the differential equation (1.1) is multiplied by Δv and integrated over Ω, it is clear that $(\Delta w, \Delta v) = (f, \Delta v)$. Note that this is formally equivalent to $(\nabla\Delta w, \nabla v) = (\nabla f, \nabla v)$ These relations do not appear to be suitable for a Galerkin procedure employing a subspace that does not incorporate the boundary values g, and we shall add two null boundary terms as follows. First, let T denote a self-adjoint second-order elliptic operator on $\partial\Omega$ with smooth coefficients. For $n=2$, it is sufficient that T be the second derivative with respect to arc length on each component of $\partial\Omega$. For $n > 2$, one reasonable choice for T would be the Laplace-Beltrami operator [15]. Assume that there exists $\alpha > 0$ such that

$$(1.5)\qquad \frac{1}{\alpha} |v|_2^2 \ge |Tv|^2 + |v|^2 \ge \alpha |v|_2^2 , \quad v \in H^2(\partial\Omega) .$$

Let B denote the bilinear form given by

$$(1.6)\qquad B(\varphi, \psi) = (\Delta\varphi, \Delta\psi) + h^{-3}\langle \varphi, \psi\rangle + h\langle T\varphi, T\psi\rangle$$

for $\varphi, \psi \in \mathcal{V}$. The H^1-Galerkin procedure for the Dirichlet problem consists in finding $W \in \mathcal{M}_h$ such that

$$(1.7)\qquad B(W, v) = B(w, v) = (f, \Delta v) + h^{-3}\langle g, v\rangle + h\langle Tg, Tv\rangle, \; v \in \mathcal{M}_h .$$

Note that (1.7) differs only in the addition of the term involving T from one of the least squares methods studied by Bramble and Schatz [6]. We shall establish an inequality for the error $w - W$ of the form

$$(1.8) \qquad \|w-W\|_{-s} + |w-W|_{-s-\frac{1}{2}} \leq C\|u\|_p h^{p+s}, \quad -2 \leq s \leq r-3, \quad \frac{5}{2} \leq p \leq r+1 .$$

The Sobolev spaces with negative indices are taken in this paper to be given by duality with respect to H^s, not H_0^s:

$$H^{-s}(\Omega) = H^s(\Omega)', \quad \|v\|_{-s} = \sup_{z \in H^s(\Omega)} \frac{(v,z)}{\|z\|_s}, \quad s \geq 0,$$

$$H^{-s}(\partial\Omega) = H^s(\partial\Omega)', \quad |v|_{-s} = \sup_{z \in H^s(\partial\Omega)} \frac{\langle v,z\rangle}{|z|_s}, \quad s \geq 0.$$

If $\dim(\Omega) \leq 3$, (1.8) implies uniform convergence, since

$$(1.9) \qquad \|w-W\|_{L_\infty(\Omega)} \leq \begin{cases} C\|u\|_p h^{p-1}, & \frac{5}{2} \leq p \leq r+1, \quad n=2, \\ C\|u\|_p h^{p-\frac{3}{2}}, & \frac{5}{2} \leq p \leq r+1, \quad n=3, \end{cases}$$

as a consequence of the two inequalities

$$\|z\|_{L_\infty(\Omega)} \leq C\|z\|_0^{\frac{1}{4}} \|z\|_1^{\frac{1}{2}} \|z\|_2^{\frac{1}{4}}, \quad n=2,$$

$$\|z\|_{L_\infty(\Omega)} \leq C\|z\|_0^{\frac{1}{12}} \|z\|_1^{\frac{1}{3}} \|z\|_2^{\frac{7}{12}}, \quad n=3.$$

An H^1-Galerkin procedure for (1.2) can be derived in the following manner. Multiply (1.2) by Δv and integrate the time derivative term by parts. Then

$$(\nabla \frac{\partial u}{\partial t}, \nabla v) + (\Delta u, \Delta v) = (f, \Delta v) + \langle \frac{\partial g}{\partial t}, \frac{\partial v}{\partial \nu} \rangle, \quad t \in J .$$

Again if we do not wish to impose the boundary conditions

on the subspace, the $(\Delta u, \Delta v)$ term should be replaced by $B(u, v)$ in the Galerkin process. Thus, we seek a differentiable map $U: J \to \mathcal{M}_h$ such that

$$(\nabla \frac{\partial U}{\partial t}, \nabla v) + B(U, v) = (\nabla \frac{\partial u}{\partial t}, \nabla v) + B(u, v), \quad t \in J, v \in \mathcal{M}_h.$$

Thus,

$$(\nabla \frac{\partial U}{\partial t}, \nabla v) + B(U, v) = (f, \Delta v) + \langle \frac{\partial g}{\partial t}, \frac{\partial v}{\partial \nu} \rangle + h^{-3} \langle g, v \rangle + h \langle Tg, Tv \rangle, \tag{1.10a}$$

$$v \in \mathcal{M}_h, \quad t \in J.$$

It is also necessary to specify $U(0)$. Let us do this by using the projection (1.7) of the initial values:

$$B(U(0), v) = B(u_0, v), \quad v \in \mathcal{M}_h. \tag{1.10b}$$

We shall show that

$$\|u - U\|_{L_\infty(J; H^s(\Omega))} \leq C(u) h^{r+1-s}, \quad 0 \leq s \leq 2, \tag{1.11}$$

for sufficiently smooth u. Obviously, (1.11) also implies an L_∞ bound on the error for $n \leq 3$.

There are several advantages to the methods (1.7) and (1.10). The boundary values are imposed weakly; thus, it is not necessary that the elements of the subspace $\mathcal{M}_h$ satisfy the boundary condition. The error estimates are valid without any so-called "inverse hypotheses" required of the subspaces $\mathcal{M}_h$. Thus, it is feasible and, in fact, probably desirable to use a subspace based on some reasonably regular grid. It is not necessary to triangulate the domain or to perturb the boundary, which eliminates the difficulty of adding considerable complication to the method in the neighborhood of the boundary in order to retain the maximal accuracy associated with $\mathcal{M}_h$ [4, 13].

Let us note some simple subspaces that satisfy the approximation requirement (1.3). Let $\delta = \{x_i, i=0, \pm 1, \pm 2, \ldots\}$, where $h_i = x_i - x_{i-1} > 0$, $\lim_{i \to \pm\infty} x_i = \pm\infty$ and $\sup_i h_i = h < \infty$.

Let $I_i = [x_{i-1}, x_i]$, and let $P_r(E)$ be the class of functions on $\mathbb{R}$ such that their restrictions to the set E are polynomials of degree not greater than r. Then let

$$M_k(r,\delta) = \{v \in C^k(\mathbb{R}) \mid v \in P_r(I_i), i = 0, \pm 1, \ldots\}, k \geq 1, r \geq 3.$$

Set

$$\tilde{\mathcal{M}}_h(k,r) = \{z = v|_\Omega, \text{ where } v \in \bigotimes_{j=1}^{n} M_k(r,\delta)\}. \tag{1.12}$$

(Obviously, we can employ a different δ in each direction.) The assumptions on $\partial\Omega$ are such that any $\varphi \in H^s(\Omega)$, $s \geq 0$, can be extended to a function $\bar{\varphi} \in H^s(\mathbb{R}^n)$ by the bounded, linear mapping given by the Calderón extension theorem [1, 7] in such a way that $\text{supp}(\bar{\varphi}) \subset \Omega_1$, a fixed domain slightly larger than Ω. It is shown in §6 that an inequality analogous to (1.3) holds for $\mathcal{M}_h = \bigotimes_{j=1}^{n} M_k(r,\delta)$ for functions in $H^p(\mathbb{R}^n)$ and with $\partial\Omega$ replaced by any smooth compact (n-1)-manifold without boundary. Consequently, the spaces $\tilde{\mathcal{M}}_h(k,r)$ defined by (1.12) can be employed in (1.7) and (1.10). They are particularly convenient choices when high order accuracy is desired.

The procedures given by (1.7) and (1.10) grew out of earlier work in a single space variable. Thomée and Wahlbin [20] introduced the H^1-Galerkin process for parabolic equations in a single space variable, and the present authors [10] extended their results in several ways under the limitation of having but a single space variable.

A specific outline of the contents is as follows. In §2 the approximation of the Dirichlet problem is analyzed. The continuous-time Galerkin method (1.10) for the heat equation is treated in §3; discretizations in time of (1.10) by the Crank-Nicolson difference method and by collocation are covered in §4 and §5, respectively. Finally, approximation properties of tensor products of piecewise-polynomial spaces are given in §6.

2. The Laplace Equation.

Let us analyze the convergence of the solution of (1.7) to that of (1.1). Let

(2.1) $$|||\varphi||| = B(\varphi,\varphi)^{\frac{1}{2}}, \quad \varphi \in \mathscr{V} .$$

Lemma 1.

There exists a constant $\beta > 0$, *independent of* h , *such that*

(2.2) $$|||\varphi||| \geq \beta \|\varphi\|_2 , \quad \varphi \in \mathscr{V} .$$

Proof.

Recall [15] that there exists $c > 0$ such that

$$\frac{1}{c} \|\varphi\|_2^2 \leq \|\Delta\varphi\|^2 + |\varphi|_{3/2}^2 \leq c\|\varphi\|_2^2 .$$

By (1.5) and standard interpolation theory [15],

$$\begin{aligned} \|\Delta\varphi\|^2 + |\varphi|_{3/2}^2 &\leq \|\Delta\varphi\|^2 + C(|\varphi|_2^{3/4}|\varphi|^{1/4})^2 \\ &\leq \|\Delta\varphi\|^2 + Ch|\varphi|_2^2 + Ch^{-3}|\varphi|^2 \\ &\leq C'|||\varphi|||^2 , \end{aligned}$$

and the lemma is demonstrated.

The next lemma is an immediate corollary of the symmetry of the form B, the projection lemma on Hilbert spaces, and Lemma 1.

Lemma 2.

There exists a unique solution $W \in \mathscr{M}_h$ *of* (1.7), *and if* w *is the solution of* (1.1), *then*

$$(2.3) \qquad \frac{1}{\beta}\|w-W\|_2 \le |||w-W||| = \inf_{\chi \in \mathcal{M}_h} |||w-\chi||| .$$

We can now demonstrate a generalization of (1.8).

Theorem 1.

Let $w \in H^p(\Omega)$, $\frac{5}{2} \le p \le r+1$, be the solution of (1.1), and let $W \in \mathcal{M}_h$ be the solution of (1.7). Then there exists a constant C, independent of h and w, such that

$$(2.4) \qquad \begin{aligned} \|w-W\|_{-s} &\le C\|w\|_p h^{p+s}, \quad -2 \le s \le r-3, \\ |w-W|_{-s-\frac{1}{2}} &\le C\|w\|_p h^{p+s}, \quad -\frac{5}{2} \le s \le r. \end{aligned}$$

Proof.

We shall derive the boundary estimate first. Let $\gamma \in H^{s+\frac{1}{2}}(\partial\Omega)$, $s \ge 0$, and let $\varphi \in H^{s+1}(\Omega)$ be the solution of the auxiliary problem

$$(2.5) \qquad \begin{aligned} &\Delta^2 \varphi = 0, \quad x \in \Omega, \\ &\Delta\varphi = 0 \text{ and } \frac{\partial}{\partial\nu}\Delta\varphi - h^{-3}\varphi = \gamma, \quad x \in \partial\Omega. \end{aligned}$$

It is obvious that the auxiliary problem reduces to a second order problem, since $\Delta\varphi$ vanishes identically; however, (2.5) conforms to $B(u, v)$, except that the term involving the operator T is missing. Clearly, $\varphi = -h^3\gamma$ on $\partial\Omega$, and standard elliptic regularity for the Dirichlet problem implies that

$$(2.6) \qquad \|\varphi\|_{s+1} \le C|\gamma|_{s+\frac{1}{2}} h^3 ,$$

with C being independent of h. If $\zeta = w-W$, the vanishing of $\Delta\varphi$ on Ω implies that

$$(2.7) \qquad \langle \zeta,\gamma\rangle = -h^{-3}\langle \zeta,\varphi\rangle = -B(\zeta,\varphi) + h\langle T\zeta, T\varphi\rangle .$$

Since $B(\zeta,\chi) = 0$ for $\chi \in \mathcal{M}_h$, (2.3) shows that

$$|B(\zeta,\varphi)| = \inf_{\chi \in \mathcal{M}_h} |B(\zeta,\varphi-\chi)| \le C|||\zeta|||\cdot|||\varphi-\chi|||$$
$$\le C\|w\|_p h^{p-2} \inf_{\chi \in \mathcal{M}_h} |||\varphi-\chi|||. \ .$$

If $\frac{3}{2} \le s \le r$ (i.e., $s+1 \le r+1$), then (1.3) and (2.6) imply that

$$\inf_{\chi \in \mathcal{M}_h} |||\varphi-\chi||| \le C\|\varphi\|_{s+1} h^{s-1} \le C|\gamma|_{s+\frac{1}{2}} h^{s+2}$$

and

$$|B(\zeta,\varphi)| \le C\|w\|_p |\gamma|_{s+\frac{1}{2}} h^{p+s}, \quad \frac{3}{2} \le s \le r .$$

Next, since T^2 is a smooth fourth-order operator on $\partial\Omega$,

$$|h\langle T\zeta, T\varphi\rangle| = |h\langle \zeta, T^2\varphi\rangle| \le h|\zeta|_{-s+\frac{7}{2}} |T^2\varphi|_{s-\frac{7}{2}}$$
$$\le Ch|\zeta|_{-s+\frac{7}{2}} |\varphi|_{s+\frac{1}{2}}$$
$$\le Ch^4|\zeta|_{-s+\frac{7}{2}} |\gamma|_{s+\frac{1}{2}} .$$

Thus, it follows that

$$|\zeta|_{-s-\frac{1}{2}} \le C\{\|w\|_p h^{p+s} + h^4|\zeta|_{-s+\frac{7}{2}}\}, \ \frac{3}{2} \le s \le r. \tag{2.8}$$

Note that (2.3) and (1.5) imply immediately that

$$|\zeta|_0 \le h^{3/2}|||\zeta||| \le C\|w\|_p h^{p-\frac{1}{2}},$$
$$|\zeta|_2 \le Ch^{-\frac{1}{2}}(|||\zeta||| + |\zeta|) \le C\|w\|_p h^{p-5/2} .$$

Interpolation gives the inequality

$$|\zeta|_q \le C\|w\|_p h^{p-q-\frac{1}{2}}, \quad 0 \le q \le 2 . \tag{2.9}$$

Since $-s+\frac{7}{2}\in[0,2]$ when $s\in[\frac{3}{2},\frac{7}{2}]$, it follows from (2.8) and (2.9) that

$$|\zeta|_{-s-\frac{1}{2}}\le C\|w\|_p\, h^{p+s}\,,\quad \frac{3}{2}\le s\le\frac{7}{2}\,,$$

and by interpolation for $-\frac{5}{2}\le s\le\frac{7}{2}$. Then (2.8) can be used as a recurrence relation to extend the range of s for which the inequality holds until s reaches the value r. Thus,

$$|\zeta|_{-s-\frac{1}{2}}\le C\|w\|_p\, h^{p+s}\,,\quad -\frac{5}{2}\le s\le r\,. \tag{2.10}$$

Note that the limitation $s\le r$ arises from the approximatic condition (1.3).

Now consider the estimate over Ω. Let $\psi\in H^s(\Omega)$, $s\ge 0$. First, we shall approximate ψ using the space introduced in (1.12). Let Δ_h denote the partition of $\mathbb{R}$ given by a uniform spacing h: $x_i = ih$, $i\in\mathbb{Z}$. Let $\mathcal{M}_h(2r\text{-}1$ denote the space of (1.12) corresponding to Δ_h. By Theore 7 below there exists $\psi_h\in\widetilde{\mathcal{M}}_h(2r\text{-}1,2r)$ such that (with C ind pendent of h)

$$\begin{aligned}\|\psi_h\|_{s+j}+|\psi_h|_{s+j-\frac{1}{2}} &\le Ch^{-j}\|\psi\|_s,\quad 0\le j\le 2r-s,\\ \|\psi-\psi_h\|_{s-j} &\le Ch^{j}\|\psi\|_s\,,\quad 0\le j\le s+2\,.\end{aligned} \tag{2.11}$$

In order to complete the proof let us change the auxiliary problem. Let $\varphi\in H^{2r+4}(\Omega)$ be the solution of

$$\begin{aligned}\Delta^2\varphi &= \psi_h\,,\quad x\in\Omega\,,\\ \Delta\varphi &= \frac{\partial}{\partial\nu}\Delta\varphi - h^{-3}\varphi = 0\,,\quad x\in\partial\Omega\,.\end{aligned} \tag{2.12}$$

Note that

$$\|\Delta\varphi\|_{p+2}\le C\|\psi_h\|_p\,,\quad p\le 2r\,, \tag{2.13}$$

and that, since $\varphi = h^3\frac{\partial}{\partial\nu}\Delta\varphi$ on $\partial\Omega$,

(2.14) $$|\varphi|_q \leq Ch^3 \|\psi_h\|_{q-\frac{1}{2}}, \quad q \leq 2r+\tfrac{1}{2},$$

with both constants being independent of h. It also follows that

(2.15) $$\begin{aligned}\|\varphi\|_{p+4} &\leq C[\|\Delta\varphi\|_{p+2} + |\varphi|_{p+\frac{7}{2}}] \\ &\leq C[\|\psi_h\|_p + h^3\|\psi_h\|_{p+3}], \quad p \leq 2r-3.\end{aligned}$$

In particular,

(2.16) $$\|\varphi\|_{s+4} \leq C[\|\psi_h\|_s + h^3\|\psi_h\|_{s+3}] \leq C\|\psi\|_s, \quad s \leq 2r-3,$$

by (2.11). It is exactly at this point that the approximation of ψ by ψ_h is critical; we cancel three differentiations by inducing a factor h^{-3}. Obviously, we could not have done this using ψ itself. Now, consider

(2.17) $$(\zeta, \psi) = (\zeta, \psi_h) + (\zeta, \psi - \psi_h).$$

First,

$$|(\zeta, \psi-\psi_h)| \leq \|\zeta\|_2 \|\psi-\psi_h\|_{-2} \leq C\|w\|_p \|\psi\|_s h^{p+s},$$

by (2.3) and (2.11). Then, for any $v \in \mathcal{M}_h$,

$$\begin{aligned}(\zeta, \psi_h) = (\zeta, \Delta^2\varphi) &= (\Delta\zeta, \Delta\varphi) + \langle \zeta, \frac{\partial}{\partial\nu}\Delta\varphi\rangle - \langle \frac{\partial\zeta}{\partial\nu}, \Delta\varphi\rangle \\ &= (\Delta\zeta, \Delta\varphi) + h^{-3}\langle\zeta, \varphi\rangle \\ &= B(\zeta, \varphi) - h\langle T\zeta, T\varphi\rangle \\ &= B(\zeta, \varphi - v) - h\langle T\zeta, T\varphi\rangle.\end{aligned}$$

Now, for $s \leq r-3$ (i.e., $s+4 \leq r+1$)

$$|B(\zeta,\varphi)| = \inf_{v \in \mathcal{M}_h} |B(\zeta,\varphi-v)| \leq C|||\zeta||| \inf_{v \in \mathcal{M}_h} |||\varphi-v|||$$

$$\leq C\|w\|_p \|\varphi\|_{s+4} h^{p+s}$$

$$\leq C\|w\|_p \|\psi\|_s h^{p+s} .$$

Also, using (2.14) with $q = s+\frac{1}{2}$ and then (2.10), we see that

$$|h\langle T\zeta, T\varphi\rangle| = |h\langle \zeta, T^2\varphi\rangle| \leq Ch|\zeta|_{-s+\frac{7}{2}} |\varphi|_{s+\frac{1}{2}}$$

$$\leq Ch^4|\zeta|_{-s+\frac{7}{2}} \|\psi\|_s$$

$$\leq C\|w\|_p \|\psi\|_s h^{p+s}, \quad \frac{3}{2} \leq s \leq r-3 .$$

Consequently,

$$\|\zeta\|_{-s} \leq C\|w\|_p h^{p+s}, \quad \frac{3}{2} \leq s \leq r-3, \quad \frac{5}{2} \leq p \leq r+1. \tag{2.18}$$

Interpolation between $s = \frac{3}{2}$ and $s = -2$ finishes the proof of (2.4), and the main result of the elliptic section has been completed.

Note that (2.4) is an optimal estimate both with respect to the exponent on h and with respect to the norm imposed on the solution w.

3. The Heat Equation.

Consider now the solution of the continuous-time H^1 Galerkin method (1.10). We shall decompose the error $\zeta = u - U$ into the sum $\zeta = \eta - \vartheta$, where $\eta = u - W$ and $\vartheta = U - W$ and where $W: J \rightarrow \mathcal{M}_h$ is given by the projection

$$B(u-W, v) = B(\eta, v) = 0 , \quad v \in \mathcal{M}_h, \ t \in J . \tag{3.1}$$

Note that (1. 10b) requires that $\vartheta(0) = 0$. It is easy to see from (1. 10) and (3. 1) that

$$(3.2) \qquad (\nabla \frac{\partial\vartheta}{\partial t}, \nabla v) + B(\vartheta, v) = (\nabla \frac{\partial\eta}{\partial t}, \nabla v), \quad v \in \mathcal{M}_h, \quad t \in J .$$

Let us obtain an H^2 estimate. Choose $v = \partial\vartheta/\partial t$. Then,

$$(3.3) \qquad \|\nabla \frac{\partial\vartheta}{\partial t}\|^2 + \tfrac{1}{2}\frac{d}{dt}B(\vartheta,\vartheta) = (\nabla \frac{\partial\eta}{\partial t}, \nabla \frac{\partial\vartheta}{\partial t}),$$

and it follows that

$$\|\frac{\partial\vartheta}{\partial t}\|_{L_2(J;H_0^1(\Omega))} + \|\vartheta\|_{L_\infty(J;\mathcal{V}_h)} \le C\|\frac{\partial\eta}{\partial t}\|_{L_2(J;H^1(\Omega))},$$

where $\|v\|_{\mathcal{V}_h} = |||v|||$. Since the defining relation for W can be differentiated with respect to t , it follows from Theorem 1 that

$$\|\frac{\partial\eta}{\partial t}\|_1 \le C\|\frac{\partial u}{\partial t}\|_p h^{p-1}, \quad \frac{5}{2} \le p \le r+1, \quad t \in J .$$

Hence,

$$(3.4) \qquad \|\vartheta\|_{L_\infty(J;\mathcal{V}_h)} \le C\|\frac{\partial u}{\partial t}\|_{L_2(J;H^p(\Omega))} h^{p-1}, \quad \frac{5}{2} \le p \le r+1.$$

Since

$$(3.5) \qquad \|\eta\|_{L_\infty(J;\mathcal{V}_h)} \le C\|u\|_{L_\infty(J;H^q(\Omega))} h^{q-2}, \quad \frac{5}{2} \le q \le r+1 ,$$

we see that

$$(3.6) \qquad \|u-U\|_{L_\infty(J;H^2(\Omega))} + h^{-3/2}\|u-U\|_{L_\infty(J;L_2(\partial\Omega))} + h^{\frac{1}{2}}\|u-U\|_{L_\infty(J;H^2(\partial\Omega))}$$

$$\leq C\left[\left\|\frac{\partial u}{\partial t}\right\|_{L_2(J;H^{q-1}(\Omega))} + \|u\|_{L_\infty(J;H^q(\Omega))}\right] h^{q-2}, \quad \frac{7}{2} \leq q \leq r+1 .$$

The inequality (3.6) is optimal both in the exponent on h and in the norms imposed on the solution u of (1.2) [12, 16].

Note that the inequality

$$(3.7) \quad \|u-U\|_{L_\infty(J;H^p(\partial\Omega))} \leq C\left[\left\|\frac{\partial u}{\partial t}\right\|_{L_2(J;H^{q-1}(\Omega))} + \|u\|_{L_\infty(J;H^q(\Omega))}\right] h^{q-p-\frac{1}{2}}, \quad \frac{7}{2} \leq q \leq r+1, \ 0 \leq p \leq 2,$$

follows immediately from (3.6) and interpolation.

Let us turn to deriving an estimate for ζ in H^1. Since

$$\|\zeta\|_1 \leq C(\|\nabla\zeta\| + |\zeta|_{\frac{1}{2}}),$$

it is sufficient to obtain an estimate for $\nabla\zeta$. Choose the test function $v = \vartheta$ in (3.2). Then, after integrating the right-hand side by parts,

$$\frac{1}{2}\frac{d}{dt}\|\nabla\vartheta\|^2 + |||\vartheta|||^2 = -\left(\frac{\partial\eta}{\partial t}, \Delta\vartheta\right) + \left\langle \frac{\partial\eta}{\partial t}, \frac{\partial\vartheta}{\partial\nu}\right\rangle$$

$$\leq \frac{1}{2}|||\vartheta|||^2 + C\left(\left\|\frac{\partial\eta}{\partial t}\right\|^2 + \left|\frac{\partial\eta}{\partial t}\right|^2_{-\frac{1}{2}}\right).$$

Hence,

$$(3.8) \quad \|\nabla\vartheta\|_{L_\infty(J;L_2(\Omega))} \leq C\left\|\frac{\partial\eta}{\partial t}\right\|_{L_2(J;L_2(\Omega)\times H^{-\frac{1}{2}}(\partial\Omega))} \leq C\left\|\frac{\partial u}{\partial t}\right\|_{L_2(J;H^q(\Omega))} h^q, \quad \frac{5}{2} \leq q \leq r+1.$$

Then note that

$$(3.9) \quad \|\nabla\eta\|_{L_\infty(J;L_2(\Omega))} \leq \|\eta\|_{L_\infty(J;H^1(\Omega))} \leq C\|u\|_{L_\infty(J;H^q(\Omega))} h^{q-1}, \quad \frac{5}{2} \leq q \leq r+1 .$$

Thus,

$$\|\nabla\zeta\|_{L_\infty(J;L_2(\Omega))} + \|\zeta\|_{L_\infty(J;H^{\frac{1}{2}}(\partial\Omega))} \tag{3.10}$$
$$\leq C\left[\|u\|_{L_\infty(J;H^q(\Omega))} + \left\|\frac{\partial u}{\partial t}\right\|_{L_2(J;H^{q-1}(\Omega))}\right]h^{q-1}$$

for $\frac{7}{2} \leq q \leq r+1$, and we have proved the following theorem.

Theorem 2.

Let u denote the solution of the boundary value problem (1.2) and let U denote the solution of the Galerkin approximation (1.10). Then, there exists a constant K, independent of h and u, such that, for $\frac{5}{2} \leq q \leq r+1$,

$$\|u-U\|_{L_\infty(J;H^p(\Omega))} \leq C(q,u)h^{q-p}, \quad 1 \leq p \leq 2,$$
$$\|u-U\|_{L_\infty(J;H^p(\partial\Omega))} \leq C(q,u)h^{q-p-\frac{1}{2}}, \quad 0 \leq p \leq 2, \tag{3.11}$$

where

$$C(q,u) = \begin{cases} K[\|u\|_{L_\infty(J;H^q(\Omega))} + \|\frac{\partial u}{\partial t}\|_{L_2(J;H^{q-1}(\Omega))}], & \frac{7}{2} \leq q \leq r+1, \\ K[\|u\|_{L_\infty(J;H^q(\Omega))} + \|\frac{\partial u}{\partial t}\|_{L_2(J;H^q(\Omega))}], & \frac{5}{2} \leq q < \frac{7}{2}. \end{cases} \tag{3.12}$$

Note that the estimates (3.11) are optimal both in the order of convergence and in the norms applied to the solution u. Next let us consider an L_2 estimate. In the case of a single space variable the authors [10] were able to derive a very precise bound for $\|\vartheta\|_0$; however, the argument used there does not seem to carry over without the imposition of additional smoothness constraints on the solution. Thus, that argument does not produce the norms to be applied to u to agree with the optimal choices that come from parabolic regularity and approximation, nor have we been able to find another. Consequently, we shall give a simple argument to obtain optimal order convergence at a cost of requiring one additional derivative in space for the $\frac{\partial u}{\partial t}$-term.

It follows easily from (3.4) and (3.8) that

$$\|\vartheta\|_{L_\infty(J;H^1(\Omega))} \le C \left\|\frac{\partial u}{\partial t}\right\|_{L_2(J;H^q(\Omega))} h^q, \quad \frac{5}{2} \le q \le r+1 .$$

Since

$$\|\eta\|_{L_\infty(J;L_2(\Omega))} \le C \|u\|_{L_\infty(J;H^q(\Omega))} h^q, \quad \frac{5}{2} \le q \le r+1 ,$$

then we have demonstrated the following theorem.

Theorem 3.

Under the same hypotheses as for Theorem 2,

$$\|u-U\|_{L_\infty(J;\, L_2(\Omega))} \le C\left\{\|u\|_{L_\infty(J;H^q(\Omega))} + \left\|\frac{\partial u}{\partial t}\right\|_{L_2(J;H^q(\Omega))}\right\} h^q \tag{3.13}$$

$$\frac{5}{2} \le q \le r+1 .$$

The additional derivative results from estimating ϑ in $H^1(\Omega)$. We have the somewhat surprising result that the boundary estimates are slightly superior to the estimates over Ω, since the optimal choice of the norms on u was applicable over the whole range $0 \le p \le 2$.

4. Discretization in Time for the Parabolic Problem by the Crank-Nicolson Difference Method.

Let $\Delta t = T_0 N^{-1}$ and $t_{n+\theta} = (n+\theta)\Delta t$, where $n \in \mathbb{Z}$ and $0 \le \theta < 1$. Then denote $f(t_n)$ by f^n, and set

(4.1) $$f^{n+\theta} = (1-\theta)f^n + \theta f^{n+1}, \quad 0 < \theta < 1 .$$

Note that $f^{n+\theta}$ is not $f(t_{n+\theta})$ for $0 < \theta < 1$.

The Crank-Nicolson discretization of (1.10) is given by the map $U:\{t_0, \ldots, t_N\} \to \mathcal{M}_h$ determined by the relations

(4.2)
$$\text{(i)}\quad B(U^0 - u_0, v) = 0, \qquad v \in \mathcal{M}_h,$$
$$\text{(ii)}\quad (\nabla\partial_t U^n, \nabla v) + B(U^{n+\frac{1}{2}}, v)$$
$$= (f^{n+\frac{1}{2}}, \Delta v) + \langle \frac{\partial g^{n+\frac{1}{2}}}{\partial t}, \frac{\partial v}{\partial \nu}\rangle + h^{-3}\langle g^{n+\frac{1}{2}}, v\rangle + \langle Tg^{n+\frac{1}{2}}, Tv\rangle,$$
$$v \in \mathcal{M}_h,$$

where $\partial_t f^n = (f^{n+1} - f^n)(\Delta t)^{-1}$. Note first that existence of U follows from uniqueness. To show uniqueness, assume that $Z^n \in \mathcal{M}_h$ satisfies

(4.3)
$$Z^0 = 0,$$
$$(\nabla\partial_t Z^n, \nabla v) + B(Z^{n+\frac{1}{2}}, v) = 0, \quad v \in \mathcal{M}_h, \quad 0 \le n \le N-1 .$$

Take the test function $v = \partial_t Z^n$. Then,

$$\|\nabla\partial_t Z^n\|^2 + \frac{1}{2\Delta t}[\,|||Z^{n+1}|||^2 - |||Z^n|||^2] = 0, \quad 0 \le n \le N-1,$$

and it is clear that $Z^n = 0$, $0 \le n \le N$. Hence, existence and uniqueness are assured for (4.2) whenever f and g are sufficiently regular that the right-hand side can be evaluated.

Let us turn to estimating the error. Let $W: J \to \mathcal{M}_h$ denote the projection given by (3.1), and let $\eta^n = u^n - W^n$ and $\vartheta^n = U^n - W^n$. It is a simple calculation to see that

$$(\nabla\partial_t W^n, \nabla v) + B(W^{n+\frac{1}{2}}, v)$$
$$= (\nabla \frac{\partial u^{n+\frac{1}{2}}}{\partial t}, \nabla v) + B(u^{n+\frac{1}{2}}, v) - (\nabla\partial_t \eta^n, \nabla v)$$
$$+ (\Delta t)^2 (\nabla \int_0^1 k(\tau) \frac{\partial^3 u}{\partial t^3}(t_n + \tau\Delta t)\, d\tau, \nabla v), \qquad v \in \mathcal{M}_h,$$

where $k(\tau) = \frac{1}{2}(\tau-\frac{1}{2})^2 - \frac{1}{8}$. It then follows that

$$(4.4)\qquad \begin{aligned}(\nabla \partial_t \vartheta^n, \nabla v) + B(\vartheta^{n+\frac{1}{2}}, v) &= (\nabla \partial_t \eta^n, \nabla v)\\ &\quad -(\Delta t)^2 (\nabla \int_0^1 k(\tau) \frac{\partial^3 u}{\partial t^3}(t_n + \tau \Delta t) d\tau, \nabla v), \quad v \in \mathcal{M}_h.\end{aligned}$$

Select the test function $v = \partial_t \vartheta^n$, and consider the two right-hand side terms separately. It will be convenient to integrate by parts in the space variables and to sum by parts in time for the first term:

$$(4.5)\qquad \begin{aligned}\sum_{n=0}^{m} (\nabla \partial_t \eta^n, \nabla \partial_t \vartheta^n)\Delta t &= \sum_{n=0}^{m} \{-(\partial_t \eta^n, \Delta \partial_t \vartheta^n) + \langle \partial_t \eta^n, \partial_t \frac{\partial \vartheta^n}{\partial \nu}\rangle\}\Delta t\\ &= -(\partial_t \eta^m, \Delta \vartheta^{m+1}) + \langle \partial_t \eta^m, \frac{\partial \vartheta^{m+1}}{\partial \nu}\rangle\\ &\quad + \sum_{n=1}^{m} \{(\partial_t^2 \eta^{n-1}, \Delta \vartheta^n) - \langle \partial_t^2 \eta^{n-1}, \frac{\partial \vartheta^n}{\partial \nu}\rangle\}\Delta t\\ &\le \tfrac{1}{2} |||\vartheta^{m+1}|||^2 + C\{\|\partial_t \eta^m\|^2 + |\partial_t \eta^m|^2_{-\frac{1}{2}}\} + \sum_{n=1}^{m} |||\vartheta^n|||^2 \Delta t\\ &\quad + C\sum_{n=1}^{m} \{\|\partial_t^2 \eta^{n-1}\|^2 + |\partial_t^2 \eta^{n-1}|^2_{-\frac{1}{2}}\}\Delta t,\end{aligned}$$

since $\vartheta^0 = 0$. The other right-hand side term can be handled directly so that we arrive at the following inequality:

$$(4.6)\qquad \begin{aligned}&\sum_{n=0}^{N-1} \|\nabla \partial_t \vartheta^n\|^2 \Delta t + \max_{0 \le n \le N} |||\vartheta^n|||^2\\ &\le C\{\|\frac{\partial \eta}{\partial t}\|^2_{L_\infty(J; L_2(\Omega) \times H^{-\frac{1}{2}}(\partial\Omega))} + \|\frac{\partial^2 \eta}{\partial t^2}\|^2_{L_2(J; L_2(\Omega) \times H^{-\frac{1}{2}}(\partial\Omega))}\\ &\qquad + (\Delta t)^4 \|\nabla \frac{\partial^3 u}{\partial t^3}\|^2_{L_2(J; L_2(\Omega))}\}\\ &\le \acute{C}\{(\|\frac{\partial u}{\partial t}\|^2_{L_\infty(J; H^p(\Omega))} + \|\frac{\partial^2 u}{\partial t^2}\|_{L_2(J; H^p(\Omega))}) h^{2p} + (\Delta t)^4 \|\frac{\partial^3 u}{\partial t^3}\|^2_{L_2(J; H^1(\Omega))}\}\end{aligned}$$

for $\frac{5}{2} \leq p \leq r+1$. Since $\|\eta\|_s \leq C\|u\|_p h^{p-s}$, $0 \leq s \leq 2$, it follows that

$$\max_{0 \leq n \leq N} \|(u-U)^n\|_s$$

$$\leq C\{(\|u\|_{L_\infty(J;H^p(\Omega))} + \|\frac{\partial u}{\partial t}\|_{L_\infty(J;H^{p-s}(\Omega))} + \|\frac{\partial^2 u}{\partial t^2}\|_{L_2(J;H^{p-s}(\Omega))})h^{p-s} + \|\frac{\partial^3 u}{\partial t^3}\|_{L_2(J;H^1(\Omega))}(\Delta t)^2\} \tag{4.7}$$

for $\frac{5}{2} + s \leq p \leq r+1$ and $0 \leq s \leq 2$. Analogous estimates hold for $\max|(u-U)^n|_s$, $0 \leq s \leq 2$, and negative norm estimates can be derived.

Theorem 4.

If U^n *denotes the solution of* (4.2), *then the optimal order estimates given by* (4.7) *for* $0 \leq s \leq 2$ *hold.*

5. Discretization in Time for the Parabolic Problem by Collocation.

Again let $\Delta t = T_0 N^{-1}$ and $t_k = k\,\Delta t$. Let $J_k = [t_{k-1}, t_k]$ and $\epsilon = \{t_0, t_1, \ldots, t_N\}$. Let

$$N_0(s, \epsilon) = \{v \in C^0(J) \,|\, v \in P_s(J_k),\ k = 1, \ldots, N\} \tag{5.1}$$

and

$$\mathcal{N}_h = \mathcal{M}_h \otimes N_0(s, \epsilon), \tag{5.2}$$

where $\mathcal{M}_h$ denotes the same subspace of $\mathcal{V}$ as used previously. The example of $\mathcal{N}_h$ that can serve as model is given by

$$\mathcal{N}_h = \tilde{\mathcal{M}}_h(k, r) \otimes N_0(s, \epsilon), \tag{5.3}$$

with

$$(5.4) \qquad r = 2s+1 , \quad \Delta t = h^2 ,$$

since it will become clear later that the selection given by (5.4) tends to minimize the total smoothness required of the solution u when we seek an approximate solution in $\mathcal{N}_h$.

Let $0 < \tau_1 < \tau_2 < \ldots < \tau_s < 1$ and $w_\ell^* > 0$, $\ell = 1,\ldots,s$, denote the Gauss points and weights, respectively, such that

$$\int_0^1 p(\tau)d\tau = \sum_{\ell=1}^{s} p(\tau_\ell)w_\ell^* , \quad p \in P_{2s-1} .$$

We shall need a Lagrange interpolation operator which was discussed in detail in §6 of [8]. The notation to be used here will differ slightly from that of [8]. Let

$$(5.5) \qquad B(t) = B_s^*(t) = \frac{1}{(2s)!} \frac{d^{s-1}}{dt^{s-1}}(t^s(t-1)^s) .$$

Then, $B \in P_{s+1}$ and $B^{(s+1)} \equiv 1$. Moreover, $B'(\tau_\ell) = 0$, $\ell = 1,\ldots,s$, and there exist simple roots of $B(\sigma) = 0$ at points $\sigma_0,\ldots,\sigma_s$ such that

$$(5.6) \qquad 0 = \sigma_0 < \sigma_1 < \ldots < \sigma_s = 1 .$$

Set

$$(5.7) \qquad \begin{aligned} \tau_{k\ell} &= t_{k-1} + \tau_\ell \Delta t, \quad k = 1,\ldots,N, \quad \ell = 1,\ldots,s , \\ \sigma_{k\ell} &= t_{k-1} + \sigma_\ell \Delta t , \quad k = 1,\ldots,N, \quad \ell = 0,\ldots,s . \end{aligned}$$

Let $A = A_\epsilon : C^0(J) \to N_0(s,\epsilon)$ be the interpolation operator given by the requirement that

$$(5.8) \qquad (A\varphi)(\sigma_{k\ell}) = \varphi(\sigma_{k\ell}), \quad k = 1,\ldots,N, \quad \ell = 0,\ldots,s.$$

(A was called $T_{s,\epsilon}^*$ in [8].) The salient facts concerning A are as follows [8]. There exist polynomials $C_q^* \in P_{s+q}$, uniformly bounded on $[0,1]$, and kernels $k_p(t,\alpha)$ having $s+p$ derivatives with respect to t such that, for $t \in J_k$,

$$(5.9)\qquad \begin{aligned} ((I-A)\varphi)(t) &= \sum_{q=1}^{p} \varphi^{(s+q)}(t_{k-\frac{1}{2}}) C_q^{*}\left(\frac{t-t_{k-1}}{\Delta t}\right)(\Delta t)^{s+q} \\ &\quad + (\Delta t)^{s+p+1} \int_0^1 k_p\left(\frac{t-t_{k-1}}{\Delta t}, \alpha\right) \varphi^{(s+p+1)}(t_{k-1} + \alpha \Delta t)\, d\alpha. \end{aligned}$$

The error expression also holds with k_0 and no leading terms. The choice of A such that $B'(\tau_\ell) = 0$ implies that $C_1^{*'}(\tau_\ell) = 0$; i.e., the principal term in the expansion of the error in the derivative vanishes at the Gauss points and the error in the derivative is $O((\Delta t)^{s+1})$ at the points $\tau_{k\ell}$.

Now, let us define the collocation discretization. We seek a function $U_h \in \mathcal{N}_h$ such that

$$(5.10)\qquad \begin{aligned} &\text{i)}\quad B(U_h(0) - u_0, v) = 0, \quad v \in \mathcal{M}_h, \\ &\text{ii)}\quad \left(\nabla \frac{\partial U_h}{\partial t}, \nabla v\right) + B(U_h, v) = (Af, \Delta v) + \left\langle A\left(\frac{\partial g}{\partial t}\right), \frac{\partial v}{\partial \nu}\right\rangle \\ &\qquad\qquad + h^{-3}\langle Ag, v\rangle + h\langle ATg, Tv\rangle, \\ &\qquad v \in \mathcal{M}_h, \ t = \tau_{k\ell}, \ k = 1, \ldots, N, \ \ell = 1, \ldots, s. \end{aligned}$$

Note that (5.10) reduces exactly to the Crank-Nicolson procedure for $s = 1$.

First, we should like to show that a solution U_h exists and is unique. Since (5.10) is linear, it is sufficient to show uniqueness, for it is a simple matter to see that the dimension of $\mathcal{N}_h$ is equal to the number of independent equations that must be satisfied. Let $Z \in \mathcal{N}_h$ be a solution of (5.10) corresponding to $u_0 = 0$, $f = 0$, $g = 0$. Then, it is clear that $Z(0) = 0$. If we take $v = \partial Z/\partial t$, then it follows that

$$(5.11)\qquad \left\|\nabla \frac{\partial Z}{\partial t}\right\|^2 + \frac{1}{2}\frac{d}{dt}|||Z|||^2 = 0, \quad t = \tau_{k\ell}.$$

Since the first term in (5.11) is a polynomial in t of degree $2s-2$ on each J_j and the second is of degree $2s-1$, Gaussian quadrature is exact on each subinterval. Thus, since $Z(0) = 0$,

$$(5.12) \quad \int_0^{t_k} \| \nabla \frac{\partial Z}{\partial t} \|^2 d\tau + |||Z(t_k)|||^2 = 0 , \quad k = 1, \dots , N .$$

If v is taken equal to Z, then it follows that

$$(5.13) \quad \|\nabla Z(t_k)\|^2 + \sum_{j=1}^{k} \sum_{\ell=1}^{s} |||Z(\tau_{j\ell})|||^2 w_\ell^* \Delta t = 0, \quad k = 1, \dots , N.$$

It is apparent that (5.12) and (5.13) imply that Z vanishes identically. In fact, we have shown more than just existence and uniqueness for U_h; we have established that U_h is determined on the interval J_k by $U_h(t_{k-1})$ and the equations (5.10) restricted to J_k; i.e., the method is local in time.

Let us consider the analysis of the error. Let $W: J \to \mathcal{M}_h$ be the elliptic projection of u given by

$$(5.14) \qquad B(W - u, v) = 0 , \quad v \in \mathcal{M}_h , \quad t \in J .$$

Let $\vartheta = U_h - AW$. Then a short calculation shows that

$$(5.15) \quad \begin{aligned} (\nabla \frac{\partial \vartheta}{\partial t}, \nabla v) + B(\vartheta, v) &= (\nabla(A \frac{\partial u}{\partial t} - \frac{\partial}{\partial t} Au), \nabla v) \\ &+ (\nabla \frac{\partial}{\partial t}(A(u-W)), \nabla v), \quad v \in \mathcal{M}_h , \quad t = \tau_{k\ell} . \end{aligned}$$

Also, $\vartheta(0) = 0$.

If $v = \partial\vartheta/\partial t$, then

$$(5.16) \quad \begin{aligned} \| \nabla \frac{\partial \vartheta}{\partial t} \|^2 + \tfrac{1}{2} \frac{d}{dt} |||\vartheta|||^2 &= (\nabla(A \frac{\partial u}{\partial t} - \frac{\partial}{\partial t} Au), \nabla \frac{\partial \vartheta}{\partial t}) \\ &+ (\nabla \frac{\partial}{\partial t}(A(u-W)), \nabla \frac{\partial \vartheta}{\partial t}), \quad t = \tau_{k\ell} . \end{aligned}$$

Now, (5.9) can be applied to $A \frac{\partial u}{\partial t} - \frac{\partial}{\partial t} Au$ to give the relation

$$(A\frac{\partial u}{\partial t}-\frac{\partial}{\partial t}Au)(\tau_{k\ell})=(\Delta t)^{s+1}\int_0^1\{\frac{\partial k_1}{\partial t}(\tau_\ell,\alpha)\frac{\partial^{s+2}u}{\partial t^{s+2}}(t_{k-1}+\alpha\Delta t)$$
(5.17)
$$-k_0(\tau_\ell,\alpha)\frac{\partial^{s+1}u}{\partial t^{s+1}}(t_{k-1}+\alpha\Delta t)\}d\alpha.$$

Hence,

$$(5.18)\qquad \|\nabla(A\frac{\partial u}{\partial t}-\frac{\partial}{\partial t}Au)(\tau_{k\ell})\|^2\le C\|u\|^2_{H^{s+2}(J_k;H^1(\Omega))}(\Delta t)^{2s+1}.$$

Also, Theorem 1 and the exactness of the Lagrange interpolation on constants imply that

$$(5.19)\qquad \|\nabla\frac{\partial}{\partial t}A(u-W)(\tau_{k\ell})\|^2\le C\|u\|^2_{H^1(J_k;H^q(\Omega))}h^{2q-2}(\Delta t)^{-1},$$
$$\frac{5}{2}\le q\le r+1.$$

Then the exactness of Gaussian quadrature for the terms on the left-hand side of (5.16) can be combined with (5.18) and (5.19) to show that

$$\int_0^{T_0}\|\nabla\frac{\partial\vartheta}{\partial t}\|^2d\tau+\max_k |||\vartheta(t_k)|||^2$$
(5.20)
$$\le C\{\|u\|^2_{H^{s+2}(J;H^1(\Omega))}(\Delta t)^{2s+2}+\|u\|^2_{H^1(J;H^q(\Omega))}h^{2q-2}\}$$

for $\frac{5}{2}\le q\le r+1$. It is immediate that

$$\max_k |||(u-U_h)(t_k)|||\le C\{(\|u\|_{L_\infty(J;H^q(\Omega))}+\|u\|_{H^1(J;H^{q-1}(\Omega))})h^{q-2}$$
(5.21)
$$+\|u\|_{H^{s+2}(J;H^1(\Omega))}(\Delta t)^{s+1}\}$$

for $\frac{7}{2}\le q\le r+1$ (and also a trivial modification holds in the range $\frac{5}{2}\le q<\frac{7}{2}$) and that

$$\max_k \|(u-U_h)(t_k)\|_1$$

$$\leq C\{\|u\|_{L_\infty(J;H^q(\Omega))} + \|u\|_{H^1(J;H^q(\Omega))})h^{q-1} + \|u\|_{H^{s+2}(J;H^1(\Omega))}(\Delta t)^{s+1}\} \tag{5.22}$$

for $\frac{5}{2} \leq q \leq r+1$.

Next, we shall consider an $L_2(\Omega)$-estimate. We shall be forced to use an $H^1(\Omega)$-estimate on ϑ, as we were in the continuous time case. Since (5.20) does not give an $O(h^{r+1})$ bound on the spatial term, we must use a different argument. Note that (5.20) does imply that

$$\max_k |\vartheta(t_k)|_{\frac{1}{2}}$$

$$\leq C\{\|u\|_{H^1(J;H^q(\Omega))} h^q + \|u\|_{H^{s+2}(J;H^q(\Omega))} h(\Delta t)^{s+1}\} \tag{5.23}$$

for $\frac{5}{2} \leq q \leq r+1$. Now, take $v = \vartheta$ in (5.15). For $t = \tau_{k\ell}$,

$$\frac{1}{2}\frac{d}{dt}\|\nabla\vartheta\|^2 + |||\vartheta|||^2$$

$$= (\nabla(A\frac{\partial u}{\partial t} - \frac{\partial}{\partial t}Au), \nabla\vartheta) + (\nabla\frac{\partial}{\partial t}A(u-W), \nabla\vartheta)$$

$$= -(A\frac{\partial u}{\partial t} - \frac{\partial}{\partial t}Au, \Delta\vartheta) + \langle A\frac{\partial u}{\partial t} - \frac{\partial}{\partial t}Au, \frac{\partial\vartheta}{\partial\nu}\rangle$$

$$- (\frac{\partial}{\partial t}A(u-W), \Delta\vartheta) + \langle \frac{\partial}{\partial t}A(u-W), \frac{\partial\vartheta}{\partial\nu}\rangle \tag{5.24}$$

$$\leq \frac{1}{2}|||\vartheta|||^2 + C\{\|A\frac{\partial u}{\partial t} - \frac{\partial}{\partial t}Au\|^2 + |A\frac{\partial u}{\partial t} - \frac{\partial}{\partial t}Au|^2_{-\frac{1}{2}}$$

$$+ \|\frac{\partial}{\partial t}A(u-W)\|^2 + |\frac{\partial}{\partial t}A(u-W)|^2_{-\frac{1}{2}}\}$$

$$\leq \frac{1}{2}|||\vartheta|||^2 + C\{\|u\|^2_{H^{s+2}(J_k;L_2(\Omega)\times H^{-\frac{1}{2}}(\partial\Omega))}(\Delta t)^{2s+2}$$

$$+ \|u\|^2_{H^1(J_k;H^q(\Omega))} h^{2q}\}(\Delta t)^{-1}, \quad \frac{5}{2} \leq q \leq r+1 .$$

Thus,

$$(5.25)\quad \max_k \|\nabla\vartheta(t_k)\| \le C\{\|u\|_{H^{s+2}(J;L_2(\Omega)\times H^{-\frac{1}{2}}(\partial\Omega))}(\Delta t)^{s+1} + \|u\|_{H^1(J;H^q(\Omega))} h^q\},\quad \frac{5}{2}\le q < r+1\ ,$$

and it follows that

$$(5.26)\quad \begin{aligned} \max_k \|\vartheta(t_k)\|_1 \le C\{&\|u\|_{H^1(J;H^q(\Omega))} h^q \\ &+ (\|u\|_{H^{s+2}(J;L_2(\Omega)\times H^{-\frac{1}{2}}(\Omega))} \\ &+ h\|u\|_{H^{s+2}(J;H^1(\Omega))})(\Delta t)^{s+1}\},\quad \frac{5}{2}\le q \le r+1\ . \end{aligned}$$

Consequently,

$$(5.27)\quad \begin{aligned} &\max_k \|(u-U_h)(t_k)\| \\ &\le C\{(\|u\|_{L_\infty(J;H^q(\Omega))} + \|u\|_{H^1(J;H^q(\Omega))})h^q \\ &\quad + (\|u\|_{H^{s+2}(J;L_2(\Omega)\times H^{-\frac{1}{2}}(\partial\Omega))} \\ &\quad\quad + h\|u\|_{H^{s+2}(J;H^1(\Omega))})(\Delta t)^{s+1}\} \end{aligned}$$

for $\frac{5}{2}\le q \le r+1$. Note that for sufficiently smooth solutions we have shown that

$$(5.28)\quad \begin{aligned} \max_k(\|(u-U_h)(t_k)\|_\alpha + h^{\frac{1}{2}}|(u-U_h)(t_k)|_\alpha) \\ \le C(u)(h^{r+1-\alpha} + (\Delta t)^{s+1}) \end{aligned}$$

for $0\le \alpha \le 2$. Also, for optimal order convergence rates the choices made in (5.4) do lead to minimal smoothness for u, at least if we take into account that the argument has tended to impose one space derivative more on the solution than is probably necessary. The results above can be summarized in the following theorem.

Theorem 5.

Let U_h denote the solution of the collocation discretization given by (5.10). Then the error $u-U_h$ can be bounded by the estimates (5.21), (5.22), and (5.27), each giving an optimal order convergence rate.

In [11] the authors showed that the collocation discretization in time of the usual Galerkin method based on the inner product on L_2 leads to superconvergence in Δt at the knots t_k; presumably, the same phenomenon occurs here, but the authors have not made the analysis required to establish that the error in $(u-U_h)(t_k)$ should involve a term of the form $O((\Delta t)^{2s})$ rather than $O((\Delta t)^{s+1})$.

6. Approximation by Tensor Products of Piecewise Polynomials.

In this section we shall verify that spaces constructed by taking tensor products of certain spaces, composed of piecewise polynomials in a single variable, satisfy the approximation assumptions of §1. The techniques which we use to show this result are reasonably standard [3, 14, 17, 18, 19]. After defining the spaces we shall use and making certain preliminary simplifications, we construct a local interpolation process and apply the Bramble-Hilbert Lemma [5].

Since functions $\varphi \in H^s(\Omega)$ can be extended to functions $\tilde{\varphi} \in H^s(\mathbb{R}^n)$ (using the elementary extension of Lions [15] in the case in which $\partial\Omega$ is smooth or the extension of Calderón [1,7] in a more general context) in such a fashion that the map $\varphi \to \tilde{\varphi}$ is continuous from $H^s(\Omega)$ to $H^s(\mathbb{R}^n)$, it is sufficient to show that we can approximate functions which belong to $H^s(\mathbb{R}^n)$ in the prescribed fashion (see also [18]).

Let $\delta = \{a_i\}_{i=-\infty}^{\infty}$ be a partition of $\mathbb{R}$ such that $h_i = a_i - a_{i-1} > 0$ and $\lim_{i\to\pm\infty} a_i = \pm\infty$; let $I_i = [a_{i-1}, a_i]$ and $h = \max h_i$. For nonnegative integers k and r such that $k < r$ let

$$M_k(r, \delta) = \{v \in C^k(\mathbb{R}): v \in P_r(I_i), \text{ all } i\}.$$

Take $k \geq 1$ and let $\mathcal{M}$ be the n-fold tensor product of the spaces $M_k(r, \delta)$. (For simplicity of notation we shall ignore the trivial refinement of allowing the partitions to be different in the different directions.) The following theorem suffices to establish the approximation properties we need for $\mathcal{M}$.

Theorem 6.

Let Γ be a compact $(n-1)$-manifold without boundary in $\mathbb{R}^n$. There is a constant C such that for $\varphi \in H^s(\mathbb{R}^n)$

$$\text{(6.1)} \quad \inf\{\sum_{j=0}^{2} h_j[\|\varphi-\chi\|_{H^j(\mathbb{R}^n)} + h^{\frac{1}{2}}\|\varphi-\chi\|_{H^j(\Gamma)}]:\chi \in \mathcal{M}\} \leq Ch^s\|\varphi\|_{H^s(\mathbb{R}^n)}$$

for $r \geq 3$ and $3 \leq s \leq r+1$.

Proof:

From Lemma 9 of [4] it follows (as in the proof of Lemma 10 of [4]) that it suffices to consider the case $s=r+1$. Also note that there is no loss of generality in assuming that $\frac{1}{3}h \leq h_i$ for all i. If this fails to hold we can pick a coarser partition from the points in δ such that each interval has length between $\frac{1}{2}h$ and $\frac{3}{2}h$; proving (6.1) for χ restricted to the subspace of $\mathcal{M}$ corresponding to this coarser mesh clearly proves (6.1). Since we decrease $\mathcal{M}$ by increasing k it suffices to demonstrate the theorem in the case $k = r-1$. It also should be noted that in the case $k = 1$ the restriction of an element of $\mathcal{M}$ is in $H^2(\Gamma)$.

Using the proof of Lemma 2.1 of [9] (or, equivalently, the quasiinterpolant of [3] or [17]) we see that in the one-dimensional case we can construct, for functions in $H^r(\mathbb{R})$, a local interpolant in $M_{r-1}(r, s)$ by matching the derivatives through order $r-1$ at each a_i for which i is an integral multiple of r. It is this Hermite-like interpolant that we shall use in the tensor product setting. To define the interpolant $\mathcal{J}w$ of a function $w \in H^{r+1}(\mathbb{R}^n)$ we need only specify the derivatives of the form

$$\left(\frac{\partial}{\partial x_1}\right)^{\alpha_1} \cdots \left(\frac{\partial}{\partial x_n}\right)^{\alpha_n} \mathcal{J}w ,$$

where each $\alpha_j \leq r-1$, at the points $x = (x_1, \ldots, x_n)$ such that each x_j is a point $a_{\ell r}$ for some integer ℓ. These cannot be set equal to the pointwise values of the corresponding derivatives of w, since for general $w \in H^{r+1}(\mathbb{R}^n)$ these derivatives may not exist; this difficulty will be easily overcome by setting the derivatives of total order greater than r to zero and by setting the others to a certain integral average of the corresponding derivatives of w.

Let $\varphi_0 : \mathbb{R} \to \mathbb{R}$ be a C^∞ function with support in $[\frac{1}{2}, 1]$ such that for polynomials p of degree less than r+1

$$p(0) = \int_{\mathbb{R}} \varphi_0(x)\, p(x) dx ;$$

such a φ_0 can be constructed by smoothly smearing a Lagrange interpolation operator. Now define $\varphi_1 : \mathbb{R}^n \to \mathbb{R}$ by

$$\varphi_1(x) = \begin{cases} 0 & , \quad x = 0 \\ |x|^{1-n} \varphi_0(|x|) |\Sigma_1| & , \quad x \neq 0 , \end{cases}$$

where $|\Sigma_1|$ is the surface area of the unit sphere in $\mathbb{R}^n$.

Now note that if $p(x)$ is a polynomial in $(x_1, \ldots, x_n)$ of total degree at most r then, for any $\epsilon > 0$,

$$p(x) = \epsilon^{-n} \int_{\mathbb{R}^n} p(x+t)\, \varphi_1(t\, \epsilon^{-1})dt .$$

Let $\mathcal{K} = \{x = (x_1, \ldots, x_n) \in \mathbb{R}^n :$ for each $j = 1, \ldots, n$ there is an integer ℓ such that $x_j = a_{r\ell}\}$. For $x \in \mathcal{K}$ we define

$$D^{\alpha} \mathcal{I} w(x) = \left(\frac{\partial}{\partial x_1}\right)^{\alpha_1} \cdots \left(\frac{\partial}{\partial x_n}\right)^{\alpha_n} \mathcal{I} w(x)$$

$$= \begin{cases} 0, & |\alpha| = \alpha_1 + \ldots + \alpha_n \geq r , \\ h^{-n} \int_{\mathbb{R}^n} \varphi_1(h^{-1}t) D^{\alpha} w(x+t)dt, & |\alpha| \leq r . \end{cases}$$

We shall now examine the error $w - \mathcal{I}w$ on the typical parallelopiped $(0, a_r)^n = D$. Let $\tau(x) = x/h$. Then τ maps the domain $D_1 = (-h, a_r+h)^n$ into the domain $\Omega_1 = (-1, r+1)^n$. Since $\mathcal{I}w$ is determined on D by the values of w in D_1 we can define a map I on $H^{r+1}(\Omega_1)$ such that, if $W \in H^{r+1}(\Omega_1)$ and $w \in H^{r+1}(\mathbb{R}^n)$ is such that $w = W \circ \tau$ on D_1, then $IW \circ \tau = \mathcal{I} w$ on D; this defines Iw uniquely on Ω_2, where $\Omega_2 = (0, a_r/h)^n = \tau(D)$. It is easily seen that the map $q: H^{r+1}(D_1) \to \mathbb{R}$ given by

$$q(W) = \sum_{|\alpha| \leq 2} \{\|D^{\alpha}(W-IW)\|_{L_2(\Omega_2)} + \|D^{\alpha}(W-IW)\|_{L_2(E)}\}$$

where $E = \tau(\bar{\Omega}_2 \cap \Gamma)$, is a continuous semi-norm on $H^{r+1}(\Omega_1)$; further, since the h_i's are bounded below by $\frac{1}{3}h$ we see that the bound is independent of the particular choice of the h_i's. It is clear from construction that q vanishes on polynomials of total degree less than $r+1$ in $x_1, \ldots, x_n$. Thus, by Bramble-Hilbert [5],

$$q(W) \leq C \sum_{|\alpha| = r+1} \|D^{\alpha} W\|_{L_2(\Omega_1)} .$$

This implies that

$$\sum_{|\alpha|\le 2} h^{|\alpha|}\{\|D^{\alpha}(w-\mathcal{I}w)\|_{L_2(D)}+h^{\frac{1}{2}}\|D^{\alpha}(w-\mathcal{I}w)\|_{L_2(\Gamma\cap\bar{D})}\}$$

$$\le Ch^{r+1}\|w\|_{H^{r+1}(\tau^{-1}(\Omega))} .$$

The inequality for cubes can then be summed to give the conclusion.

A corresponding theorem can be proved for $k = 0$; the $H^2(\mathbb{R}^n)$ and $H^2(\Gamma)$ terms should be omitted and the range on s is $2 \le s \le r+1$.

Theorem 6 shows that the tensor product spaces $\tilde{\mathcal{M}}_h(k,r)$ satisfy the assumptions made on the trial spaces $\mathcal{M}_h$; however, we need to be able to approximate well in $H^{-s}(\Omega)$ at one point in the proof of Theorem 1, and interpolation does not provide quite the result we want. The following technical lemma suffices.

Theorem 7.

Let $\psi \in H^s(\Omega)$, $0 \le s \le r-3$. Then there exists $\psi_h \in H^{2r}(\Omega)$, $0 < h \le 1$, such that

$$\text{(6.2)}\quad \begin{array}{ll}\text{(i)} & \|\psi_h\|_{s+j} \le Ch^{-j}\|\psi\|_s , \quad 0 \le j \le 2r-s ,\\ \text{(ii)} & \|\psi-\psi_h\|_{s-j} \le Ch^{j}\|\psi\|_s , \quad 0 \le j \le 2r-s+1 .\end{array}$$

Proof:

Let $\mathcal{R}_h = \tilde{\mathcal{M}}_h(2r-1, 2r)$, where the partition of $\mathbb{R}$ is the uniform one of subinterval length h . Let

$$(\varphi,\psi)_q = (\varphi,\psi)_{H^q(\Omega)} = \sum_{|\alpha|\le q} (D^{\alpha}\varphi, D^{\alpha}\psi).$$

Let $\psi_h \in \mathcal{R}_h$ be determined by the minimization problem

$$\|\psi-\psi_h\|_s^2 + h^{2(2r-s)}\|\psi_h\|_{2r}^2 \tag{6.3}$$
$$= \inf_{\chi\in\mathfrak{R}_h} \{\|\psi-\chi\|_s^2 + h^{2(2r-s)}\|\chi\|_{2r}^2\} ;$$

i.e., ψ_h satisfies the Galerkin equations

$$(\psi-\psi_h, v)_s + h^{2(2r-s)}(\psi_h, v)_{2r} = 0 , \quad v \in \mathfrak{R}_h . \tag{6.4}$$

The trivial choice $\chi = 0$ shows that

$$\|\psi-\psi_h\|_s + h^{2r-s}\|\psi_h\|_{2r} \leq 2\|\psi\|_s ; \tag{6.5}$$

thus $\|\psi_h\|_s \leq 3\|\psi\|_s$, and interpolation shows that (6.2.i) is valid.

Next, let $\eta\in H^p(\Omega)$, $0 \leq p \leq 2r-2s$, and let $\varphi \in H^{2s+p}(\varphi)$ be defined by

$$(\varphi, v)_s = (\eta, v)_0 , \quad v \in H^s(\Omega) .$$

Note that

$$\|\varphi\|_{2s+p} \leq C\|\eta\|_p .$$

If $\zeta = \psi-\psi_h$, then

$$(\zeta, \eta) = (\zeta, \varphi)_s = (\zeta, \varphi-\chi)_s - (\psi_h, \chi)_{2r} h^{2(2r-s)}$$

for any $\chi\in \mathfrak{R}_h$. First extend φ to a function in $H^s(\mathbb{R}^n)$ and then apply the construction in the proof of Theorem 6 to obtain an interpolant of φ such that, if $p \leq 2(r-s)+1$,

$$\|\varphi-\chi\|_s \leq C\|\varphi\|_{2s+p} h^{s+p} \leq C\|\eta\|_p h^{s+p} ,$$

$$\|\chi\|_{2r} \leq Ch^{2s+p-2r}\|\varphi\|_{2s+p} \leq Ch^{2s+p-2r}\|\eta\|_p ;$$

we have used homogeneity in h to trade derivatives of χ against powers of h^{-1}. Then, (6.5) shows that

$$|(\zeta, \eta)| \leq C\|\zeta\|_s \|\eta\|_p h^{p+s} + C\|\psi_h\|_{2r} h^{2s+p-2r} \|\eta\|_p h^{2(2r-s}$$

$$\leq C\|\psi\|_s \|\eta\|_p h^{p+s} ,$$

and

$$\|\zeta\|_{-p} \leq C\|\psi\|_s h^{p+s} , \quad 0 \leq p \leq 2(r-s) + 1.$$

Interpolation finishes the proof of (6.2.ii).

The choice of $\mathfrak{R}_h$ was made more or less arbitrarily. The ranges of validity for the bounds in (6.2) can be made as large as desired by increasing the degree of the smooth spline space $\mathfrak{R}_h$. We needed only to be able to bound $\|\psi_h\|_r$ and to approximate well in $H^{-2}(\Omega)$; it would have been sufficient to take $\mathfrak{R}_h = \widetilde{\mathcal{M}}_h(2r-5, 2r-4)$, but it was notationally simpler to use $\mathfrak{R}_h$ as it was chosen.

References.

1. S. Agmon, Lectures on Elliptic Boundary Value Problems, Van Nostrand, Princeton, 1965.

2. G. A. Baker, Simplified proofs of error estimates for the least squares method for Dirichlet's problem, Math. Comp. 27 (1973), 229-235.

3. C. de Boor and G. J. Fix, Spline approximation by quasiinterpolants, J. Approx. Theory 8 (1973), 19-45.

4. J. H. Bramble, T. Dupont, and V. Thomée, Projection methods for Dirichlet's problem in approximating polygonal domains with boundary value corrections, Math. Comp. 26 (1972), 869-879.

5. J. H. Bramble and S. Hilbert, Bounds for a class of linear functionals with application to Hermite interpolation, Numer. Math. 16 (1971), 362-369.

6. J. H. Bramble and A. H. Schatz, Rayleigh-Ritz-Galerkin methods for Dirichlet's problem using subspaces without boundary conditions, Comm. Pure Appl. Math. 23 (1970), 653-675.

7. A. P. Calderón, Lebesgue spaces of differentiable functions and distributions, Partial Differential Equations, Proc. Symp. Pure Math., Vol. 4, A. M. S., 1961.

8. J. Douglas, Jr., and T. Dupont, Collocation methods for parabolic equations in a single space variable based on C^1-piecewise-polynomial spaces, Lecture Notes in Mathematics no. 385, Springer-Verlag, Heidelberg, 1974.

9. J. Douglas, Jr., T. Dupont, and L. Wahlbin, Optimal L_∞ error estimates for Galerkin approximations to solutions of two point boundary value problems, to appear in Math. Comp.

10. J. Douglas, Jr., T. Dupont, and M. F. Wheeler, Some superconvergence results for an H^1-Galerkin procedure for the heat equation, Proceedings of a symposium on Méthodes de calcul scientifique et technique, to appear in the Springer Lecture Note Series.

11. J. Douglas, Jr., T. Dupont, and M. F. Wheeler, A quasi-projection approximation method applied to Galerkin procedures for parabolic and hyperbolic equations, to appear.

12. T. Dupont, Some L^2 error estimates for parabolic Galerkin methods, The Mathematical Foundations of the Finite Element Method with Applications to Partial Differential Equations, A. K. Aziz (ed.), Academic Press, New York, 1972.

13. T. Dupont, L^2 error estimates for projection methods for parabolic equations in approximating domains, these Proceedings.

14. S. Hilbert, A mollifier useful for approximations in Sobolev spaces and some applications to approximating solutions of differential equations, Math. Comp. 27 (1973), 81-89.

15. J.-L. Lions and E. Magenes, Problèmes aux limites non homogènes et applications, Vol. 1, Dunod, Paris, 1968.

16. J.-L. Lions and E. Magenes, ibid, Vol. 2.

17. T. Lyche and L. L. Schumaker, Local spline approximation methods, MRC Technical Summary Report #1417, 1974.

18. M. H. Schultz, L^2-multivariate approximation theory, SIAM J. Numer. Anal. 6 (1969), 184-209.

19. G. Strang, Approximation in the finite element method, Numer. Math. 19 (1972), 81-98.

20. V. Thomée and L. Wahlbin, To appear.

Jim Douglas, Jr.
Mathematics Research Center
University of Wisconsin
and
Department of Mathematics
University of Chicago,
Chicago, Illinois 60637.

Todd Dupont
Department of Mathematics
University of Chicago
Chicago, Illinois 60637

Mary F. Wheeler
Department of Mathematics
Rice University
Houston, Texas 77001

Subject Index

Q

R

S

T

U

V

W

A 4
B 5
C 6
D 7
E 8
F 9
G 0
H 1
I 2
J 3